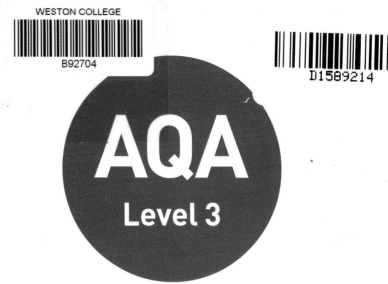

AQA
Level 3

Mathematical Studies
(Core Maths)

Approval message from AQA

This textbook has been approved by AQA for use with our qualification. This means that we have checked that it broadly covers the specification and we are satisfied with the overall quality. Full details of our approval process can be found on our website.

We approve textbooks because we know how important it is for teachers and students to have the right resources to support their teaching and learning. However, the publisher is ultimately responsible for the editorial control and quality of this book.

Please note that when teaching the *AQA Level 3 Mathematical Studies* course, you must refer to AQA's specification as your definitive source of information. While this book has been written to match the specification, it cannot provide complete coverage of every aspect of the course.

A wide range of other useful resources can be found on the relevant subject pages of our website: www.aqa.org.uk.

Authors
Anne Haworth

Steven Lomax

Elaine Lambert

David Bowman

Ruth Gibson

Deborah McCarthy

Series editor
Heather Davis

AN HACHETTE UK COMPANY

Every effort has been made to trace all copyright holders, but if any have been inadvertently overlooked, the Publisher will be pleased to make the necessary arrangements at the earliest opportunity.

Although every effort has been made to ensure that website addresses are correct at time of going to press, Hodder Education cannot be held responsible for the content of any website mentioned in this book. It is sometimes possible to find a relocated web page by typing in the address of the home page for a website in the URL window of your browser.

Hachette UK's policy is to use papers that are natural, renewable and recyclable products and made from wood grown in sustainable forests. The logging and manufacturing processes are expected to conform to the environmental regulations of the country of origin.

Orders: please contact Bookpoint Ltd, 130 Milton Park, Abingdon, Oxon OX14 4SB.

Telephone: +44 (0)1235 827720. Fax: +44 (0)1235 400454. Lines are open 9.00a.m.–5.00p.m., Monday to Saturday, with a 24-hour message answering service. Visit our website at www.hoddereducation.co.uk

© Anne Haworth, Steven Lomax, Elaine Lambert, David Bowman, Ruth Jones, Deborah McCarthy and Heather Davis 2017

First published in 2017 by

Hodder Education,

An Hachette UK Company

Carmelite House

50 Victoria Embankment

London EC4Y 0DZ

Impression number 10 9 8 7 6 5 4 3 2 1

Year 2021 2020 2019 2018 2017

Cover photo © Thinkstock/Getty Images/iStockphoto

Typeset in Bembo Std, 11/13 pts by Integra Software Services Pvt. Ltd, Pondicherry, India

Printed in Italy

A catalogue record for this title is available from the British Library

ISBN 978 1471 863752

Acknowledgements

The Publisher would like to thank the following for permission to reproduce copyright material.

p.23 Extracts from www.iaaf.org/disciplines/throws/javelin-throw Used with permission from IAAF; **p.39** Graph of population sizes: *Sustainable Energy – without the hot air*, by David JC MacKay, published by UIT: www.uit.co.uk/sustainable. Also available free to download for personal non-commercial use from www.withouthotair.com; p.45 Formulae: Used with permission from GLOBALRPh; **p.64:** Table of data, Heights of 18 year olds: From APPLIED LINEAR REGRESSION by Sanford Weisberg. Reproduced with permission of John Wiley & Sons Ltd via PLSclear. **p.76** Table of data, Underweight/overweight children: Global and regional trends by UN Regions, 1990–2025, Overweight: 1990–2015, Global and regional trends by UN Regions, 1990–2025, Underweight: 1990–2015. The original data is available at http://apps.who.int/gho/data/view.main.NUTUNOVERWEIGHTv; **p.90** Table of data, Women's heptathlon results: This work is licensed under Creative Commons ShareAlike 3.0 Unported (CC BY-SA 3.0). It is attributed to 'Athletics at the 2012 Summer Olympics – Women's heptathlon' by Wikipedia; **p.139** Text extract, The Guardian: Copyright Guardian News & Media Ltd 2016; **p.153** Alcohol facts table: From Statistics on Alcohol, March 2016 published by Alcohol Concern. Used with permission of Alcohol Concern; **p.156** Alcohol recommendations: Used with permission of Alcohol Concern. **p.163** Pie chart of government spending: © MailOnline; **p.163** Chart of the rising costs of renewable energy: © The Sun/News Syndication; **p.163** Bar chart of level of government investment: Contains public sector information licensed under the Open Government Licence v3.0. It is attributed to 'National Infrastructure Plan 2013' and the original version can be found at www.gov.uk/government/uploads/system/uploads/attachment_data/file/263159/national_infrastructure_plan_2013.pdf; **p.191** Graph, Heights and weights: from http://wiki.stat.ucla.edu/socr/index.php/SOCR_Data_Dinov_020108_HeightsWeights; **p.192** Graph, iPhone sales vs falling down stairs: Covered under Creative Commons. It is attributed to http://tylervigen.com/page?page=3; **p.194** Graphs drawn using BMI data: From http://wiki.stat.ucla.edu/socr/index.php/SOCR_Data_Dinov_020108_HeightsWeights; **p.227** Text extract, Quote from MoneySavingExpert.com: Martin Lewis, founder of MoneySavingExpert.com; **p.232** Table data, Average salary data, plus student loan data: Contains public sector information licensed under the Open Government Licence v3.0. and the original source can be found at www.ons.gov.uk/employmentandlabourmarket/peopleinwork/employmentandemployeetypes/articles/graduatesintheuklabourmarket/2013-11-19; **p.256** Table of data, Number of internet users: Used with permission from Internet Live Stars; **p.259** Table of data, Number of transistors: Covered under Creative Commons. From Transistor Count by Wikipedia; **p.262** Headline, The mathematical equation that caused the banks to crash: Copyright Guardian News & Media Ltd 2017; **p.266** Graph, Models of world population: Covered by Creative Commons. Based on https://en.wikipedia.org/wiki/world_population, UN data; **p.270** Graph of concentration of a drug: Information from www.intmath.com/blog/mathematics/math-of-drugs-and-bodies-pharmacokinetics-4098; **p.374** Table of data, CPI over the 10 years: ©InflationData.com; **p.389** Graph, Number of medals: Craig Nevill-Manning, medalspercapita.com.

Photo credits:

p 16 © Claudio Divizia/Shutterstock.com; **p 17** *tl* © Tom Gowanlock/Shutterstock.com; **p 18** © pergo70/Fotolia; **p 19** *mr* © dzain/Fotolia; **p 20** *tr* © 1997 Doug Menuez/Photodisc/Getty Images/Lifestyles Today 45; **p 21** © Actionplus sports images/Topfoto; **p 22** *ml* © Choon Kheong Gui/Alamy Stock Photo; **p 22** *b* © Imagestate Media (John Foxx)/F1rst V3071 ; **p 25** *tl* © Imagestate Media (John Foxx)/Vol 12 Nature & Animals 2; **p 28** © Lana Smirnova/Shutterstock.com; **p 29** *bl* © C Squared Studios/Photodisc/Getty Images/Travel Vacation Icons OS23; **p 29** *bm* © imstock/Shutterstock.com **p 29** *br* © Stockbyte/Getty Images Ltd/ Fast Food SD175; **p 29** *br* © Stockcreations/Shutterstock.com; **p 30** *mr* © photka/Shutterstock.com; **p 31** *tr* © photka/Shutterstock.com; **p 33** *ml* © dft.gov.uk; **p 33** *mr*| © Mitsubishi Motors; **p 35** *tl* © Lena Pan/Shutterstock.com; **p 39** *tl* © 1xpert /123RF; **p 42** *ml* © Nito/Fotolia; **p 43** *ml* © Freefly/Fotolia; **p 43** *bl* © Haveseen/Fotolia; **p 47** © Rob Walls/Alamy Stock Photo; **p 47** *mr* © World History Archive/TopFoto; **p 48** *tl* © Bilderstoeckchen/Fotolia; **p 51** © Imagestate Media (John Foxx)/

Acknowledgements

Vol 07 Business & Industry 3; **p 52** *ml* © Getty Images/Image Source - OurThreatened Environment IS236; **p 53** *tr* © DayOwl/Shutterstock.com; **p 55** © David38/Fotolia; **p 56** *mr* © Imagestate Media (John Foxx)/ London V3037; **p 60** *ml* © muhammad_karim/iStock/Thinkstock; **p 63** © Digital Vision/Getty Images/ Groups Children and Teenagers; **p 66** *tr* © Fotolia; **p 66** *mr* © Monkey Business/Fotolia; **p 69** © Monkey Business/Fotolia; **p 71** *bl* © Georgejmclittle/Shutterstock.com; **p 73** *mr* © Imagestate Media (John Foxx)/ F1rst V3071; **p 77** © Stockbyte/Getty Images/Everyday Medicine 194; **p 79** *tl* © Yuri Arcurs/Fotolia; **p 80** *bl* © Boggy/Fotolia; **p 84** © Imagestate Media (John Foxx)/Gold Disc 3 SS21; **p 85** *br* © Simone Werner/Ney/ Fotolia; **p 87** *tr* © Gina Sanders/Fotolia; **p 89** © Pete Saloutos/Fotolia; **p 90** *ml* © Jupiterimages/Pixland/ Thinkstock; **p 91** *ml* © Stockbyte/Getty Images/Sport: The Will to Win CD93; **p 94** *t* © lenets_tan/Fotolia; **p 94** *b* © peepo/iStock; **p 95** *tl* © BasPhoto/Fotolia; **p 97** *tr* © Jale Ibrak/Fotolia; **p 98** © Craig Prentis/ Stringer/Getty Images; **p 99** *bl* © overthehill/Fotolia; **p 102** *ml* © John Brueske/Shutterstock.com; **p 105** © Photodisc/Getty Images/ Business Today 35; **p 107** *br* © Leafy/Fotolia; **p 108** *br* © Design Pics Inc./ Alamy/Education and Learning 2 CD00E90; **p 110** © ttatty/iStock/Fotolia; **p 111** *br* © Imagestate Media (John Foxx)/Store V3042; **p 113** *tr* © Sean Gladwell/Fotolia; **p 119** © Jose Luis Pelaez, Inc./Blend Images/ Photolibrary.com; **p 120** *br* © michaelfitz/Fotolia; **p 121** *bl* © Kzenon/Fotolia; **p 124** © Syda Productions/ Fotolia; **p 126** *mr* © Svetography/Shutterstock.com; **p 128** *mr* © Imagestate Media (John Foxx)/Vol 05 Concepts & Backgrounds; **p 131** © Gina Sanders/Fotolia; **p 132** *ml* © LuckyImages/Fotolia; **p 134** *bl* © Dušan Zidar/Fotolia; **p 137** © Weyo/Fotolia; **p 140** *bl* © Heather Davis; **p 142** © Fazon/Fotolia; **p 144** *mr* © Ping han/Fotolia; **p 145** *ml* © Imagestate Media (John Foxx)/Super Business Mix SS71; **p 150** © Luftbildfotograf/Fotolia; **p 153** *tl* © Stockbyte/Getty Images/Everyday Medicine 194; **p 157** *tl* © Yuri Arcurs/ Fotolia; **p 160** © afateev/Fotolia; **p 162** *tr* © ArnoMassee/iStock; **p 162** *mr* © atoss/iStock/Thinkstock; **p 165** © Elaine Lambert; **p 168** *tr* © Nikolay Grigoryev/Fotolia; **p 170** *br* © Paulcraven/iStock/Thinkstock; **p 171** © Cathy Yeulet/123RF; **p 172** *tl* © Piotr Adamowicz/Fotolia; **p 173** *tl* © BrandX/Getty Images/ Food and Textures CD X025; **p 175** *mr* © Eric Isselée/Fotolia; **p 177** © Michael Brown/Fotolia; **p 178** *br* © Imagestate Media (John Foxx)/Vol 03 Nature & Animals; **p 180** *tl* © Stefano Buttafoco/Shutterstock. com; **p 182** © Graja/Fotolia; **p 189** *bl* © itsallgood/Fotolia; **p 189** *mr* © Jérôme Rommé/Fotolia; **p 190** © C Squared Studios/Photodisc/Getty Images/Moments in Life OS36; **p 191** *bl* © Cambo photography/Fotolia; **p 192** *br* © Fotocrisis/Shutterstock.com; **p 195**© Yuri Arcurs/Fotolia; **p 197** *mr* © Pink Badger/Fotolia; **p 198** *br*© Maciej Olszewski/Fotolia; **p 200** © Kadmy/Fotolia; **p 204** *m* © Tim Large/Shutterstock.com; **p 205** *tl* © benjaminnolte/Fotolia; **p 210** © Joe Gough/Fotolia; **p 211** *ml* © Joe Gough/Shutterstock.com; **p 212** *tl* © Gail Philpott/Alamy Stock Photo; **p 215** © Mariusz Blach/Fotolia; **p 217** *tl* © Krzysiek z Poczty/ Fotolia; **p 217** *ml* © sveta/Fotolia; **p 221** © Yuri Arcurs/Fotolia; **p 222** *tl* © Denis Makarenko/Shutterstock. com; **p 224** *mr* © Ded Pixto/Fotolia; **p 227** © Stockr/Fotolia; **p 228** *bl* © Sean Gladwell/Fotolia; **p 229** *tr* © gemphotography/Fotolia; **p 231** © CandyBox Images/Fotolia; **p 232** *tl* © Ludmila Smite/Fotolia; **p 232** *ml* © Moodboard/Thinkstock; **p 233** *tl* © Christophe Fouquin/Fotolia; **p 235** © Purestock/119 Best Friends; **p 237** *tr* © Adam Radosavljevic/Fotolia; **p 238** *bl* © Skogas/Fotolia; **p 241** © Tiffany Schoepp/ Blend Images/Getty Images/Small Business BLDV035; **p 242** *ml* © a4stockphotos/Fotolia; **p 242** *bl* © Ben Schonewille/Shutterstock.com; **p 247** © Brenda Carson/Fotolia; **p 252** *tl* © Lucielang/Fotolia; **p 253** *ml* © Loren Rodgers/Fotolia; **p 255** © agsandrew/Fotolia; **p 258** *tr* © aey/Fotolia; **p 259** *mr* © Andrew Barker/Fotolia; **p 262** © STILLFX/Shutterstock.com; **p 264** *tl* © Luis Viegas/Fotolia; **p 266** *tl* © Arthimedes/ Shutterstock.com; **p 268** © Imagestate Media (John Foxx) / Vol 10 Travel & Transportation; **p 270** *tl* © AZP Worldwide/Fotolia; **p 270** *br* © Marco Saracco/Fotolia; **p 273** © Bitter/Fotolia; **p 276** *bl* © Vladitto/ Fotolia; **p 279** *tl* © Paul Liu/Fotolia; **p 282** © Pixel 4 Images/Shutterstock.com; **p 285** *b* © Dean Pennala/ Fotolia; **p 287** © Grebcha/Fotolia; **p 291** *ml* © Imagestate Media (John Foxx)/Industry SS10; **p 291** *br* © Photodisc/Getty Images/Business & Industry 1; **p 293** © www.BibleLandPictures.com/Alamy Stock Photo; **p 293** *mr* © Emilio Segre Visual Archives/American Institute Of Physics/Science Photo Library; **p 295** *tr* © Brooke Becker/Fotolia; **p 296** *ml* © volff/Fotolia; **p 315** *mr* © February Destu/Shutterstock.com; **p 376** *mr* © Heather Davis;

t = top, *b* = bottom, *l* = left, *r* = right, *m* = middle

CONTENTS

Contents

Contents

Paper 2C: Graphical techniques

Section 3: Maths help

Paper 1: Compulsory content

Paper 2: Options

Index

All answers can be found at www.hoddereducation.co.uk/AQALevel3CertMathStudiesanswers

HOW TO GET THE MOST FROM THIS BOOK

Sections 1 and 2 of the book cover the content of the AQA Level 3 Certificate in Mathematical Studies. Section 3 gives you extra explanation, examples and practice for key techniques.

› Section 1: Getting started – talking about maths

This section:

› revisits GCSE knowledge and skills

› develops collaborative working skills

› develops investigational skills

› develops criticality and evaluative skills

› should be done first and worked through in order.

Each Section 1 chapter starts with a scenario that you discuss and investigate in order to explore the content. There are also some closed questions to test your understanding and questions for practice. The reflect questions are to help you see mathematics differently and recognise the skills you are developing.

There is a list of the process skills developed in the chapter and the maths help box gives references to relevant chapters in Section 3.

The questions marked with a blue bullet have answers available online, the ones with a green bullet are to answer but there are many possible responses to them. Answers can be found at www.hoddereducation.co.uk/AQALevel3CertMathStudiesanswers

The stars for the maths level give you a rough idea of how challenging the maths is!

❯ Section 2: Learning the new stuff

This section:

> ❯ develops fluency with the compulsory content and each of the options
> ❯ continues to develop skills in collaborative working and investigation
> ❯ develops modelling skills and continues to develop criticality

Chapter 2.1 to 2.4 cover estimation and should be done in order.

Chapters 2.5 to 2.10 cover data handling and should be done in order.

Chapters 2.11 to 2.19 cover personal finance and should be done in order.

Chapters 2.20 and 2.21 cover critical analysis and should be done in order.

Chapters 2.22 to 2.43 cover the three options.

Each Section 2 chapter starts with a scenario that you discuss and investigate in order to learn the content. There are also closed questions to test your understanding which have red bullets and answers online, worked examples and questions for practice.

The green bullets are questions to work on to deepen your understanding and they have many possible outcomes. The maths help box gives references to relevant chapters in Section 3. Answers are not supplied for these questions.

The practice questions are on the topics in the specification listed in the contents. Those marked with an **E** are to support the assessment of students' progress.

The stars for the maths level give you a rough idea of how challenging the maths is!

❯ Section 3: Maths help

This section:

> ❯ is a reference section for techniques
> ❯ gives additional worked examples and explanations
> ❯ gives additional practice questions
> ❯ is to be dipped into when you need it.

❯ Answers

Answers can be found at www.hoddereducation.co.uk/AQALevel3CertMath Studiesanswers

Answers are provided for all the practice questions and questions in the text with blue (Section 1) and red (Section 2) bullets. There are no answers provided for questions with green bullets.

SPECIFICATION REFERENCES

Chapter 1.1	*	Use of spreadsheets and tables Knowledge and use of the GCSE mathematical content of analysis of data
Chapter 1.2	*	Knowledge and use of the GCSE mathematical content of maths for personal finance elements
Chapter 1.3	*	Knowledge and use of the formulae for the circumference and the area of a circle
Chapter 1.4	*	Knowledge and use of the formulae for the perimeter of two-dimensional shapes, their areas and for calculating fractional areas of circles and composite shapes
Chapter 1.5	*	Calculate surface areas of spheres, cones, pyramids and composite solids Apply the concepts of similarity, including lengths in similar figures, and Pythagoras' theorem applied to two- and three dimensional figures
Chapter 1.6	*	Knowledge and use of the formula $y = mx + c$ Find the gradient of a straight line connecting two different points
Chapter 2.1	*	E1.1 Representing a situation mathematically, making assumptions and simplifications; students will engage in the tackling of 'open' mathematical problem-solving where there may not be a clear single approach or 'correct' answer E1.2 Selecting and using appropriate mathematical techniques for problems and situations E1.3 Interpreting results in the context of a given problem E1.4 Evaluating methods and solutions including how they may have been affected by assumptions made
Chapter 2.2	**	E1.1 Representing a situation mathematically, making assumptions and simplifications; tackling of 'open' mathematical problem-solving where there may not be a clear single approach or 'correct' answer E1.2 Selecting and using appropriate mathematical techniques for problems and situations E1.3 Interpreting results in the context of a given problem E1.4 Evaluating methods and solutions including how they may have been affected by assumptions made
Chapter 2.3	*	E2.1 Making fast, rough estimates of quantities that are either difficult or impossible to measure directly
Chapter 2.4	*	E2.1 Making fast, rough estimates of quantities that are either difficult or impossible to measure directly
Chapter 2.5	*	D1.1 Appreciating the difference between qualitative and quantitative data, including the difference between discrete and continuous quantitative data D4.1 Constructing and interpreting diagrams for grouped discrete data, knowing their appropriate use and reaching conclusions based on these diagrams, including box and whisker plots and stem-and-leaf diagrams (including back-to-back) D3.1 Calculating/identifying mean, median, mode, quartiles, either from raw data or from stem-and-leaf diagrams or box plots D3.2 Interpreting these numerical measures and reaching conclusions based on these measures

Chapter 2.6	*	D1.2 Appreciating the difference between primary and secondary data D1.3 Collecting quantitative and qualitative primary and secondary data D4.1 Constructing and interpreting diagrams for grouped continuous data, knowing their appropriate use and reaching conclusions based on these diagrams including cumulative frequency graphs
Chapter 2.7	**	D3.1 Calculating/identifying the (mean, median, mode, quartiles), percentiles, range, interquartile range whether from raw data or from cumulative frequency diagrams, stem-and-leaf diagrams or box plots D3.2 Interpreting these numerical measures and reaching conclusions based on these measures
Chapter 2.8	**	D4.1 Constructing and interpreting diagrams for grouped continuous data, knowing their appropriate use and reaching conclusions based on these diagrams, including histograms with unequal class intervals
Chapter 2.9	**	D2.1 Inferring properties of populations or distributions from a sample, whilst knowing the limitations of sampling D2.2 Appreciating the strengths and limitations of random, cluster, stratified and quota sampling methods and applying this understanding when designing sampling strategies
Chapter 2.10	**	D3.1 Calculating/identifying the mean and standard deviation, whether from raw data or from stem-and-leaf diagrams D3.2 Interpreting these numerical measures and reaching conclusions based on these measures
Chapter 2.11	**	F4.1 Student loans F2.5 Solving problems involving percentage increase
Chapter 2.12	**	F3.1 Simple and compound interest, Annual Equivalent Rate (AER) F3.2 Savings and investments F2.5 Solving problems involving simple and compound interest F7.2 Setting up, solving and interpreting the solutions to financial problems, including those that involve compound interest using iterative methods F1.3 Applying and interpreting limits of accuracy, specifying simple error intervals due to truncation or rounding
Chapter 2.13	**	F4.1 Annual Percentage Rate (APR) F1.1 Substituting numerical values into formulae and financial expressions
Chapter 2.14	**	F4.1 Mortgages and Annual Percentage Rate F5.1 Graphical representation, interpreting results from graphs in financial contexts F1.1 Substituting numerical values into formulae, spreadsheets and financial expressions F1.2 Using conventional notation for priority of operations, including brackets, powers, roots and reciprocals
Chapter 2.15	**	F6.1 Income tax, National Insurance F1.2 Using conventional notation for priority of operations, including brackets, powers, roots and reciprocals
Chapter 2.16	**	F2.1 Interpreting percentages and percentage changes as a fraction or a decimal and interpreting these multiplicatively F2.4 Working with percentages over 100% F2.5 Solving problems involving percentage change, including percentage increase/decrease and original value problems F6.1 Value Added Tax (VAT)
Chapter 2.17	**	F7.1 The effect of inflation, Retail Price Index (RPI), Consumer Price Index (CPI) F2.2 Expressing one quantity as a percentage of another F2.3 Comparing two quantities using percentages F2.5 Solving problems involving percentage change, including percentage increase/decrease

Chapter 2.18	**	F7.3 Currency exchange rates including commission F1.1 Substituting numerical values into (formulae), spreadsheets (and financial expressions) F5.1 Graphical representation percentage change
Chapter 2.19	*	F7.4 Budgeting F1.1 Substituting numerical values into formulae, spreadsheets and financial expressions, including bank accounts F1.4 Finding approximate solutions to problems in financial contexts
Chapter 2.20	*	C1.1 Criticising the arguments of others C2.1 Summarising and report writing C3.1 Comparing results from a model with real data C3.2 Critical analysis of data quoted in media, political campaigns, marketing etc
Chapter 2.21	*	C3.2 Critical analysis of data quoted in media, political campaigns, marketing and so on. Knowing and using the mathematical content of analysis of data C1.1 Criticising the arguments of others C2.1 Summarising and report writing
Chapter 2.22	**	S1.1 Knowledge that the normal distribution is a symmetrical distribution and that the area underneath the normal 'bell' shaped curve represents probability. Knowledge that approximately 2/3 of observations lie within 1 standard deviation of the mean and that approximately 95% of observations lie within 2 standard deviations of the mean
Chapter 2.23	***	S2.1 Use of the notation $N(\mu, \sigma^2)$ to describe a normal distribution in terms of mean and standard deviation and use of the notation $N(0, 1)$ for the standardised normal distribution with mean = 0 and standard deviation = 1 S3.1 Using a calculator or tables to find probabilities for normally distributed data with known mean and standard deviation
Chapter 2.24	**	S4.1 Understanding what is meant by the term 'population' in statistical terms. S4.2 Developing ideas of sampling to include the concept of a simple random sample from a population and a systematic sample
Chapter 2.25	***	S5.1 Knowing that the mean of a sample is called a 'point estimate' for the mean of the population; appreciating that accuracy is likely to be improved by increasing the sample size S6.1 Confidence intervals for the mean of a normally distributed population of known variance using σ/n; confidence intervals will always be symmetrical, the confidence level required and the sample size will always be stated
Chapter 2.26	**	S7.1 Recognising when pairs of data are uncorrelated, correlated, strongly correlated, positively correlated and negatively correlated S7.2 Appreciating that correlation does not necessarily imply causation S7.3 Understanding the idea of an outlier; identifying and understanding outliers and making decisions whether or not to include them when drawing a line of best fit
Chapter 2.27	***	S8.1 Understanding that the strength of correlation is given by the pmcc S8.2 Understanding that the pmcc always has a value in the range −1 to 1 S8.3 Appreciating the significance of a positive, zero or negative value of pmcc in terms of correlation of data S10.1 Using a calculator to calculate the pmcc of raw data
Chapter 2.28	***	S9.1 Plotting data pairs on scatter diagrams and drawing, by eye, a line of best fit through the mean point S9.2 Understanding the concept of a regression line S9.3 Plotting a regression line from its equation S9.4 Using interpolation with regression lines to make predictions S9.5 Understanding the potential problems of extrapolation S10.1 Use a calculator to calculate the equation of the regression line

Chapter 2.29	***	R1.1 Representing compound projects by activity networks R1.2 Activity-on-node representation will be used R2.1 Using early time and late time algorithms to identify critical activities and find the critical path(s) R3.1 Using Gantt charts (cascade diagrams) to present project activities
Chapter 2.30	***	R4.1 Understanding that uncertain outcomes can be modelled as random events with estimated probabilities; knowing that the probabilities of an exhaustive set of outcomes sum to one R4.2 Applying ideas of randomness, fairness and equally likely events to calculate expected outcomes
Chapter 2.31	**	R5.1 Understanding and applying Venn diagrams and simple tree diagrams; understanding the P(A), P(A'), P(A∪B) and P(A∩B) notation R6.1 Calculating the probability of combined events: both A and B; neither A nor B; either A or B (or both); to include independent and dependent events
Chapter 2.32	**	R7.1 Calculating the expected value of quantities such as financial loss or gain
Chapter 2.33	*	R8.1 Understanding that many decisions have to be made when outcomes cannot be predicted with certainty
Chapter 2.34	**	R9.1 Understanding that the actions that can be taken to reduce or prevent specific risks may have their own costs; including the costs and benefits of insurance
Chapter 2.35	**	R10.1 Using probabilities to calculate expected values of costs and benefits of decisions; other factors must be considered, for example: the regulatory framework (e.g. compulsory insurance) minimising the maximum possible loss R10.2 Understanding that calculating an expected value is an important part of such decision making
Chapter 2.36	**	G1.1 Sketching and plotting curves defined by simple equations
Chapter 2.37	**	G2.1 Plotting and interpreting graphs (including exponential graphs) in real contexts, to find approximate solutions to problems; including understanding the potential problems of extrapolation G2.2 Interpreting the solutions of equations as the intersection points of graphs and vice versa
Chapter 2.38	**	G3.1 Interpreting the gradient of a straight line graph as a rate of change G3.3 Estimating rates of change for functions from their graphs
Chapter 2.39	***	G3.2 Interpreting the gradient at a point on a curve as an instantaneous rate of change G3.3 Estimating rates of change for functions from their graphs
Chapter 2.40	**	G4.1 Knowing that the average speed of an object during a particular period of time is given by distance travelled/time taken G5.1 Knowing that the gradient of a distance–time graph represents speed and that the gradient of a velocity–time graph represents acceleration
Chapter 2.41	***	G7.1 Understanding that e has been chosen as the standard base for exponential functions; knowing that the gradient at any point on the graph of $y = e^x$ is equal to the y value of that point
Chapter 2.42	***	G6.1 Using a calculator to find values of a^x; the laws of logarithms will not be required G6.2 Using a calculator log function to solve equations of the form $a^x = b$ and $e^{kx} = b$ G8.1 Formulating and using equations of the form $y = Ce^x$ and $y = Ce^{kx}$ G8.2 Using exponential functions to model growth in various contexts
Chapter 2.43	***	G6.1 Using a calculator to find values of the function a^x; the laws of logarithms will not be required G6.2 Using a calculator log function to solve equations of the form $a^x = b$ and $e^{kx} = b$ G8.1 Formulating and using equations of the form $y = Ce^x$ and $y = Ce^{kx}$ G8.2 Using exponential functions to model growth and decay in various contexts

CHAPTER

> **1.1**

VOTING

Do you understand this information?

Does it explain 'first past the post'?

> ## Are elections fair?

This is what happened in the 2015 UK General Election:

What questions do you need to ask?

Do these results seem fair?

First Past the Post
Explained

UKIP 3.8 m votes = 1 MP

Greens 1.1 m votes = 1 MP

SNP 1.5 m votes = 56 MPs

How could you show whether these results are fair or not?

WHAT YOU NEED TO KNOW ALREADY

→ Using fractions and percentages

→ Interpreting tables, charts and diagrams

→ Calculating averages and range

It is easy to say that results are unfair when we disagree with them, but harder to back up our statement when we are challenged. Processing the data helps to make the case, but can also be used to make whatever case you want it to!

MATHS HELP

p304 Chapter 3.2 Spreadsheet formulae

p322 Chapter 3.8 Representing and analysing data

p359 Chapter 3.15 Percentages

PROCESS SKILLS

→ Researching

→ Using spreadsheets

→ Representing data

→ Interpreting data

Investigate

More detail about the results of the 2015 General Election is available on line – search for it!

❭ Most of the parties could be voted for in each of the four countries of the UK – which ones? Which parties could you vote for in only one of the UK's countries? Is that fair?

❭ Set up a spreadsheet to calculate, for each of the parties, how many votes were cast for each seat. You may wish to focus on your own region.

❭ Draw some charts and graphs using the options on the spreadsheet. Do they help to communicate the data?

❭ Calculate the mean and the median number of votes cast per seat. Which do you think best represents the data?

Discuss

❭ What do your charts show? What sort of chart do you think shows the data most clearly – and why?

❭ What is meant by 'first past the post' and 'proportional representation' voting systems? Which is fairer?

❭ Would this data encourage you to consider proportional representation rather than 'first past the post' to elect your Members of Parliament?

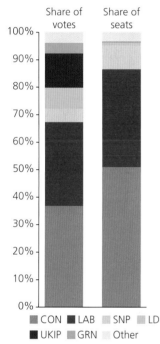

CON   LAB   SNP   LD
UKIP   GRN   Other

QUESTIONS

1 Set up a spreadsheet to draw charts to compare the number of seats actually won by each party in 2015 with the number of seats they would have won if the seats had been distributed in proportion to the number of votes cast. Would a proportional system have affected the outcome of the election?

2 If proportional representation had been used for the data shown, what are the possibilities for a coalition government?

Reflect

Did your research help you understand the problem?

Did you have new ideas?

Were you able to use a spreadsheet?

Were you able to interpret the data in different ways?

CHAPTER

❯ 1.2

RUNNING YOUR OWN CAR

I'll need insurance

❯ Now I've passed my test I want my own car!

Where can I buy a cheap car?

How will I pay for it?

Having your own car means that you have much more freedom over where you go and when, without having to rely on others. There are lots of benefits to having your own car but there are plenty of costs too. This chapter looks at the financial aspects of running a car.

WHAT YOU NEED TO KNOW ALREADY

→ Calculating with money

MATHS HELP

p304 Chapter 3.2 Spreadsheet formulae
p316 Chapter 3.6 Personal finance
p352 Chapter 3.13 Formulae and calculation

PROCESS SKILLS

→ Estimating costs
→ Researching costs
→ Making decisions
→ Evaluating decisions

Buying a car

It is easy to find a car to buy in newspapers (still), online or at a local showroom. The difficult bit is finding the money to pay for it.

Discuss

> What make and model of car do you want?

> How much can you afford to pay?

> What factors may affect what car you buy?

Investigate

Research the cars for sale in your local area.

What is the price range of cars that are for sale? What does the price depend on? What kind of things make the car more expensive? What brings the price down?

Discuss

How can you pay for the car you want? Do you have any savings? Would a family member help you out? You have to wait until you are 18 to get a loan. Do you have a part-time job?

You may have the skills to do up an old car, which would reduce the cost substantially.

Running a car

Discuss

> What are the costs involved in running a car? How much are they likely to be for the car you have your eye on?

You will have to pay road tax for the car. At the time of writing, that costs more for cars with more powerful engines. You need to have insurance and this is more expensive for young drivers. You can choose comprehensive insurance or save by just having third party insurance.

> What is the difference between them?

The car needs an MOT certificate once it is three years old, to show it is roadworthy, and you should have some money available for repairs and maintenance. The cost of fuel depends on how many miles you do and the size and efficiency of the engine.

Investigate

Work out how much the likely costs will be for a car that you have identified, and which you can reduce if you need to. You may wish to work this out monthly rather than yearly. You can pay insurance and the tax by instalments, so you do not need to find all of the money in one go. You could set up a spreadsheet for this so that you can quickly see the effect of reducing or increasing a cost.

Funding it

You may have had help buying the car, but it is likely that you will have to fund the running costs yourself. What income do you have? Can you get a part-time job? How much is it likely to pay? Will your friends pay towards the petrol if you give them lifts? How much would you ask for?

Investigate

Research opportunities locally for part-time work to generate some income. You should also consider how the time spent might affect your studies. Some jobs may want you all through the year, others are seasonal so need you at Christmas or during the summer. Remember to include likely contributions from friends and family. Again, a spreadsheet may be helpful.

QUESTIONS

1 Jan buys a car for £500 and insurance for £700. The car tax costs £30. She allows £200 for repairs and anticipates that she will spend £50 per month on petrol. She hopes to pay off the car, having borrowed money from her parents, in the first year. Her parents have said she can have the equivalent of her bus fares for college (£14 per week) towards running costs.

   How much extra does she need to find each month?

2 Jen decides she needs a part-time job to fund running a car. She can get a job at the local superstore earning £3.95 an hour. They have 12 hours of work each week to offer her. If her costs are similar to Jan's in question 1, will she be able to run the car?

Reflect

Did your research contribute to the discussion?

Were you able to use mathematics to solve problems?

Were you able to see different approaches to the problem?

Were you able to adjust your ideas?

RUNNING SMART

How does the staggering work?

Why did Kelly Holmes run in lane 2?

> Which lane is best?

Dame Kelly Holmes won two gold medals at the 2004 Olympic Games

Lane 1 is the shortest distance

Why is the track that shape?

To win the 800 m, she completed nearly the whole of the final lap in lane 2

Geometrical shapes are used in most sports to create the pitch or area that they are played on. You have probably never thought about the calculations and measurements that are involved. Perhaps they are to make the game fair, or there may be other reasons behind the measurements used.

WHAT YOU NEED TO KNOW ALREADY

→ Calculating perimeters and areas of compound shapes

PROCESS SKILLS

→ Representing a problem
→ Conjecturing about decisions
→ Deducing from information
→ Making assumptions

MATHS HELP

p306 Chapter 3.3 Perimeter and area

Investigate

You may know the answers to some of these already but there may be more than one answer.

> What shape is a running track?

> How is this structured as a combination of shapes?

> How long is each section of the track? Is there a rule?

> What do you need to know to find the area and perimeter of a running track?

> How wide is each lane on a running track?

Discuss

The lanes are different lengths so how is this managed in middle-distance running races?

> What about the 800 m specifically?

Investigate

What other activities take place in the centre of a running track during an athletics competition?

> How much space do these activities require?

> Can you find out why running tracks are usually 400 m?

Discuss

> What is the problem with running the whole of the last lap of a race in lane 2, when everyone else is running in lane 1?

> In which sections of the track is this most challenging?

> Why might an athlete choose to run in lane 2 throughout most of the final lap?

> Is it better to have a track with longer straights or longer bends? What would be the advantages and disadvantages of each design? Think about the competitors as well as other factors that may affect the design.

> If you know the lengths of the straights of a track, how can you work out the lengths of the bends? Or the distance between the straights?

QUESTIONS

1 One stadium designs their 400 m track like this:

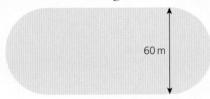

60 m

a What is the total distance round the two bends on this track?

HINT

What shape do the two bends make when put together?

b One lap is 400 m. How long is each straight of the track?

c The lanes on this track are 1.22 m wide.

Work out the difference between the distance round this track in lane 2 and lane 1.

What assumptions did you make in order to work this out?

d Work out the area of the ground contained within the track.

2 Another stadium designs a track with a width of 65 m. The lanes are 1.22 m wide.

a What is the difference between the distance round this track in lane 2 and lane 1?

b How far down the track does an athlete running a lap in lane 8 start to ensure their lap is the same distance as an athlete running in lane 1?

What assumptions did you make in order to work this out?

c What is the area of the ground contained within the track?

3 A stadium designer produces a track that has bends and straights each 100 m long.

a What is the distance between the straights so that each of the sections of the track is 100 m long?

b What is the area of ground inside the track, for field events?

What assumptions did you make in order to work this out?

4 In 1986 the men's javelin was redesigned; its centre of gravity was moved forward by four centimetres. This shortened throwing distances by approximately 10 per cent. This was done because the men, following a world record of 104.80 m by East Germany's Uwe Hohn in 1984, were in danger of throwing the javelin beyond the space available in normal stadiums.

Source: http://www.iaaf.org/disciplines/throws/javelin-throw

A javelin thrower takes their run-up across the track. What are the dimensions of the track if there is enough space for a throw of 104.8 m?

What assumptions did you make in order to work this out?

Reflect

Were you able to see different points of view?

How has your research helped you with your designs?

Could you use your mathematical knowledge to solve these problems?

Did you think of any ideas?

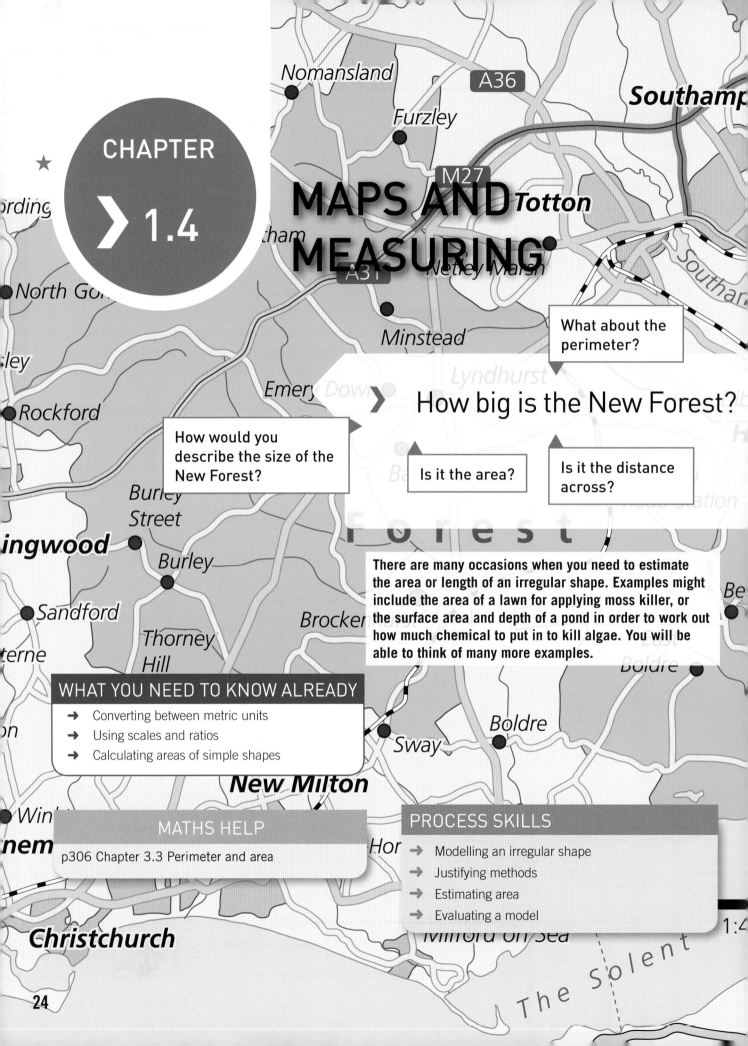

MAPS AND MEASURING

How big is the New Forest?

> **What about the perimeter?**

> **How would you describe the size of the New Forest?**

> **Is it the area?**

> **Is it the distance across?**

There are many occasions when you need to estimate the area or length of an irregular shape. Examples might include the area of a lawn for applying moss killer, or the surface area and depth of a pond in order to work out how much chemical to put in to kill algae. You will be able to think of many more examples.

WHAT YOU NEED TO KNOW ALREADY

→ Converting between metric units
→ Using scales and ratios
→ Calculating areas of simple shapes

MATHS HELP

p306 Chapter 3.3 Perimeter and area

PROCESS SKILLS

→ Modelling an irregular shape
→ Justifying methods
→ Estimating area
→ Evaluating a model

HINT

Because the reproduction of the map on the next page is not the same size as the original, the scale 1:400000 will not be correct.

You can, however, use the marked 0__ 2.5__ 5km line to determine the scale.

Discuss

How could you work out these measurements for the New Forest?

Which shapes could you use to approximate the shape of the forest on the map?

To find the area of the forest, you could start by approximating its shape with a very simple shape, such as the red triangle shown here.

❯ How good an approximation do you think the triangle is for the shape of the forest?

❯ Do you think it is an underestimate or an overestimate?

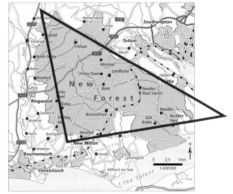

You may know two ways to find the area of a triangle:

– Calculate half the base times the perpendicular height $\left(\frac{1}{2}bh\right)$

– Calculate half the product of two sides multiplied by the sine of the angle between them $\left(\frac{1}{2}ab\sin C\right)$

In either case, you have to do some measuring on the diagram, and use the scale to get the lengths you need for the calculations.

HINT

You will need a ruler.

Discuss

❯ How can you use the scale on the map to get the real-life measurements of the base and height of the triangle?

The area of the triangle works out as approximately 320 km^2.

Can you explain to someone else in the group how to get that answer (approximately)?

Discuss

❯ What other simple shapes could you draw on the map to approximate the area of the New Forest?

Investigate

❯ Try drawing a rectangle to approximate the area and calculate this area using base multiplied by height.

〉 Is the area you get bigger or smaller than that given by the triangle approximation?

〉 Which do you think is closer to the actual area of the forest?

〉 What other shapes could you try?

Whatever shape you draw, you need to be able to find its area.

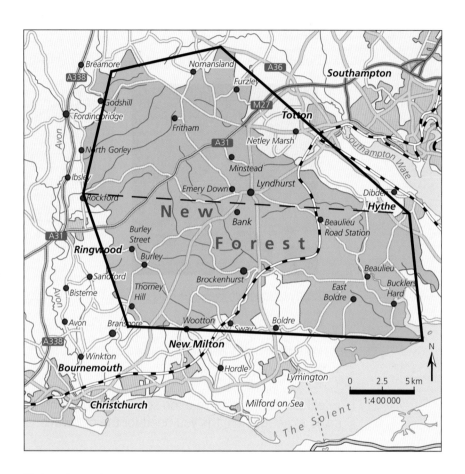

If the shape has straight sides, it can always be split into triangles whose areas can be found, so you could, for example, use the black hexagon shown here.

The dashed line shows the hexagon split into two quadrilaterals, one of which is a trapezium.

Investigate

〉 Use the internet to research a more accurate figure for the area of the forest.

〉 Use perimeter as your measure and estimate the perimeter of the forest.

HINT

The formula for the area of a trapezium is half the sum of the parallel sides multiplied by the distance between them.

QUESTIONS

1 There are $1\,000\,000$ m^2 in 1 km^2. Explain why.

2 Use a different method to find the area of the red triangle on the New Forest map.

3 Draw a circle that approximates to the New Forest and use it to calculate the area of the forest using the formula πr^2.

4 This diagram shows a simplified version of the map of a field, drawn to a 1:10 000 scale (which means that every centimetre on the map represents 10 000 cm in real life).

   a Find the perimeter of the field.

   b Calculate the area of the field in hectares. (A hectare is 10 000 m^2.)

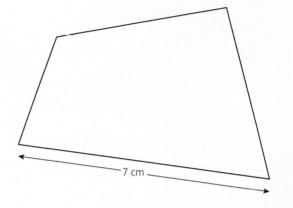

7 cm

Reflect

Were you able to see different approaches?

Were you able to use maths to solve the problem?

Were you able to find helpful information on the internet?

Were you able to think of new ideas?

ICE CREAM CONES

There are different types of wafer

It's handy if it stands up

> ## What is the optimal size and shape of an ice cream cone?

If it's too small at the top the ice cream falls off!

My daughter drops hers if the cone is too fat

There is a huge range of cones that ice cream sellers can buy. But how do they decide? What factors do they consider? Where do they start? What would you choose? Why? Would your choice be the most popular choice for the majority of customers? Many problems that need to be solved are affected by several factors. It is important to make sure you have considered them all in order to decide which ones are the priority for further investigation.

WHAT YOU NEED TO KNOW ALREADY

→ Finding surface area of cylinders, cones and frustums
→ Substituting into formulae
→ Using similarity and ratio
→ Using Pythagoras' theorem

PROCESS SKILLS

→ Modelling a shape
→ Evaluating results
→ Justifying decisions
→ Criticising designs

MATHS HELP

p309 Chapter 3.4 Surface areas
p312 Chapter 3.5 Triangles and similarity

The image on the previous page shows a variety of 'cones'. Although they are called cones they are not all cone-shaped! To avoid confusion in this chapter:

- **Cornet** means the hand-held container for ice cream made of wafer
- **Cone** means the mathematical shape.

Investigate

> By eating ice cream, or by researching on the internet, draw up a list of images representing all the different shapes of cornet that are available. Describe each shape. Try to include the words cylinder, sphere, cone and frustum in your descriptions.

Discuss

> Draw up a list of **factors** that an ice cream seller might want to consider when deciding what cornets to buy.

Here are just a few factors you might consider.

> **HINT**
>
> When there are a lot of factors, it is a good idea to change one factor at a time.

The height of the cornet

Discuss

> What factors do you need to consider when deciding the optimum height for a cornet?

For now, assume the height of the cornet is 11 cm.

The size of the opening that holds the ice cream ball

Discuss

> The ice cream can sit on the top, inside the cornet or somewhere in between. Which is best?

A   B   C

Safety of the ball is part of the answer, but taste is also a factor. The safest option is C above, but would the customer taste too much wafer in proportion to the amount of ice cream?

For now, fix the size of the opening so that the ball sits in a similar way to the vanilla ice cream in B above.

Investigate

> Draw a large circle to represent the **ball of ice cream**. Draw a line across the ball (a **chord**) to show where you think the top of the cornet should be to fit snuggly, as in example B. Here is a diagram showing four possible positions for the top of the cornet.

> Measure the diameter of your circle and the length of your chosen chord.

> Now use similarity to calculate the length this chord would be on a ball of diameter 5 cm. Use this as the diameter at the top of your cornet. Call it d.

The shape of the cornet

Some cornets are simple shapes like the cornet in example A above. Others are 'compound' shapes like some of these. For now, consider only cornets that are simple shapes:

- a cone

- a cylinder

- a frustum.

The ratio of ice cream to wafer

We have now narrowed our investigation down to three cornets to investigate the taste factor.

Investigate

> Make drawings showing the dimensions of each of your cornets. There is one dimension missing that you will need for the frustum. Call it w. It is the diameter of the base.

> Choose a value for w. Give reasons for your choice.

In order to work out the ratio of ice cream to wafer, you will need to work out the volume of ice cream in a ball and the volume of wafer in each cornet.

The volume of a sphere is $\frac{4}{3}\pi r^3$ where r is its radius.

The volume of wafer = surface area of cornet × thickness of wafer.

Assume the thickness of wafer to be 2 mm.

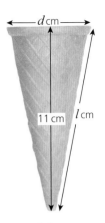

HINT

You know the height of the **cone** is 11 cm, but to calculate the surface area of a cone you need the length of the side of the cone. You can use Pythagoras' theorem to find this.

Investigate

> Calculate the volume of ice cream in a ball.

> Calculate the volume of wafer in both the **cone** and cylinder you drew previously.

> Work out the ratio of ice cream to wafer in both of these cases.

To calculate the surface area of the frustum first find the surface area of the cone that the frustum is part of. This means you need to find the full height of the cone.

You can find the height, h, of the cone using similarity. The large beige and green cone is mathematically similar to the green cone.

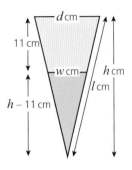

This means the ratio of top width (d) to height (h) is the same for both cones.

$d:h$ is the same ratio as $w:h-11$

HINT

You will need to subtract the surface area of one cone from the other to find the surface area of the frustum.

Investigate

> You know d and w. Find h.

> Now find l using Pythagoras' theorem.

> Find the surface area of the frustum.

> Work out the ratio of ice cream to wafer in this case.

The taste test

This is key. Experimentation is the only way to discover what the right ratio is. But an ice cream seller would want to choose a cornet that achieved a ratio close to it.

Discuss

> By considering your three designs, and judging by your past experience of eating ice cream, what do you think is the optimum ratio of ice cream to wafer?

> What cornet shape and size do you think might achieve this ratio?

Investigate

Find out how ice cream cornets are made.

QUESTIONS

1 Change the dimensions of the cone-shaped ice cream cornet to increase the ratio of ice cream to wafer. It should still be designed to take a 5 cm ball of ice cream.

a State the dimensions of your cone. Justify your choices for each dimension.

b Calculate the ratio of ice cream to wafer for your design.

2 Choose a cornet that is a compound shape.

a Provide a sketch of your design, including dimensions.

b You could estimate the surface area of your design by splitting it into different sections and estimating the area of each section. Explain what shapes you would use for each section and how you would find the surface area of each.

c Estimate the surface area of your cornet.

3 Design three ice cream cornets of your own that are not yet on the market.

a Provide a sketch of your designs including dimensions.

b What are the advantages and disadvantages of each of your cornets?

c Calculate the surface area of wafer in each design.

Reflect

Were you able to think of different approaches?

Did you have any good ideas?

Were you able to use the ideas suggested?

How have you been able to use your mathematical understanding to solve a real-life problem?

LINES MAKING LOGOS

> ## A picture paints a thousand words

How can I store a design on my computer?

I want to be able to rotate it

How can I make different sizes of it?

MITSUBISHI MOTORS

WHAT YOU NEED TO KNOW ALREADY

→ Plotting and reading coordinates
→ Identifying parallel lines and lines of symmetry

Coding a design so that it can be stored and manipulated in a computer has many uses. A simple icon can be coloured and sized differently or shown from a different perspective. A landscape in a computer game can be coded so that it can be viewed in different ways.

PROCESS SKILLS

→ Representing an image using algebraic equations
→ Deducing equations
→ Comparing representations
→ Creating designs

MATHS HELP

p301 Chapter 3.1 Straight line graphs

A designer is making a logo for a company's new range of products.

The designer produces this image and will add the product name across the top.

Discuss

> What are the advantages and disadvantages of using only two colours for a logo?

> What are the advantages and disadvantages of using only straight lines for a logo?

> How do companies ensure that they choose a logo that can be replicated and transformed easily?

> How do printers and other manufacturers ensure they can produce an identical logo each time?

> How can we convert the designer's image into a mathematical object that a computer can understand?

Investigate

Here is the image without the shading in the original picture:

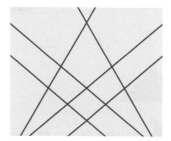

And here is the same picture shown on a coordinate grid:

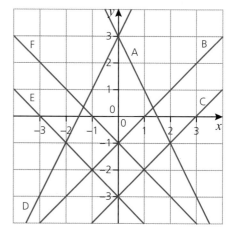

HINT

The gradient of a line is the steepness.

We can find it by choosing two points on the line and calculating

change in y-coordinate

change in x-coordinate

> How many lines make up the logo design when you strip out the shading?

> How do the lines relate to each other?

> Can you find the gradient of each line in the design? Which ones are the same or related to each other?

> Can you find the y-intercept of each line A–F in the design?

Discuss

> Could you predict where the y-axis is positioned by looking at the original image?

> What do you notice about the gradients of the lines that are parallel?

> What do you notice about the y-intercepts of the lines?

> How is the symmetry of the design shown in the gradients and y-intercepts of the lines?

> Can you predict the gradient of a line if you know the gradient of its mirror image?

Investigate

> Find or design a logo using only straight lines. Use graphing software to draw it and experiment with different colourways.

> Try repeating the logo either by translating it (sliding it), reflecting it or rotating it.

> Experiment with different ways of using the logo.

QUESTIONS

1 **a** Give the equations of the six lines that make up our design in the form $y = mx + c$.

   b The designer adds a line, G, with equation $y = x + 1$, and a line, H, that is its mirror image when reflected in the y-axis.

   Give the equation of the line, H, that is the mirror image of G when reflected in the y-axis.

2 Here is another design for a logo, shown on a coordinate grid:

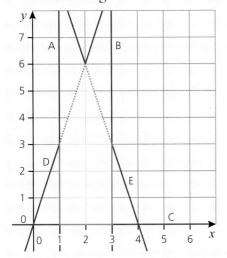

Form a list of equations to produce this design.

Give your equations in the form $y = mx + c$.

HINT

We cannot see line E's y-intercept but we can work it out!

3 A computer programmer for a textile company has input the following equations for a new design:

$y = 0.5x + 3$

$y = 2x - 3$

The programmer wants the design to be symmetrical about the y-axis.

   a State the equations of the other two lines that will be required to make a symmetrical image.

   b Draw the image produced by all four lines on coordinate axes.

Reflect

Were you able to use different approaches?

Were you able to use equations of lines to solve the problem?

Did you have any ideas?

Have you found it easy to create a design?

★

CHAPTER

〉2.1

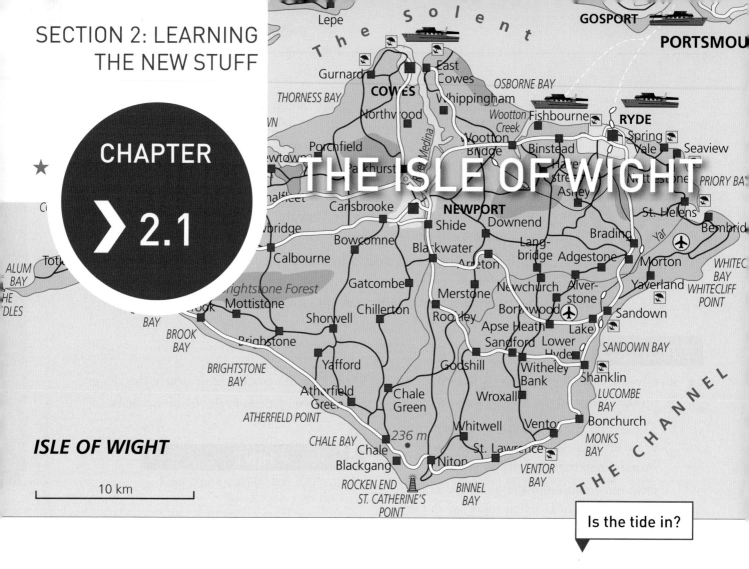

THE ISLE OF WIGHT

ISLE OF WIGHT

|— 10 km —|

Is the tide in?

WHAT YOU NEED TO KNOW ALREADY

→ Working with large numbers
→ Estimating amounts

MATHS HELP

p306 Chapter 3.3 Perimeter and area
p376 Chapter 3.20 Estimating

PROCESS SKILLS

→ Modelling the problem
→ Checking by recalculating in a different way
→ Estimating quantities
→ Evaluating the method and answer

> ## Can everyone in the world fit on the Isle of Wight?

What could 'fit on the island' mean?

Are they all standing up?

Making an estimate requires mathematical modelling. You need to find numbers, measurements and shapes that you can use to answer the question. You have to make some assumptions, such as using approximate measurements to simplify the working, which may affect the validity of your answer. The last step is to look back and review your method and answer to see if the assumptions were reasonable.

> What is your gut feeling about this question?

> How could you work out whether people would fit?

> What do you need to know?

> What mathematics could you use?

Note down your ideas and questions.

The area of the Isle of Wight is 380 km². There are 7.4 billion people in the world.

Investigate

> How big is a square kilometre? Can you compare it with an area you are familiar with?

> How could you model the area of ground needed for each person? Circle? Rectangle? Hexagon? What assumptions do you have to make?

> Do you get the same answer when you use different ways of working out whether everyone will fit?

> What if everyone in the world would fit on the island? Would there be space left over for more people?

> What if they wouldn't fit? Can you find an island that they *could* stand on?

> Is your answer sensible? Were the assumptions reasonable?

The modelling cycle

You have used the modelling cycle to decide whether the population of the world would fit on the Isle of Wight.

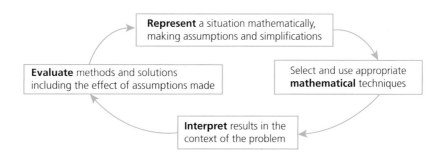

You **represented** the Isle of Wight using 2-D shapes, and the amount of space taken up by one person using a measure of area.

You **used mathematics** to calculate the area of the shape, and to divide by the area needed by one person.

You compared your answer with an estimate of the world's population and **interpreted** what you found to decide whether there is enough space.

You **evaluated** your assumptions and decided whether to try to improve your answer.

WORKED EXAMPLE

People come to Trafalgar Square to celebrate the New Year.

Estimate how many people can stand in Trafalgar Square.

The distance between the hand rails on the steps, the darker

lines extending the full length of the steps, is 65 feet.

You should make clear any assumptions you make and evaluate them.

SOLUTION

We need the area available and the amount of space taken up by a person.

Estimating the area:

In Chapter 1.4 you estimated the area of the New Forest by representing it by familiar 2-D shapes. Here the area is approximately a rectangle.

We need the length and width of the rectangle so must estimate them from — **Representing**
the aerial view using the fact that the distance across the steps is 65 feet.

Width, including the trees, is about 6 times the width of the steps.

Length, including the area at the top of the steps, is also about 6 times the width of the steps.

Area of rectangle = 6 × 65 × 6 × 65 square feet — **Using mathematics**
= 152 100 square feet

People won't be allowed to stand in the pools or on the plinth of the statue.

We subtract those areas from our estimate.

The pools are approximately squares of side 70 feet and the plinth is — **Representing**
approximately a square of side 60 feet.

Estimate = $152\,100 - 2 \times 70^2 - 60^2$ — **Using mathematics**
≈ 140 000 square feet

The area taken up by one person standing is approximately 1 square foot.

So the number of people who can fit on Trafalgar Square is about 140 000. — **Interpreting**

Assumptions: — **Evaluating**

Trafalgar Square is a rectangle – reasonable as the extra bits compensate for the missing bits if you draw a rectangle over the image.

Other shapes are squares – reasonable as they are designed to be close to squares in shape.

Distances used – these are likely to be accurate to 1 significant figure. The 65 feet is measured but the scaling involved is an estimate so 1 significant figure accuracy is reasonable.

A person takes up 1 square foot – obviously this varies but it is a reasonable average amount to 1 significant figure.

Overall it might be better to give the estimate to 1 significant figure so 100 000 people and it is probably an under-estimate as it was rounded down.

Investigate

Whether or not everyone in the world can fit on the Isle of Wight, they certainly couldn't live there!

> How much area do people need to live and work on?

Investigate

> What does the diagram below show?
> What does the position of Greenland on the graph tell us about it?
> Compare England and Scotland using the information on the graph.

Using information in graphs to make estimates

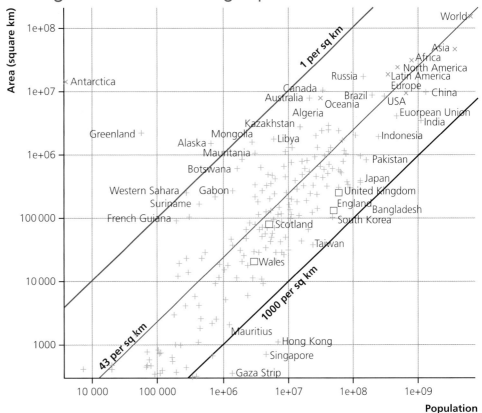

Reading the labels on the axes of a graph is the first step. You can see that the area of each country is plotted against its population to create a scatter graph.

Look at the scales on the axes and notice that they are non-linear. The numbers are marked at equal intervals but each is 10 times the previous number. Scales which use a multiplier like this are called logarithmic scales. Why is this scale used?

In this case the magnitude of the numbers involved varies so much that you would not be able to see the patterns in the data with a linear scale.

Notice that Greenland has a population between 10 000 and 100 000 and an area between 1 million and 10 million square kilometres. It has less than 0.1 person per square kilometre (on average) and so is very sparsely populated. You could deduce that from its position, low on the population axis and high on the area axis. It is the least densely populated country in the world after Antarctica.

Notice that England has a larger area than Scotland and a much larger population which makes its population density greater.

Discuss

This diagram gives you some information about the countries labelled. Do you think their position on the graph tells you whether they could support their own population? Use the results of your previous investigation, as well as finding out more about the countries, to help you decide.

WORKED EXAMPLE

A possible estimate of the area needed to support one person, eating a mixed diet, is 0.5 hectares of usable land. Obviously, this figure increases if the conditions are difficult and decreases if they are favourable or the person has a meat-free diet.

A hectare = 10 000 m^2.

a   Work out how many people can be supported by a square kilometre of land using this estimate.

b   Use the information on the graph to select one example of a country that you estimate:

   (i)   should be able to support their population

   (ii) is unlikely to be able to support their population.

SOLUTION

a   1 hectare = 10 000 m^2

   $$1 \text{ km}^2 = 1000 \times 1000 \text{ m}^2$$
   $$= 100 \times 10\ 000 \text{ m}^2$$
   $$= 100 \text{ hectares}$$

   So number of people supported = 100 ÷ 0.5
   = 200 people

b   The graph shows lines for 1 person per km^2, 43 people per km^2 and 1000 people per km^2.

   You could add lines for 10 people per km^2 and 100 people per km^2 as they pass through the intersections on the grid.

   (i)   Scotland should be able to support its own population as it lies between the 43 and 100 people per km^2 lines. Any country above the 100 line could be selected here. However, the graph does not tell the whole story as geographical characteristics need to be taken into account as well.
   Some countries close to the 100 people per km^2 line, but below it, such as Wales, China and the European Union, could also be selected as likely to be around 200 people per km^2.

   (ii) Singapore is unlikely to be able to support its population as it lies below the 1000 people per km^2 line and is close to the 10 000 people per km^2 line. Countries such as England, Japan and Mauritius could also be selected here.
   Again, the graph does not tell the whole story as human and physical geographical issues affect the answer.

Making an estimate gives you an idea of what to do next. The next step depends on the context. In the example above you may wish to investigate some of the countries further to see if they do support their population. Perhaps Singapore manages through careful use of land and a diet that does not rely on meat. Perhaps instead they rely on importing food or many of the people go hungry. Further work is needed to find out.

Estimating the number of people who can gather in Trafalgar Square gives you an idea of how many police to have in attendance. Knowing whether this is an under or over-estimate is useful in managing the risk of so many people in one place at a time.

QUESTIONS

1 Find Hong Kong on the graph on page 39.

Estimate how many people can live there based on the area needed to feed and house the population.

2 Find Australia on the graph on page 39.

Estimate how many people can live there based on the area needed to feed and house the population.

Research the potential land usage in Australia. Does that affect your answer?

St Mary's, the largest of the Scilly Isles

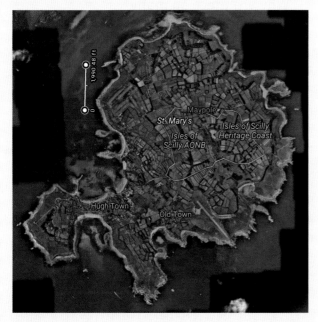

E 3 Use the aerial view of St Mary's in the Scilly Isles.

Estimate the number of people that can be supported on St Mary's.

You may assume each person needs 0.25 hectares of land.

Show details of any other assumptions you make, and your calculations.

E 4 Use the map above of St Mary's in the Scilly Isles.

Estimate the time to walk around the coast of the island.

You may assume that someone walks at a speed of 3 feet per second.

Show details of any other assumptions you make, and your calculations.

CHAPTER
> 2.2

SUN CREAM

What affects the amount of sun cream you use?

> ## Are you using enough sun cream?

What does 'SPF 30' actually mean?

Why are there so many types of sun cream?

Why do we use sun cream?

Mathematical modelling is involved in any estimation work. In this task, the assumptions are about how much of a product is used, and thus how long a bottle of it will last. The questions suggest different approaches, including the use of various formulae as a model. Reviewing your method and answer to see if the assumptions were reasonable is an important step in the modelling process.

WHAT YOU NEED TO KNOW ALREADY

→ Substituting values into given formulae
→ Converting between metric units
→ Calculating percentage error

MATHS HELP

p309 Chapter 3.4 Surface areas
p352 Chapter 3.13 Formulae and calculation
p359 Chapter 3.15 Percentages
p376 Chapter 3.20 Estimating

PROCESS SKILLS

→ Modelling a problem using mathematics
→ Representing a real-life situation mathematically
→ Interpreting results in context
→ Evaluating the methods, solutions and assumptions

Investigate

NHS guidelines suggest that in order to achieve the SPF (Sun Protection Factor) level advertised on a sun cream bottle, the sun cream needs to be applied at a thickness of 2 mg of product for each cm^2 of skin.

> What are the NHS guidelines for how regularly you should apply sun cream?

Investigate

> Since sun cream is a liquid it tends to be sold in millilitres. How many millilitres (ml) are there in one milligram (mg)?

> The average body surface area (BSA) for men and for women is taken to be:

Women	$1.6 \, m^2$
Men	$1.9 \, m^2$

What is the BSA for a man in cm^2?

What is the BSA for a woman in cm^2?

> NHS guidelines suggest that sun cream needs to be applied at a thickness of 2 mg per cm^2 of skin. Using the BSA figures in cm^2 for men and women above, calculate how many millilitres of sun cream should be used for each application of sun cream.

> Using your previous answers work out how long this 200 ml bottle of sun cream will last.

> What assumptions have you made?

Discuss

> Were the results for how long a 200 ml bottle of sun cream will last what you expected?

> Are your results an overestimate or an underestimate?

> How could you adjust your calculations to improve your results?

> How regularly do you apply sun cream while on holiday?

> How many bottles of sun cream do you take on holiday with you?

> How does the type of holiday affect the amount of sun cream you apply?

WORKED EXAMPLE

Using your knowledge of the surface area of 3-D shapes such as spheres, cylinders and cuboids estimate the BSA of a volunteer in the class.

How could you improve this estimate?

> Surface area of a sphere, radius r, $= 4\pi r^2$
>
> and
>
> curved surface area of a cylinder, radius r, height h, $= 2\pi rh$

SOLUTION

There are many ways to model the human body using mathematical shapes. You may like to try another way to do it.

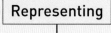 **Representing**

Using mathematics

Head, sphere radius 10 cm, surface area $= 4 \times \pi \times 10^2$
$$\approx 1200 \text{ cm}^2$$

Body, cuboid 25 cm by 15 cm by 60 cm, surface area $= 2 \times (25 \times 15 + 15 \times 60 + 60 \times 25)$
$$\approx 5500 \text{ cm}^2$$

Legs, 2 cylinders, radius 6 cm, length 90 cm, curved surface area $= 2 \times 2 \times \pi \times 6 \times 90 \text{ cm}^2$
$$\approx 6500 \text{ cm}^2$$

Arms, 2 cylinders, radius 4 cm, length 60 cm, curved surface area $= 2 \times 2 \times \pi \times 4 \times 60 \text{ cm}^2$
$$\approx 3000 \text{ cm}^2$$

Surface area of the whole body $\approx 16\,000 \text{ cm}^2$ — **Interpreting**

The estimate could be improved by breaking the body down into more pieces; using a more accurate value for π; using more accurate measurements. — **Evaluating and refining**

Estimating using formulae

The previous method gives an estimate which is close to the average given earlier in the chapter. It does, however, need quite a few measurements! Human bodies are broadly similar in shape and so some formulae have been devised to calculate surface area using two measurements, height and weight. They are estimates because human bodies do vary in shape as you can tell by looking at your classmates! The formulae are also models of a human body, algebraic ones rather than geometrical ones.

WORKED EXAMPLE

Here are four suggested formulae for calculating the BSA of the human body:

DuBois & DuBois: $A = 0.007184 \times H^{0.725} \times W^{0.425}$

Gehan: $A = 0.0235 \times H^{0.42246} \times W^{0.51456}$

Haycock: $A = 0.024265 \times H^{0.3964} \times W^{0.5378}$

Mosteller: $A = \sqrt{\dfrac{H \times W}{3600}}$

a   Use these formulae to calculate your BSA.

Height, H (cm)	Mass, W (kg)	DuBois & DuBois	Gehan	Haycock	Mosteller

b   How do these results relate to your earlier estimations? Calculate the percentage error between your estimated BSA and the BSA you have calculated using these formulae.

c   What are the advantages of using a formula, rather than modelling the body as a series of standard components?

d   Which formula is the easiest to use? Why do you think that?

SOLUTION

a

Height, H (cm)	Mass, W (kg)	DuBois & DuBois	Gehan	Haycock	Mosteller
165	70	1.77	1.81	1.80	1.79

b   Dubois & Dubois percentage error $= \dfrac{1.62 - 1.77}{1.77} \times 100 = -8.47\%$

Gehan percentage error $= \dfrac{1.62 - 1.81}{1.81} \times 100 = -10.5\%$

Haycock percentage error $= \dfrac{1.62 - 1.80}{1.80} \times 100 = -10\%$

Mosteller percentage error $= \dfrac{1.62 - 1.79}{1.79} \times 100 = -9.50\%$

c   When using a formula, fewer measurements are needed; only the height and mass of an individual needs to be measured, this reduces the risk of calculation and measurement errors and therefore the results should be more accurate. The results from using the formulae are quite consistent.

d   The Mosteller one is easiest to use and gives a result, in this case, that is within the range of the others.

Discuss

Human BSA has many important applications in medical practice. What are these applications?

QUESTIONS

1 The distance, in km, to the horizon, from a height of h metres above sea level can be estimated using the formula $d = 1.6 \times \sqrt{5h}$

Estimate how far you could see from the top of a light-house.

2 Grains of rice are piled up on a tennis court. Estimate how many grains of rice there are in the pile.

You should list the assumptions you make.

3 A human cell is about 50 micrometres wide. Estimate how many are in your hand.

Remember to list the assumptions you make.

4 Estimate how many jelly beans would fit in a double decker bus.

Give one way that you could improve the accuracy of your estimate.

E 5 The bar graph below shows the number of people who died in road accidents in 2014 and 2015.

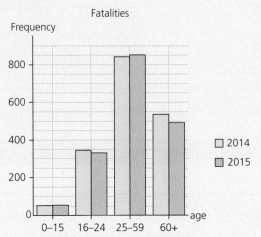

Estimate the total number of years of life lost by those who died in road accidents in 2015.

Show details of your assumptions and calculations.

Give one way that you could improve the accuracy of your estimate.

PIANO TUNERS

What about schools and concert halls?

> ## How many piano tuners are there in Chicago?

I've no idea!

Not many – who has a piano these days?

Quick estimates of numbers or quantities, such as the amount of food wasted in the USA each year, are often called Fermi estimates after the Italian physicist Enrico Fermi (1901–1954), who gained a Nobel Prize in 1938 for his work on neutrons.

Fermi was very good at making 'back of the envelope' calculations, such as finding the number of piano tuners in Chicago – the classic example of a Fermi estimate. Such estimates can be surprisingly accurate and can have scientific and economic importance.

WHAT YOU NEED TO KNOW ALREADY

→ Calculating with powers of 10

PROCESS SKILLS

→ Modelling a problem using powers of 10
→ Estimating quantities
→ Conjecturing about quantities
→ Evaluating solutions

MATHS HELP

p376 Chapter 3.20 Estimating

The method depends on making sensible estimates about the numbers or quantities involved. This can be quite difficult, but one simple method is to estimate using powers of 10. For example, if we need an estimate for the height of a building, we can choose a power-of-10 answer: is the building 1 metre tall, 10 metres tall, 100 metres tall or 1000 metres tall? It is quite easy to make a choice from these, even if you don't have an accurate idea of how tall the building is.

Discuss

Make 'power-of-10' estimates of:

> the number of people in the UK

> the number of people in the world

> the area of a football pitch

> the average household income in the UK.

You may know more accurate values for some of the quantities involved – that's fine and you can use these values in making estimates, but, if you don't know, a power-of-10 estimate is a good way forward. When such estimates are used in calculations, the overestimates are likely to compensate for the underestimates, so the result can be quite accurate, or at least of the right order of size.

WORKED EXAMPLE

How many piano tuners are there in Chicago?

SOLUTION

How many people are there in Chicago? Is it ten thousand, a hundred thousand, a million or ten million? A million seems the best of these.

What proportion of these people own a piano? One-tenth, one hundredth, one thousandth, one ten-thousandth? Perhaps one hundredth.

So our estimate of the number of pianos in Chicago is one million divided by 100, that is 10 000.

How often do pianos need tuning? Every year, every 10 years, every 100 years? Every year is good.

How many pianos can a tuner tune in a year? One, 10, 100, 1000? A good choice may be 100.

So, our estimate of the number of piano tuners in Chicago is $10\,000 \div 100 = 100$.

Apparently the number of piano tuners in Chicago is in fact approximately 80 – so our quick estimate is certainly in the right order of size.

You may have made different choices for some of the estimates, but the final answer could easily end up being the same. It may not matter whether you got 10 or 1000 instead, but you probably didn't estimate 10 000 or more.

Discuss

When is a power-of-10 estimate sufficiently close? Can you think of some situations when a power-of-10 estimate is fine and some where you would want a more accurate estimate?

Investigate

Use a similar method to estimate the number of schools in the UK.

You will need to make estimates of:

> the number of people in the UK

> the proportion of the population who are of school age

> the average number of pupils per school.

Use power-of-10 estimates unless you have more accurate knowledge of any of these.

What if you can get closer than a power of 10?

You may know some of the numbers you need in your calculation more accurately than the nearest power of ten. Calendar or other time facts such as the number of days in a year or seconds in a minute are better used as 365 and 60 respectively. You may know the population of the UK or of the world to a greater accuracy than a power of ten. The radius of the earth or distance of the Earth from the Sun are also facts you may have committed to memory. In those cases it is appropriate to use the more accurate number in your estimate.

WORKED EXAMPLE

Estimate how far you walk in a lifetime. State any assumptions you make.

SOLUTION

10 000 steps is considered a target amount for people wishing to get fit.

So, assume that someone takes 5 000 steps in an average day.

One step is roughly a metre so one day's worth of walking is 5 000 m = 5 km

One year = 365 days, life expectancy is approximately 80 years.

In a lifetime, distance walked $\approx 5 \times 365 \times 80$

$$\approx 146\,000 \text{ km}$$
$$\approx 150\,000 \text{ km}$$

Notice that the basis for this solution is 10 000 steps. You could base your estimate on how long someone spends walking in a day. A reasonable walking speed is about 5 km/h although most people will walk more slowly than that when at work. They will also take shorter strides than 1 metre.

QUESTIONS

1 Make power of 10 estimates of:

   a the number of dogs in the UK

   b the number of dentists in Bristol

   c the number of football pitches in the UK

   Show your working clearly.

2 Estimate the amount of money spent on food in the UK each week. State your assumptions clearly.

3 Estimate the number of pizzas eaten in your home town each year.

4 Estimate the value of 10p coins, laid end to end, that would be needed to go around the Earth at the equator. Show clearly how you work out your estimate.

E 5 Estimate how long it would take one person to paint the hull of the Titanic. State the assumptions you make. You **must** show working to justify your answer.

STOCKING A SUPERMARKET

> ## How many lorries are needed to stock a supermarket?

How many aisles of produce would you fit in one lorry?

The supermarket lorries are very big!

No two supermarkets are the same size

Maybe 20?

WHAT YOU NEED TO KNOW ALREADY

→ Calculating volumes
→ Comparing lengths

MATHS HELP

p376 Chapter 3.20 Estimating

PROCESS SKILLS

→ Modelling the problem
→ Selecting the maths to use
→ Interpreting results
→ Evaluating results

If you have ever been involved in stocking a new supermarket, you might know the answer to this question. Otherwise, maybe it is a question you ask yourself whenever you see a lorry delivering to your local supermarket. Most people will need to answer this question using lots of estimates. It is another Fermi question like those you met in Chapter 2.3. The questions in Chapter 2.3 used strategies of powers of 10 (and a mixture of known quantities and ball park figures). This chapter introduces two further useful strategies:

— **Break the problem down then calculate each step**

— **Compare elements of the problem with things you know about.**

In order to answer this question, there are two key quantities we need to determine:

- the volume of a supermarket lorry (it is probably fair to assume a lorry can be loaded with goods without leaving any space)

- the total space for goods inside a supermarket.

In this chapter we will work in metres.

The volume of a supermarket lorry

So how big is a supermarket lorry? It might be easier to work this out if you try to compare it with something you know better. What about a car?

Discuss

> How many car lengths do you think the container part of a supermarket lorry is?

> How many car heights do you think the container part of a supermarket lorry is?

> How much wider than a car do you think a lorry is?

> How long is a car?

> How high is a car?

> How wide is a car?

What are the disadvantages of comparing a lorry with a car?

Investigate

> Here is a calculation for the volume of the lorry. It uses possible estimates for the six measurements above. Which estimate do you think goes with each question?

$$2 \times 3.5 \times 2.5 \times 1.6 \times (0.5 + 1.5) = 56 \text{ m}^3$$

> Using your own estimates for the six measurements above, estimate the volume of the lorry.

The total space for goods inside a supermarket

How big is a supermarket? No two supermarkets are exactly the same size. In order to work out the total space, some clarification is necessary:

WORKED EXAMPLE

What is the volume of stock in a supermarket? How many lorries does it fill?

SOLUTION

We assume that the type of supermarket you are going to use for the estimates is 'A superstore but without clothes, toys, DVDs, kitchen equipment and electrical goods'. You might think of a better definition!

What space is actually used for products? Goods are usually arranged in aisles.

Number of aisles = 8 say

An aisle is 15 m say

There is shelving on each side of the aisle. Roughly 6 shelves on average and 0.5 m deep.

Stock is piled roughly 30 cm high on average.

Amount = $8 \times 15 \times 2 \times 6 \times 0.5 \times 0.3$ m^3

   = 216 m^3

Number of lorries = 216 ÷ 56

   = 3.86 or round up to 4 lorries

Discuss

> Were the assumptions made likely to give an overestimate or an underestimate?

> Can you think of some other ways to approach this problem? Try working them through to see how closely your results match the previous answer.

Using statistics to estimate

Statistics are widely available from government agencies and these can be used to make estimates. You still need to make your own estimates for some of the calculation.

WORKED EXAMPLE

Use the information given to estimate how much electricity you use in one year.

Assume you live in a one bedroomed flat.

	Per hour	Per month
Heater	1.5 kW	
Water heating		400 kWh
Cooking	2kW	
Fridge		40 kWh
Freezer		90 kWh
TV	0.2 kW	
Gaming	0.15 kW	
Lights, per light	0.015 kW	
Laundry per load	4 kW	

SOLUTION

Heater, average 7 hours a day for 6 months of the year $= 1.5 \times 7 \times 180$ kW $= 1890$ kWh

Water, fridge and freezer for 1 year $= (400 + 40 + 90) \times 12$ kWh $= 6360$ kWh

Cooking, 4 hours per week average $= 2 \times 4 \times 52 = 416$ kWh

Laundry, 3 loads per week on average $= 3 \times 4 \times 52 = 624$ kWh

TV for 3 hours per day, gaming for 2 hours per day $= (3 \times 0.2 + 2 \times 0.15) \times 365 = 329$ kWh

Lights, 2 on average 5 hours per day $= 0.015 \times 2 \times 5 \times 365 = 55$ kWh

Total $= 1890 + 6360 + 416 + 624 + 329 + 55 = 9674$ kWh

Round to 10 000 kWh to account for other items.

QUESTIONS

1 How much water does your household use each week?

2 How many bricks are used to build a house?

3 A city is organising a parade with a mile-long route. About one hundred and fifty organisations have expressed interest in being in the parade. For how long will the streets need to be closed along the route?

4 How many drops of water are needed to fill a 50-metre swimming pool?

5 When you buy contents insurance for your house you need to estimate their value.

Item	Cost
Bed and bedding	£400
Wardrobe	£250
Clothes	£20 each on average
Kitchen appliances	£250 each
Kitchen equipment	£200
Sofa	£400
Table and chairs	£300
IT equipment	£1000

Use the information in the table to estimate the value of the contents of a one bedroom flat.

Show details of your assumptions and calculations.

CAMPAIGNING FOR CHANGE

I know I'm right

How can I prove it?

> ## How can you effectively make your point heard?

How can I make them listen?

Ashingford News
Overcrowding on local commuter trains

A commuter organisation says that 21% of trains are overcrowded but the rail company says the average number of passengers per train arriving in Ashingford is 239. There are 320 seats per train so there is plenty of space.

WHAT YOU NEED TO KNOW ALREADY

→ Calculating the mean and median for raw data
→ Using tally marks to create a grouped frequency table

MATHS HELP

p320 Chapter 3.7 Statistical terms
p322 Chapter 3.8 Representing and analysing data
p333 Chapter 3.10 Calculating statistics
p337 Chapter 3.11 Diagrams for grouped data

PROCESS SKILLS

→ Modelling the data using diagrams
→ Interpreting statistical diagrams
→ Representing data graphically and numerically
→ Summarising a statistical argument

Things are not always how we would like them to be. When things are challenged, a statistic is often quoted as justification. This is often an average or a percentage. In many cases the statistic is quoted without the context that it needs to be valid. To get things changed, put forward a case based on real and appropriate data. This chapter explores how to present data clearly when the data is quantitative and discrete.

In this chapter we are dealing with **discrete** data, which tends to be (but is not solely) whole numbers that result from counting things such as numbers of passengers.

The trouble with quoting single statistics such as the mean, median, or a percentage is that it does not allow you to get a full picture of the data. A diagram showing the distribution of the data itself can be very useful. Possibilities include:

– Grouped frequency tables

– Stem-and-leaf diagrams

– Box and whisker diagrams.

Here are the number of passengers on every train arriving at Ashingford in one particular week in May.

4	7	9	25	159	256	55	56	57	59
102	104	111	111	112	143	115	118	118	123
200	299	281	207	208	280	214	280	219	224
311	261	268	337	340	345	349	289	364	378
269	271	445	460	467	294	147	153	77	250
513	537	544	567	574	587	186	197	198	65
639	646	653	77	80	84	240	243	243	68
164	164	177	184	244	114	434	440	125	140
217	243	324	337	256	213	153	155	228	234
361	165	161	163	38	51	203	204	388	167

Discuss

> Do you think there were overcrowded trains the week this data was recorded?

> How many passengers do you think a train needs to have in order to be classified as overcrowded?

> How many trains do you think were overcrowded?

> When do you think these were?

> What problems would you have calculating the mean of this data?

It would be easier to understand the data if it was organised in some way.

The data has been collected into a **grouped frequency table.**

No. of passengers (p)	Tally	Frequency (f)
$0 \leqslant p < 50$	ℍℍ	5
$50 \leqslant p < 100$	ℍℍ ℍℍ I	11
$100 \leqslant p < 150$	ℍℍ ℍℍ IIII	14
$150 \leqslant p < 200$	ℍℍ ℍℍ ℍℍ	15
$200 \leqslant p < 250$	ℍℍ ℍℍ ℍℍ III	18
$250 \leqslant p < 300$	ℍℍ ℍℍ II	12

No. of passengers (p)	Tally	Frequency (f)
$300 \leqslant p < 350$	ⵘ II	7
$350 \leqslant p < 400$	IIII	4
$400 \leqslant p < 450$	III	3
$450 \leqslant p < 500$	II	2
$500 \leqslant p < 550$	III	3
$550 \leqslant p < 600$	III	3
$600 \leqslant p < 650$	II	2
$p \geqslant 650$	I	1

HINT

A train with 350 passengers would be included in the group $350 \leqslant p < 400$. A train with 400 passengers would be included in the next group $400 \leqslant p < 450$. When choosing groups make sure every number is included in one and only one group.

Discuss

> In the grouped frequency table the same data has been grouped into group sizes of 50. How can you tell that from the table?

> There are four items of data in the group $350 \leqslant p < 400$. What are their values?

> What do these values represent?

A good way to work out the mean is from a grouped frequency table.

WORKED EXAMPLE

Estimate the mean number of passengers in May.

SOLUTION

No. of passengers (p)	Frequency (f)	Midpoint (m)	$m \times f$
$0 \leqslant p < 50$	5	25	125
$50 \leqslant p < 100$	11	75	825
$100 \leqslant p < 150$	14	125	1750
$150 \leqslant p < 200$	15	175	2625
$200 \leqslant p < 250$	18	225	4050
$250 \leqslant p < 300$	12	275	3300
$300 \leqslant p < 350$	7	325	2275
$350 \leqslant p < 400$	4	375	1500
$400 \leqslant p < 450$	3	425	1275
$450 \leqslant p < 500$	2	475	950
$500 \leqslant p < 550$	3	525	1575
$550 \leqslant p < 600$	3	575	1725
$600 \leqslant p < 650$	2	625	1250
$p \geqslant 650$	1	700	700
Total	**100**		**23 925**

$$mean = \frac{fm}{f}$$

$$Mean = \frac{23\ 925}{100} = 239.25$$

When you put data into a grouped frequency table, the actual values are lost and it is assumed that all data takes the value of the midpoint of the group. So the total number of passengers on the four trains in the group $350 \leqslant p < 400$ is estimated at 1500, whereas the actual total was 1491. For this reason, any calculations will only give an estimate.

Alternatively, the raw data can be entered in a spreadsheet. This makes it easy to calculate an accurate mean of 237.2.

Discuss

> Explain where all the numbers in the grouped table above come from.

> How different is the estimate from the accurate mean?

> Do you think this matters?

The data in the chart is organised in a **stem-and-leaf diagram.**

It shows the numbers of passengers recorded on the trains arriving at Ashingford for one week in August.

The digits to the left of the vertical line are called **stems** and those to the right are the **leaves**.

Notice that the leaves are arranged in ascending order.

Stem							
0	4	6	7	8	9		
1	0	1	5	7	8	8	
2	4	5	6				
3	2	2	4	6	7	8	9
4	3	4	5	6	7	7	
5	0	2	6				
6	5	6					
7	5	6	8				
8	0	1	5				
9	3	5	6				
10	4	5	6	6			
11	0	0	1	5			
12	3	4	6	8	8		
13	0	1	2	4	6		
14	4	5	5	7	7	8	
15	3	3	4	5	6	7	8
16	4	5	6	6	7	8	8
17	0	1	2	3	4	5	5	7
18	0	0	2	4	4	5	
19	1	4	7				
20	2	5					
21	3	5					

Key 11 | 5 means 115

Discuss

> Look at the key. One train had 115 passengers, represented by a stem of 11 and a leaf of 5. Find this train within the table.

> Which of the following numbers of passengers were also recorded as part of the data?

105 15 5 115 150 250

Investigate

> Display this data as a grouped frequency table. Find an estimate for the mean using your grouped frequency table.

> Compare the data collected in August with the data collected in May. What do you notice? Give a possible reason for these differences.

DID YOU KNOW?
The stems could be 'doubled up' to simplify the diagram. The line with a stem of 7 would become: 7|5 6 8 10 11 15

HINT
A stem and a leaf can represent different amounts. For example, 34|5 can represent 345, 34.5 and 3.45. This is why having a key is vital.

HINT
Stem-and-leaf diagrams and grouped frequency tables tend to be most useful when there are between 5 and 15 stems or groups.

Discuss

> > How many stems would there be if you represented the May data as a stem-and-leaf diagram?

Another diagram used to represent data is a **box and whisker** diagram (also known as a **box plot**). This uses the median and quartiles to give a diagrammatic representation of the shape of the distribution of the data.

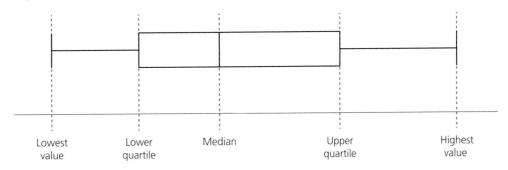

| Lowest value | Lower quartile | Median | Upper quartile | Highest value |

The median and quartiles are most easily found from an ordered set of data or from a stem-and-leaf diagram.

WORKED EXAMPLE

Find the median and quartiles of the data below. This is the passenger data collected in May, arranged in order.

4	7	9	25	38	51	55	56	57	59	65	68	77	77	80	84	102	104	111	111
112	114	115	118	118	123	125	140	143	147	153	153	155	159	161	163	164	164	165	167
177	184	186	197	198	200	203	204	207	208	213	214	217	219	224	228	234	240	243	243
243	244	250	256	256	261	268	269	271	280	280	281	289	294	299	311	324	337	337	340
345	349	361	364	378	388	434	440	445	460	467	513	537	544	567	574	587	639	646	653

SOLUTION

The **median** item of 100 pieces of data is the $\frac{1+100}{2} = 50.5$th item, so the median is the mean of the 50th and 51st items. It is halfway through the data.

So median $= \frac{208 + 213}{2} = \frac{421}{2} = 210.5$

The **lower quartile** is the middle item of the first half of the data, the $\frac{1+50}{2} = 25.5$th item, so the lower quartile is the mean of the 25th and 26th items. It is a quarter of the way through the data.

So lower quartile $= \frac{118 + 123}{2} = \frac{241}{2} = 120.5$

The **upper quartile** is the middle item of the second half of the data, the $\frac{51+100}{2} = 75.5$th item, so the upper quartile is the mean of the 75th and 76th items. It is three-quarters of the way through the data.

So upper quartile $= \frac{299 + 311}{2} = \frac{610}{2} = 305$

The box and whisker diagram for May trains at Ashingford is drawn to scale as shown:

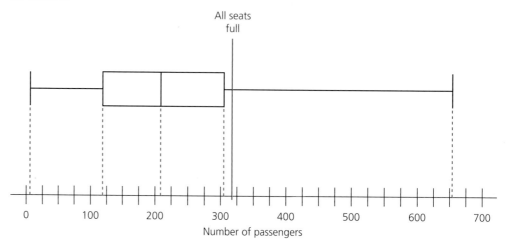

All seats full

Number of passengers

By adding the line that shows when all the seats are full, we can see that almost a quarter of the trains have more passengers than seats.
Box and whisker diagrams are particularly useful when comparing two sets of data.

Investigate

❯ Draw a box and whisker diagram for the passenger data collected in August (on page 58). Make a note of the differences between the two box and whisker diagrams.

QUESTIONS

1
> Subjects studied; number of people studying each subject; exam marks; exam grades; holiday destinations; holiday prices; how much people earn.

   a Which of these are quantitative data?

   b Which are qualitative data?

   c List more examples of qualitative and quantitative data.

2 A website designer wants to demonstrate the effect she has had on the number of hits received by a customer's website. She records the number of hits each day in the month before she makes the changes and each day for a month after she has made all the changes.

The data obtained is recorded in the back-to-back stem-and-leaf diagram.

→

Number of website hits

Before										Stem	After								
			8	8	5	3	0	0		0	2	2	3	7	7				
9	7	7	5	4	4	4	1	0		1	4	7	7	7	8	8	9		
		8	7	5	5	3	2	1		2	2	4	6	7	7	8	8	8	9
			9	8	7	7	6	6		3	5	6	7	7	8	8	9	9	
						2	1	0		4									
										5	6								

Key   6 | 3 | means 36

a Calculate the mean and median number of hits both before and after her changes.

b What evidence is there that she has increased the number of hits?

c What evidence is there that she has **not** increased the number of hits?

3 The villagers of Trowford are campaigning for a village bypass. For one complete week, they count the number of lorries passing the village church each hour between 7 a.m. and 7 p.m. They record the following data.

6	12	14	26	35	23	36	23	32	26
12	3	5	13	24	23	34	31	31	28
10	7	7	6	11	24	34	38	32	34
23	34	32	22	12	12	35	32	41	21
32	37	27	23	25	24	23	7	9	11
17	26	23	27	32	32	26	14	13	10
6	8	10	11	12	12	10	14	12	4
3	2	0	0	0	0	2	1	2	1
0	0	0	0						

a Use either a stem-and-leaf diagram or a grouped frequency table to summarise this data. Justify your choice of diagram.

b Calculate the mean and median number of lorries per hour over the week.

c What advice would you give the villagers in order for their campaign to have the greatest impact?

4 A bird conservation group is investigating whether there has been any growth in guillemot breeding along a particular piece of cliff. They photograph the cliff at midday every day in May in two consecutive years and then count the number of birds they can see.

The results are shown in this back-to-back stem-and-leaf diagram.

Year 1							Stem	Year 2								
						3	18									
					7	3	19	2								
		6	5	4	0		20	3	5	6						
		7	6	2	2		21	4	5							
		9	8	6	5		22	5	7	9						
7	6	4	2	1			23	5	6	7	6	8	9	9		
8	7	5	3	0			24	3	4	5	6	8	8			
		3	3	2			25	0	1	3	5					
			1	0			26	2	4	5	7					

Key | 26 | 2 means   262

a Looking at the stem-and-leaf diagram, do you think there is any evidence that the number of guillemots on the cliff has increased in year 2? Explain your answer.

b Draw a box and whisker diagram for year 1 and year 2 using the same scale.

c Compare your two box and whisker diagrams to quantify the differences between the two years.

HINT

Quantify means you have to give some numbers to back up your statements.

5 Peter is a fisherman. During the lobster season he catches crabs and lobsters in his lobster pots. He records his daily catch each day of the season and the data is summarised in the box plots below.

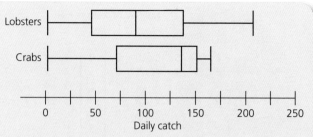

Compare his daily catch of crabs and lobsters.

IS IT REALLY TRUE?

> Why should men be taller?

> ## Are males really taller than females?

Is the tallest person in the picture a man?

Why should men be taller?

WHAT YOU NEED TO KNOW ALREADY

→ Methods for collecting data
→ Compiling tables of data
→ Constructing and interpreting frequency diagrams

MATHS HELP

p320 Chapter 3.7 Statistical terms
p322 Chapter 3.8 Representing and analysing data
p333 Chapter 3.10 Calculating statistics
p337 Chapter 3.11 Diagrams for grouped data
p345 Chapter 3.12 Estimating statistics from graphs

PROCESS SKILLS

→ Comparing data
→ Representing data
→ Evaluating representations
→ Summarising
→ Comparing representations

There are plenty of instances where people say something without evidence to base it on, such as the frequently repeated 'fact' that the EU will not allow the sale of bananas that are not straight enough. Such falsities can becomes accepted as true. They can even lead to prejudice against groups in society, such as unfounded assumptions about immigrants being criminals or that they are only in this country for benefit payments. It is important to challenge these assertions in order to achieve a fairer world. Statistics have a role here, gathering unbiased data and representing it in ways that communicate the patterns in it clearly so that people become better informed. This chapter introduces you to some simple statistics and what they show.

The table shows the heights in centimetres (rounded to the nearest cm) of a sample of male and female 18-year-olds.

Discuss

> What can you tell from the table?

> Are the males taller than the females?

Heights are **continuous** data and are rounded to the nearest centimetre for the table.

A height rounded to the nearest centimetre can be up to half a centimetre out either way. So, the number 179 in the table represents a measurement between 178.5 cm and 179.5 cm.

DID YOU KNOW?

Continuous data is data that can take all the possible values in an interval, even if you only record it to a few significant figures. It usually comes from measuring something.

Height (cm)	
Male	**Female**
179	170
195	167
184	157
179	181
172	158
182	166
173	167
175	157
190	168
174	165
173	164
185	174
178	169
178	170
184	164
178	176
177	171
173	169
188	172
169	163
180	155
189	173
182	176
186	167
181	164
179	162

Heights of a sample of 26 males and 26 females

Discuss

The diagram shows the same height data presented as two overlapping frequency graphs.

> What can you tell from the diagram?

> What is the same and what is different about the graph for males and the graph for females?

> What are the advantages and disadvantages of the table and the diagram?

> What other diagrams could you use to represent the data?

Grouping data

To draw the frequency graphs, the data has to be grouped.

If x cm is a (rounded) height included in the 150 cm to 155 cm group, that means that $150 \leqslant x < 155$. A measurement of 155 cm is put in the 155 cm to 160 cm group. Strictly speaking, the grade boundaries should be $149.5 \leqslant x < 154.5$, but the difference is too small in comparison to the heights to matter.

Note that the horizontal axis is marked with a continuous scale and the bars have no gaps between them.

Discuss

What statements can you make about male and female heights as shown in the graphs?

> For example, are these statements true or false?
>
> 'All the males are taller than all the females'
>
> 'The tallest male is taller than the tallest female'

> Write some more statements about the male and female height data and decide whether they are true or false.

Investigate

> Collect your own height data, group it and draw the overlapping frequency graphs.

When collecting your data:

> What decisions did you have to make?

> What difficulties did you encounter?

> How did you record and present the data?

> How does your data compare with the data given above?

> What statements can you make about the male and female heights in your data set?

Discuss

> Can you explain the differences between the different types of data?

> Can you use examples of primary, secondary, qualitative, quantitative, discrete and continuous data in your explanations?

> Which of these are continuous data?

> Which are discrete data?

> Heights; weights; foot lengths; shoe sizes; ages; prices; wages; exam marks; numbers of students in different classes; proportions of faulty components in a batch.

> Think of more examples of continuous and discrete data.

DID YOU KNOW?

Data that is collected for a specific investigation is **primary data**. It may actually be collected by you or another person.

Data that you get from elsewhere that was collected for some other investigation is **secondary data**.

DID YOU KNOW?

Continuous data, such as measurements of height, can have any numerical value in a range in contrast with **discrete data**, such as the number of children in a family, which can only have particular values in a range.

WORKED EXAMPLE

Jane is researching what courses students in her college are taking and comparing this with national data.

She considers the following data:

- results of a survey she conducts in her year at college
- information on the internet
- data sent from a neighbouring college.

Which of these is primary data and which is secondary data?

Explain your decisions.

SOLUTION

The survey that Jane conducts generates primary data as she has collected it herself for this purpose.

The other information is secondary data as she has not collected it herself, and it has been collected for another purpose.

How else could the height data be presented?

The table and the graph that follows show the **cumulative frequency** of the male heights.

Heights (cm)	Male frequency	Female frequency	Male cumulative frequency	Female cumulative frequency
$150 \leqslant x < 155$	0	0	0	0
$155 \leqslant x < 160$	0	3	0	3
$160 \leqslant x < 165$	0	4	0	7
$165 \leqslant x < 170$	1	8	1	15
$170 \leqslant x < 175$	5	6	6	21
$175 \leqslant x < 180$	8	3	14	24
$180 \leqslant x < 185$	6	2	20	26
$185 \leqslant x < 190$	4	0	24	26
$190 \leqslant x < 195$	1	0	25	26
$195 \leqslant x < 200$	1	0	26	26
$200 \leqslant x < 205$	0	0	26	26

❯ Can you see how cumulative frequencies in the table have been worked out?

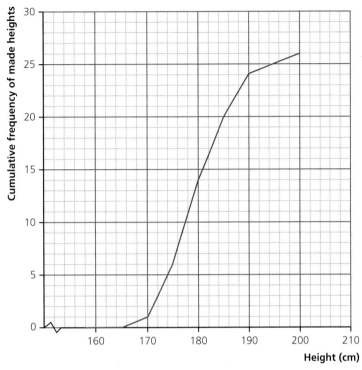

The cumulative frequencies are the running total of the frequencies. Notice that they are plotted at the end of the interval. This is because it is only at the end of the interval that you know you have included all the data for the group.

❯ Work out the cumulative frequencies for the female heights and draw the cumulative frequency graph.

❯ Compare the male and female graphs. What is the same and what is different?

The data can also be presented as box and whisker diagrams, which you met in the last chapter. The whiskers show the range of the data and the box shows the interquartile range. They are ways of showing how spread out, or variable, the data is.

DID YOU KNOW?

The **range** is the difference between the largest and the smallest items of data.

The **interquartile range** is the difference between the upper and lower quartiles.

QUESTIONS

1 Write each of these in the appropriate cell of a two-way table like the one on the right.

a temperature readings from a chemistry experiment you carry out

b traffic passing a junction at different times of day from a council survey

c your class register

d types of car you've spotted during a traffic jam

	Primary	Secondary
Qualitative		
Quantitative		

e summary attendance data for all classes in mathematics

Make up three more examples so that there are two in each cell.

2 The table shows the runners' finishing times for the marathon in the 2012 Olympics in hours and minutes.

	2:00→	2:10→	2:20→	2:30→	2:40→	2:50→	3:00→	3:10→	3:30→
Women	0	0	28	49	22	3	3	0	1
Men	3	47	27	4	2	1	0	0	0

a Draw overlapping frequency block graphs to show both sets of results.

b Compare the performance of the athletes in the two events.

c Criticise the choice of graph. Is there a better way of showing the results?

Are they achieving this based on the data provided?

> **HINT**
>
> First construct a cumulative frequency graph from the table provided.

3 The table below shows the amount of time customers waited to have their call answered at an energy provider, company A.

Waiting time, t (seconds)	Frequency
$0 \leqslant t < 10$	5
$10 \leqslant t < 20$	12
$20 \leqslant t < 30$	11
$30 \leqslant t < 60$	19
$60 \leqslant t < 120$	33

The cumulative frequency graph shows the amount of time customers waited to have their call answered at a rival energy provider, company B.

a Which company has a longer customer waiting time on average?

b Which company is more consistent in its customer waiting time?

c The companies both have a target to answer 80% of calls within 90 seconds.

E 4 Use the data in the table to draw, on the same diagram, box and whisker diagrams for the male and female heights.

	Male	Female
Minimum	169	155
Lower quartile	175	164
Median	179	167
Upper quartile	184	171
Maximum	195	181

Compare the male and female heights.

5 Draw cumulative frequency graphs for the data in question 4.

Comment on whether it is better presented in this way.

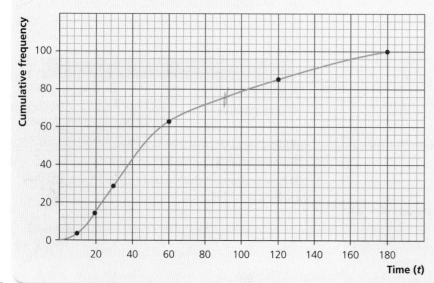

HOW FAIR IS OUR SOCIETY?

What is fair pay?

> 'Women will get equal pay ... in 118 years' – *Guardian* 18/11/2015

Is it equal pay for equal work?

How could you judge that?

WHAT YOU NEED TO KNOW ALREADY

→ Calculating the mean, median and mode from a set of data
→ Calculating the lower and upper quartiles
→ Calculating the range for a set of data
→ Constructing a stem-and-leaf diagram, box plot and cumulative frequency diagram

MATHS HELP

p322 Chapter 3.8 Representing and analysing data
p333 Chapter 3.10 Calculating statistics
p337 Chapter 3.11 Diagrams for grouped data
p345 Chapter 3.12 Estimating statistics from graphs

PROCESS SKILLS

→ Representing data
→ Comparing related data sets
→ Interpreting charts and calculations

Gathering data to test whether something is actually true or whether people just think it is true is essential in a modern organisation. Separating rumour from fact is possible using statistics. We all know that statistical graphs can be interpreted, and misinterpreted, in a number of ways. Averages are also open to abuse, with people, perhaps deliberately, misunderstanding what they are – suggesting that everyone can be above average! Statistics are more often used to support a view rather than to clarify understanding of a situation or phenomenon. It would be more helpful to include a measure of spread to show how consistent the data is, but this rarely gets into the headline. This chapter is about comparing data using both averages and measures of spread.

Following a newspaper story that women are discriminated against at work in their organisation, a company conducts a review of the wages of its male and female employees in the last tax year.

The data is shown in the table and the box plots:

	Females	Males
Mean earnings (£)	29 280	31 475

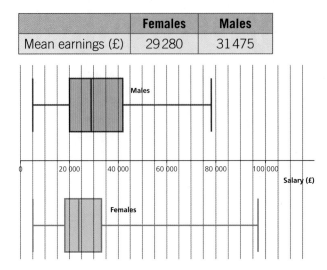

The company produces this summary of their analysis:

- The average salaries of females and males are within £2200, indicating that there is not a significant difference in the earnings of the different genders.
- The width of the box plots shows that there is less variation in the salaries of the male employees than the female employees.

Therefore, there is no specific concern about the variation in salaries between different genders.

Investigate

Reading the report, you may think the company is being fair, or you may be concerned that they are trying to mislead, or cover up poor practice in the workplace.

> Compare the mean earnings of the males and females. Does this support the company's first bullet point?

> How can you use a box plot to give you a different measure of the average? Does this measure (median) support the company's first bullet point?

> What story lies behind the mean and median averages being different?

> Compare the range of wages for the males and females using the box plots. Does this support the company's second bullet point?

> The company measured the spread of data by looking at the width of the box plot. You can use a box plot to give you an alternative measure of the spread. Does this measure (interquartile range) support the company's second bullet point?

> What explanation could there be for the ranges and interquartile ranges being different?

> Do you agree with the company's conclusion that there is no significant difference between the wages of its female and male employees?

Look at each of the box plots above and write down approximate values for:

> the median

> the lowest value

> the highest value

> the range of the data

> the lower quartile

> the upper quartile

> the interquartile range.

Now look at the wages of the board of directors:

£97 000   £78 000   £75 000   £82 000   £78 000   £78 000   £77 000

For this data, work out:

> the mean

> the median

> the mode

> the range of the data

> which of the salaries must be for a female employee.

As we have seen, we can measure the average of a set of data and the spread of a set of data in more than one way to give us more information.

DID YOU KNOW?

Percentiles split the data into hundredths in the same way as quartiles split it into quarters.

DID YOU KNOW?

The **lower quartile** can be thought of as the 25th percentile and the **upper quartile** can be thought of as the 75th percentile.

DID YOU KNOW?

The **mode** is the most common data value in the data set. The mode is useful for discrete data where there is significant replication of data values, for example, shoe sizes.

WORKED EXAMPLE

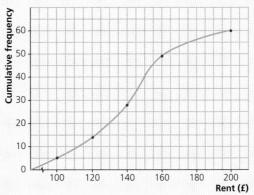

a   The data shows the amount of rent paid per week by students living away from home:

£123	£135	£84	£95	£101	£105	£120	£118	£123	£124	£98	£121	£115

Calculate the median and interquartile range of the data.

b   The cumulative frequency graph shows the rent paid per week by young professionals:
Calculate the median and interquartile range of the data.

c   Compare and contrast the rent paid by students and young professionals.

SOLUTION

a   Place the data in order:

| £84 | £95 | £98 | £101 | £105 | £115 | £118 | £120 | £121 | £123 | £123 | £124 | £135 |

As you know from Chapter 2.5, the median is the middle value, which is the 7th value as there are 13 pieces of data, so median = £118.

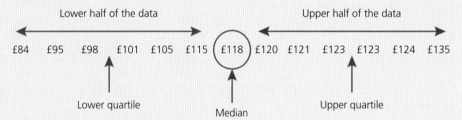

Lower half of the data   Upper half of the data

£84   £95   £98   £101   £105   £115   (£118)   £120   £121   £123   £123   £124   £135

Lower quartile   Upper quartile

Median

The lower quartile lies between £98 and £101. Therefore, lower quartile = $\frac{98+101}{2}$ = £99.50.

The upper quartile lies between £123 and £123. Therefore, upper quartile = £123.

Therefore, the interquartile range = £123 − £99.50 = £23.50.

b   The graph goes up to a cumulative frequency of 60.

So the median will be at a cumulative frequency of $\frac{60}{2}$ = 30, and the lower and upper quartiles

will be found at a cumulative frequency of $\frac{60}{4}$ = 15 and $\frac{3 \times 60}{4}$ = 45 respectively.

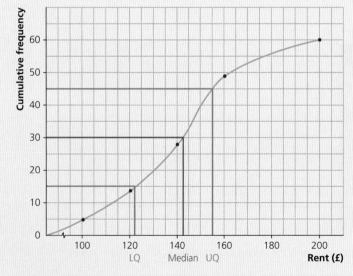

Reading across the graph, the median is at £142.

The interquartile range = £155 − £122 = £33.

HINT

Mentioning the two medians that are used to work out £24 gives a better idea of the context.

c When we compare the data, we need to make a comment about the average and a comment about the spread. So we could say:

On average, young professionals pay £24 more for their rent than students (£142 − £118).

There is more variation in the rent paid by young professionals than students (or the students' rent was more consistent).

Sometimes you cannot find the median and quartiles immediately, so you have to use another representation.

WORKED EXAMPLE

The table shows the times of runners in the 100 m heats at the World Championships.

Time (seconds)	Frequency
$9.5 \leq x < 9.6$	1
$9.6 \leq x < 9.8$	3
$9.8 \leq x < 10.0$	12
$10.0 \leq x < 10.2$	14
$10.2 \leq x < 10.4$	7
$10.4 \leq x < 10.7$	3

The box plot shows the times of the runners in the semi-finals of the same competition.

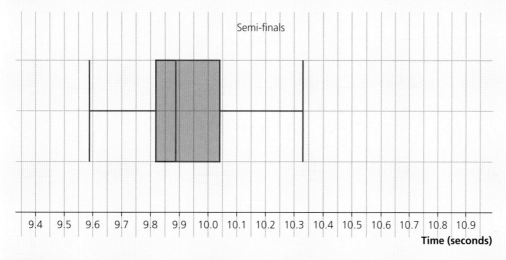

Compare the race times of the heats and semi-finals.

SOLUTION

We need to compare the average and the spread of both data sets using the median and interquartile range (as we cannot easily find the mean from a box plot).

So we will draw a cumulative frequency diagram from the frequency table to help us find these values.

We start by producing a cumulative frequency table for the data:

Time (seconds)	Cumulative frequency
$9.5 \leqslant x < 9.6$	1
$9.5 \leqslant x < 9.8$	4
$9.5 \leqslant x < 10.0$	16
$9.5 \leqslant x < 10.2$	30
$9.5 \leqslant x < 10.4$	37
$9.5 \leqslant x < 10.7$	40

Then we plot the graph and find the median and quartiles.

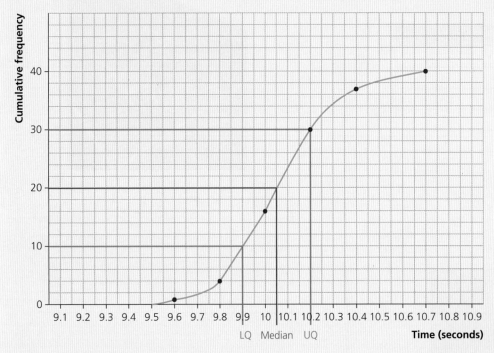

So, for the heats:

Median = 10.05 seconds.

Interquartile range = 10.2 − 9.9 = 0.3 seconds.

We now look at the box plot to find the same statistics for the semi-finals:

Median = 9.89 seconds.

Interquartile range = 10.04 − 9.82 = 0.22 seconds.

Therefore, on average the runners ran 0.16 seconds faster in the semi-finals than in the heats (by comparing the medians).

The times that the runners produced in the semi-finals were more consistent than those in the heats (or the times in the heats were more varied).

QUESTIONS

1 The box plot shows the scores of the French gymnastics team at a competition.

Here are the scores of the Great Britain gymnastics team across four pieces of apparatus (to 1 decimal place):

| 14.1 | 15.3 | 15.6 | 14.9 | 15.9 | 16.0 | 15.3 | 15.3 |
| 15.8 | 14.8 | 15.1 | 13.8 | 14.6 | 14.9 | 15.5 | 15.7 |

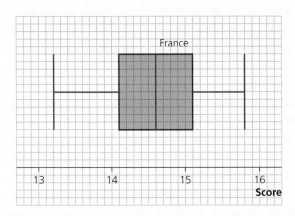

France

Score

a Construct a box plot for the Great Britain data.

b Compare the average scores of Great Britain and France.

c Compare the spread of scores of Great Britain and France.

2 Look at the stem-and-leaf diagram below, which shows the delay (in minutes) to every flight on a European route for two competitor airlines.

```
        Airline A              Airline B
  9  5  1  0  0  0  │ 0 │ 5  5  8
           5  5  0  │ 1 │ 2  8
                    │ 2 │ 0  0  4  9
              5  3  │ 3 │ 2  3  5  9  9
                 1  │ 4 │ 5
                 4  │ 5 │ 1  1
           5  2  0  │ 6 │ 0
              2  0  │ 7 │ 5
```

Key   0 | 6 | 0   means   60

a Complete this table of results for both airlines:

	Airline A	Airline B
Maximum		
Minimum		
Median		
Mode		
Range		
Lower quartile		
Upper quartile		
Interquartile range		

b Use your findings in the table to compare the performance of both airlines.

c Write a short report to summarise your findings.

3 The table shows the distribution of internet spending by item value in the UK in 2015:

Item value	Frequency (as % of the whole sample)
$v < £50$	12
$£50 \leq v < £100$	11
$£100 \leq v < £500$	42
$£500 \leq v < £1000$	13
$£1000 \leq v < £2000$	11
$£2000 \leq v < £3000$	9

a Construct a cumulative frequency graph for this data.

b Hence estimate:

   i the median item value

   ii the interquartile range of the values.

c An internet shop claims that '30% of the items that people purchase cost under £200'. Use your graph to find an estimate of the 30th percentile to see if the claim is correct.

d A forecaster predicts that, in 2025, the median item value of internet purchases will be £250 and the interquartile range will be £1000.

Compare and contrast the current distribution of purchasing with the 2025 forecast.

4 The scoring statistics in a season for two basketball players are shown:

	Median score	Lowest score	Highest score	Lower quartile	Upper quartile
Player A	37	0	61	12	44
Player B	31	0	58	26	39

Which player would you hire? Justify your choice.

5 The table below shows the prevalence of being underweight and overweight as a child.

Region	% Underweight	% Overweight
Africa	15.9	5.7
Asia	16.9	20.1
Latin America and Caribbean	3.0	3.9
Oceania	18.4	0.1
North America and Europe	–	–
Global average	**13.9**	**41.6**

a Compare the proportion of children who are underweight and overweight by region and with the global averages provided in the table.

b The data for North America and for Europe is not provided but is part of the global average. Suggest how the percentage of children who are underweight and overweight in North America and Europe will compare with the global averages provided.

E 6 Anna is a chef in a large hotel.

She needs to buy a large number of eggs to serve at breakfast.

She needs the eggs to all be of similar weight to create consistent dishes.

She can buy from supplier A or supplier B.

Anna samples eggs from supplier A and measures their weight in grams.

Here are her results:

62	65	71	55	62	60	59	59	65	63
58	56	56	61	62	68	60	59	59	62

Anna researches supplier B and finds this information on their website about the weights of a sample of eggs.

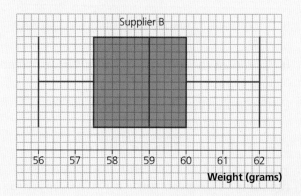

Supplier B

Weight (grams)

Compare the eggs from the two suppliers.

SHOULD I TAKE A DAY OFF?

> ## How do you know if you are unwell?

Is my temperature normal?

Am I feeling sick?

What other symptoms do I have?

WHAT YOU NEED TO KNOW ALREADY

→ Drawing box plots and cumulative frequency curves

→ Calculating the mean from a grouped frequency table

MATHS HELP

p320 Chapter 3.7 Statistical terms
p322 Chapter 3.8 Representing and analysing data
p333 Chapter 3.10 Calculating statistics
p337 Chapter 3.11 Diagrams for grouped data
p345 Chapter 3.12 Estimating statistics from graphs

PROCESS SKILLS

→ Interpreting data
→ Representing data
→ Comparing data
→ Estimating

One of the principal uses of statistics is in decision making. It does not guarantee that you make the correct decision, but it helps you to use the data available in a sensible way. Drawing a diagram is a way of both organising the data and representing it in a way that you can readily understand. That means you can make a better-informed decision. This chapter continues to look at statistical diagrams and introduces histograms as a particularly powerful way of representing data.

A common way to decide if we are ill is to take our temperature.

Recent medical evidence suggests that the assumed 'normal' body temperature might be outdated.

Investigate

> What is normal body temperature?

> Why was this particular temperature suggested?

> What factors will affect an individual's temperature?

Discuss

The first thing to do is take a quick look at the data in the table. Perhaps you can begin to form an idea from that. It is noticeable that there is a greater range for the female data than for the male data. Why do you suppose that happens? How does the data support the notion of a normal body temperature?

The next step is to analyse the data by calculating some statistics and to represent it using appropriate graphs.

DID YOU KNOW?

The mode is the most frequently occurring item; the **modal class** is the class with the largest frequency when the class widths are equal.

Males

Temperature, t (°C)	Frequency
$35.7 \leqslant t < 35.9$	1
$35.9 \leqslant t < 36.1$	1
$36.1 \leqslant t < 36.3$	6
$36.3 \leqslant t < 36.5$	10
$36.5 \leqslant t < 36.7$	7
$36.7 \leqslant t < 36.9$	14
$36.9 \leqslant t < 37.1$	12
$37.1 \leqslant t < 37.3$	9
$37.3 \leqslant t < 37.5$	4
$37.5 \leqslant t < 37.7$	1

Females

Temperature, t (°C)	Frequency
$35.8 \leqslant t < 36.0$	2
$36.0 \leqslant t < 36.2$	1
$36.2 \leqslant t < 36.4$	3
$36.4 \leqslant t < 36.6$	3
$36.6 \leqslant t < 36.8$	12
$36.8 \leqslant t < 37.0$	15
$37.0 \leqslant t < 37.2$	17
$37.2 \leqslant t < 37.4$	7
$37.4 \leqslant t < 37.6$	2
$37.6 \leqslant t < 37.8$	1
$37.8 \leqslant t < 38.0$	1
$38.0 \leqslant t < 38.2$	0
$38.2 \leqslant t < 38.4$	1

> Calculate the mean male and female body temperatures from the tables.

> Identify the modal class for the male and female body temperatures from the tables.

> Calculate the median male and female body temperatures from the tables.

> Draw cumulative frequency curves to find the quartiles and interquartile range for both data sets.

> Draw box plots to compare the statistics for male and female body temperatures.

An outlier may be identified as any data point that is either more than $1.5 \times$ the IQR below the lower quartile or above the upper quartile.

> Why might outliers be useful in identifying whether you are unwell?

> How many males/females might be unwell based on the data set provided?

Looking at your results for the five tasks on page 78, which do you think was the most informative average?

Histograms

Histograms are another useful way in which to represent grouped data. They provide a visual representation of grouped data similar to bar charts except that with a histogram area is used to represent frequency. The scale on the vertical axis is described as **frequency density**.

WORKED EXAMPLE

A call centre is looking to improve waiting times for customers. The target is to answer 80% of calls in 2 minutes or less.

The table shows the waiting times, in seconds, for a sample of 150 calls.

From this sample comment on how far the call centre is from meeting its target.

Time, t (seconds)	Frequency
$10 \leqslant t < 50$	24
$50 \leqslant t < 100$	34
$100 \leqslant t < 130$	39
$130 \leqslant t < 150$	26
$150 \leqslant t < 180$	27

SOLUTION

To calculate frequency density, divide the frequency by the difference between the upper and lower limits of the group interval.

Time, t (seconds)	Frequency	Frequency density
$10 \leqslant t < 50$	24	$24 \div 40 = 0.6$
$50 \leqslant t < 100$	34	$34 \div 50 = 0.68$
$100 \leqslant t < 130$	39	$39 \div 30 = 1.3$
$130 \leqslant t < 150$	26	$26 \div 20 = 1.3$
$150 \leqslant t < 180$	27	$27 \div 30 = 0.9$

The target for the call centre is for 80% of calls to be answered in 2 minutes (120 seconds) or less.

150 calls have been sampled.

80% of 150 = 120 calls.

To meet the target, 120 calls need to have been answered in 120 seconds or less.

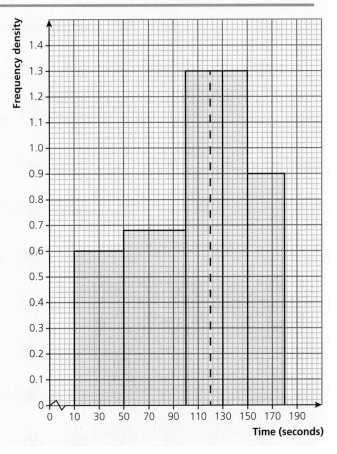

Area of the bars up to and including 120 seconds:

first bar = 40 × 0.6 = 24

second bar = 50 × 0.68 = 34

part of third bar, 20 × 1.3 = 26

So 24 + 34 + 26 = 88.

In total 88 calls were answered in 2 minutes or less. Therefore, from this given sample, the call centre is a long way (32 calls) from meeting its target.

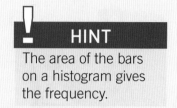

HINT

The area of the bars on a histogram gives the frequency.

DID YOU KNOW?

The **modal class** is the one with the highest frequency density when the classes are unequal in width.

Discuss

Identify the modal class in the previous worked example from the frequency table, and then from the histogram. What do you notice? Which of the two is more representative of the data? Try to justify your answer.

Discuss

› How would you find the median from a histogram?

› How would you work out the quartiles?

WORKED EXAMPLE

The grouped frequency table and histogram provide some of the details for the finishing times of the top 40 runners in the 2016 London Marathon.

Finishing time (hours:mins)	Number of runners	Frequency density
2:03 ≤ t < 2:07		45
2:07 ≤ t < 2:10	3	
2:10 ≤ t < 2:20	14	84
2:20 ≤ t < 2:25	6	72
2:25 ≤ t < 2:28	3	60
2:28 ≤ t < 2:30	3	
2:30 ≤ t < 2:35		72
2:35 ≤ t < 2:40		24

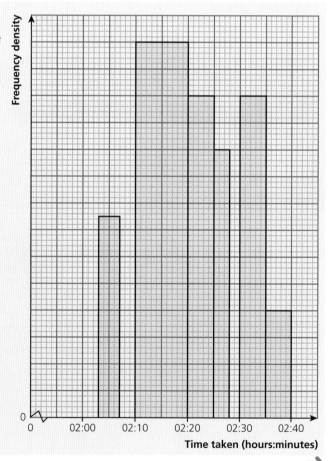

a   Use the information provided to complete the table and corresponding histogram.

b   Which is the modal class?

c   The mean finish time for all runners for the 2016 London marathon was 4 hours 30 minutes. Estimate how many of these 40 runners finished in less than half the mean finishing time for all runners.

SOLUTION

a

Finishing time (hours:mins)	Number of runners	Frequency density
$2{:}03 \leqslant t < 2{:}07$	3	45
$2{:}07 \leqslant t < 2{:}10$	3	60
$2{:}10 \leqslant t < 2{:}20$	14	84
$2{:}20 \leqslant t < 2{:}25$	6	72
$2{:}25 \leqslant t < 2{:}28$	3	60
$2{:}28 \leqslant t < 2{:}30$	3	90
$2{:}30 \leqslant t < 2{:}35$	6	72
$2{:}35 \leqslant t < 2{:}40$	2	24

> ### HINT
> The class widths are in fractions of an hour, so $\dfrac{4}{60}$ for the first group.

b   The modal class is $2{:}28 \leqslant t < 2{:}30$ as its frequency density is the greatest.

c   The mean finish time for all runners was 4 hours and 30 minutes.
Half of this time would be 2 hours 15 minutes.
We need to estimate how many runners in the top 40 finished in under 2 hours and 15 minutes.
To this we need to find the area of the bars up to and including 2 hours 15 minutes on the histogram.
Area of bars:

$$\frac{4}{60} \times 45 = 3$$

$$\frac{3}{60} \times 60 = 3$$

$$\frac{5}{60} \times 84 = 7$$

So $3 + 3 + 7 = 13$

In total 13 runners completed the 2016 London marathon in less than half the mean finishing time for all runners.

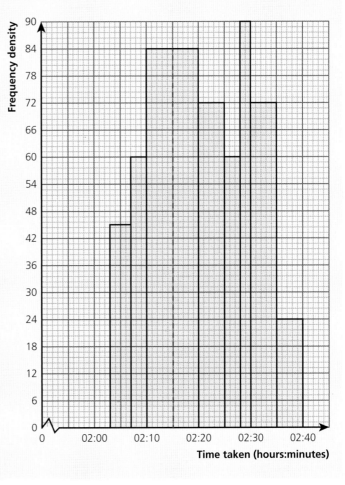

81

QUESTIONS

1 The temperature of a supermarket fridge is regularly checked to ensure that it is working correctly. Over a period of four months the temperature (measured in degrees Celsius) is checked 600 times. These temperatures are displayed in the table below:

Temperature, t (degrees Celsius)	$3.0 \leqslant t \leqslant 3.3$	$3.3 < t \leqslant 3.6$	$3.6 < t \leqslant 4.2$	$4.2 < t \leqslant 4.8$	$4.8 < t \leqslant 5$
Frequency	98	46	247	167	42

The recommended temperature for the supermarket fridge is 4°C or cooler. It should never go above 5°C and should only be above 4°C for a maximum of 30% of the time. Is this fridge functioning correctly?

Use a suitable statistical diagram or statistical analysis to support your argument.

2 A taxi driver operates from a taxi rank at a mainline railway station in London. During one particular week he makes 120 journeys, the lengths of which are summarised in the table below:

Length (x miles)	$0 < x \leqslant 1$	$1 < x \leqslant 2$	$2 < x \leqslant 3$	$3 < x \leqslant 4$	$4 < x \leqslant 6$	$6 < x \leqslant 10$
Number of journeys	38	30	21	14	9	8

a Draw a histogram to represent this information.

b During a different week the taxi driver makes fewer journeys as shown in the histogram:

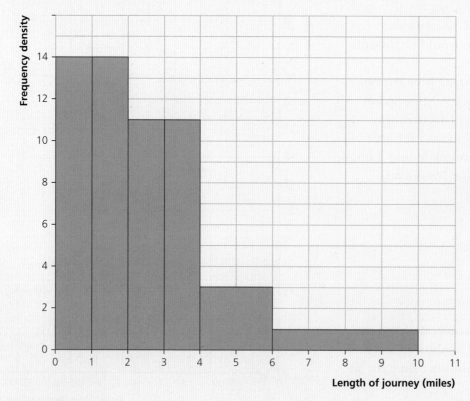

He claims that a quarter of all his journeys were one mile or under. Is this claim justified?

E **3** The histogram shows the distribution of ages of members of a local swimming club.

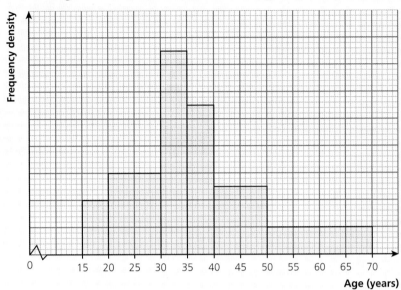

There are 300 members in total.

The bar between 20 and 30 represents the 60 members aged 20 or over but less than 30.

The table shows the monthly membership fees for different age categories.

Age	Monthly fee
15 and under	£23.50
16–64	£31.00
65 and over	£23.50

Estimate the total amount of membership fees that are paid in a month.

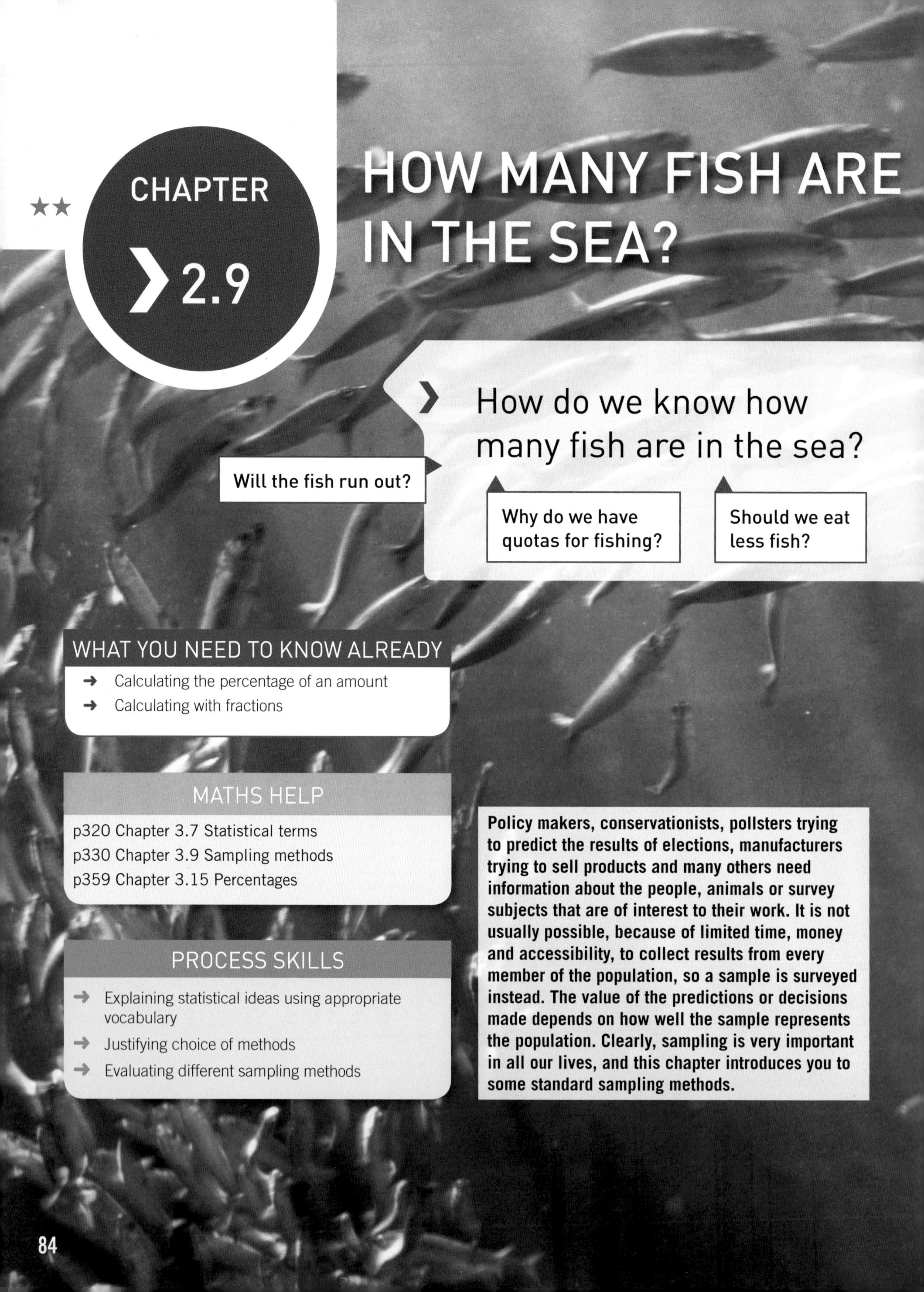

CHAPTER

> 2.9

HOW MANY FISH ARE IN THE SEA?

> ## How do we know how many fish are in the sea?

Will the fish run out?

Why do we have quotas for fishing?

Should we eat less fish?

WHAT YOU NEED TO KNOW ALREADY

→ Calculating the percentage of an amount
→ Calculating with fractions

MATHS HELP

p320 Chapter 3.7 Statistical terms
p330 Chapter 3.9 Sampling methods
p359 Chapter 3.15 Percentages

PROCESS SKILLS

→ Explaining statistical ideas using appropriate vocabulary
→ Justifying choice of methods
→ Evaluating different sampling methods

Policy makers, conservationists, pollsters trying to predict the results of elections, manufacturers trying to sell products and many others need information about the people, animals or survey subjects that are of interest to their work. It is not usually possible, because of limited time, money and accessibility, to collect results from every member of the population, so a sample is surveyed instead. The value of the predictions or decisions made depends on how well the sample represents the population. Clearly, sampling is very important in all our lives, and this chapter introduces you to some standard sampling methods.

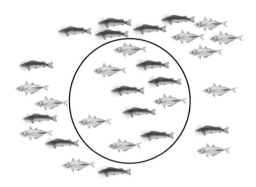

How many fish in a pond?

Ecologists catch a sample of 20 fish from a pond, tag them (harmlessly!) and put them back in the pond. The next day, they catch a sample of 15 fish and note that 3 of them are tagged. How can this information be used to estimate the number of fish in the pond?

Assume that the proportion of tagged fish in the second sample is the same as the proportion of fish in the whole pond.

So $\dfrac{3}{15} = \dfrac{20}{N}$

where N is an estimate of the number of fish in the pond.

$$N = \frac{20 \times 15}{3}$$

So $N = 100$.

Discuss

> Why is this an estimate rather than the exact number of fish in the pond?

> Discuss the limitations of this method (which is called capture/recapture). How could the accuracy of the estimate be improved?

Try capture/recapture for yourself

Put a large number of objects (counters, matchsticks, Multilink pieces or slips of paper, for example) in a box or bag. Use capture/recapture to estimate the number of objects in the bag. Collect different sizes and numbers of samples and calculate the corresponding estimates – then check their accuracy at the end when you count the number of objects.

More sampling

A common method for determining the number of plants in a field is to take samples using a quadrat. This is a frame 1 metre by 1 metre.

WORKED EXAMPLE

A researcher puts a quadrat in a random place in a 0.5 hectare field (a hectare is $10\,000\,\text{m}^2$) and counts the number of dandelions inside the square metre. To increase the accuracy, the researcher takes a total of ten samples.

The numbers of dandelions in the ten samples are 4, 1, 2, 4, 3, 1, 6, 0, 3, 5.

Estimate the number of dandelions in the field.

SOLUTION

The median number of dandelions per square metre is 3. So an estimate of the number of dandelions in the field is $3 \times 5000 = 15\,000$.

❯ Could you work out an estimate for the number of dandelions in a different way?

Why and how to use sampling

You can only be completely accurate about the number and characteristics of a population if you collect information from every member of the population – as in a census.

Discuss

What are the advantages and disadvantages of conducting a census?

How could you conduct a census for the fish and dandelion examples above?

In many circumstances, a census is impractical and a representative sample of the population is taken instead. The planning and conducting of appropriate sampling is an important part of all sorts of research.

There are several different types of samples, including random, stratified, quota and cluster.

What do you know already about these?

In **random sampling**, each member of the population has an equal chance of being chosen. For example, if we want a 10% sample from a population of 100 people, we could put a named slip of paper for each person into a hat and pick out 10 slips at random.

Alternatively, random numbers can be used to choose participants from a numbered list of the members of the population.

Stratified sampling is about making sure that the sample represents one or more characteristics of the population. For example, if the population consists of 60 males and 40 females, the sample should consist of males and females in the ratio 3:2. Within both parts of the population, random sampling is used to choose members of the sample.

A variation of stratified sampling is **quota sampling**. The population is split into categories as above, but maybe the researcher wants to interview twice as many males as females, so the sample would be chosen to reflect this.

Quotas may be obtained by a researcher, perhaps working in a town centre, choosing to ask people of a particular age range or other category as required for the survey. This method may be subject to **sample bias.**

❯ What do you think is meant by the term sample bias?

In **cluster sampling**, a random selection of groups is chosen to represent the population. For example, if a 10% sample of the pupils in a school is required, 10% of the form groups could be chosen.

WORKED EXAMPLE

A UK company wants to know about people's shopping habits. Do they shop once a week or daily? Do they go to a large supermarket, the local corner shop or order online?

For each of the sampling methods, describe how it could be used to choose a sample of 1000 and decide which you would recommend and why.

SOLUTIONS

(Note that there are many alternative solutions.)

Random sample

Use a list of all the landline numbers and use a random number generator to pick the ones for the sample.

Cluster sample

Choose a town and take a random sample from the electoral register.

Stratified sample

Decide whether particular groups need to be represented in your sample, for example, adults who are in work and adults who are not working. Choose a random sample in the right proportions from the electoral roll.

Quota sample

Decide how many young, middle-aged and older men and women you wish to ask. Define young as 18 to 35, middle-aged as 36 to 55 and older as over 55.

Either telephone or do street interviews to get the data.

I would recommend a quota sample. All of the methods should give a reasonably representative sample. If you used landline numbers you would include more people than if you used the electoral roll, but you would be choosing households rather than individuals. Since shopping is done on behalf of a household that would be valid. However, the other methods are more time-consuming, and so more costly, than the quota sample.

You would have to take care that the sample was taken from a variety of locations.

There are ways to improve the sample to make it more representative, such as increasing the sample size and careful planning to avoid bias in the sample. Both of these are likely to increase the time taken and cost of the survey. Getting a representative sample is very difficult, and a compromise is inevitable between cost and accuracy.

QUESTIONS

1 Researchers capture a sample of 25 birds in a forest, tag them, and release them back into the forest. Two days later, they capture a second sample of 25 and note that 4 of them are tagged. Estimate the number of birds in the forest.

2 The numbers of wild garlic plants in five quadrat samples in a 2400 m^2 field is:

0, 2, 1, 2, 7

Estimate the number of wild garlic plants in the field.

3 Three different moss species were found in a quadrat sample of a 100 m^2 lawn. Explain why it is not correct to say that there are an estimated 300 different species of moss in the whole lawn.

4 A workplace has 1250 females and 750 males on its staff. A quota sample of 60 people, made up of twice as many females as males, is taken for research into working hours. Who has the greater probability of being chosen, a female or a male?

5 Give an example of cluster sampling and discuss its advantages and disadvantages.

E 6 A sample of 150 pupils, stratified by year group, is needed for a survey.

The numbers of pupils in each year are shown in the table.

Year	7	8	9	10	11	12	13
Number of pupils	150	167	148	170	162	90	75

Work out how many pupils are from Year 9.

CHAPTER

> 2.10

WINNING THE HEPTATHLON

Is it better to be good at field events or track events?

> ## Is the scoring system for the heptathlon fair?

How do they decide how the points work?

How can you compare performance in different events?

Comparing data using an average can give an idea about whether one set is 'better' than another. However, as we have seen in previous chapters, a measure of spread is useful in that it gives information about the variability of the data. The range and the interquartile range are only calculated using particular values in the data, and so are only sensitive to changes in these values. The same applies to the median and mode. The mean, however, is affected by all data values.

A measure of spread that is affected by all values is standard deviation. The mean and standard deviation are widely used to analyse data for this reason. We shall look at some applications in this chapter.

WHAT YOU NEED TO KNOW ALREADY

→ Calculating the mean and mode from a set of data

→ Constructing a stem-and-leaf diagram

MATHS HELP

p322 Chapter 3.8 Representing and analysing data
p333 Chapter 3.10 Calculating statistics
p337 Chapter 3.11 Diagrams for grouped data

PROCESS SKILLS

→ Representing data
→ Comparing related data sets
→ Interpreting charts and calculations
→ Drawing conclusions

The data gives the points scored by the first ten athletes in the women's heptathlon in the 2012 Olympics.

	Athlete	Total points	100 m hurdles	High jump	Shot put	200 m	Long jump	Javelin throw	800 m
1	Jessica Ennis	6955	1195	1054	813	1096	1001	812	984
2	Lilli Shwarzkopf	6649	1086	1016	845	908	943	894	957
3	Tatyana Chernova	6628	1053	978	805	1013	1020	788	971
4	Austra Skujyte	6599	978	1132	1016	848	927	882	816
5	Antoinette Nana Djimou Ida	6576	1130	978	811	913	890	974	912
6	Jessica Zulinka	6480	1178	830	848	1047	822	778	977
7	Kristina Savitskaya	6452	1069	1016	845	937	915	738	932
8	Laura Ikauniece	6414	1020	1016	704	965	890	885	934
9	Hanna Melnychenko	6392	1077	978	725	972	975	742	923
10	Brianne Theisen	6383	1080	1016	720	947	853	792	975
	Mean	6553	1087	1001	813	965	924	829	938
	Standard deviation	172.8	66.5	75.8	90.0	72.3	63.2	76.7	49.8

The mean and standard deviation have been worked out for the points in each event and for the total for the ten athletes listed in the table. There were more who took part and so the actual mean is lower and the actual standard deviation is greater.

❯ Why is this?

Discuss

❯ What do you notice about the results?

❯ Which is the easiest event? How do you know?

❯ Which event is the most challenging? How do you know?

❯ In which event is it best to improve your performance? Justify your choice.

The standard deviation is a calculation to measure the average (standard) distance (deviation) from the mean of the data. It tells you the zone around the mean where most of the data lies. A small standard deviation means that most of the data lies in a small zone around the mean; a larger standard deviation means that most of the data lies in a larger zone around the mean, so is more spread out. Another way to say it is more spread out is to say it is more variable, has greater variability or is less consistent.

There are several ways to calculate standard deviation. The results in the table above were calculated on a spreadsheet.

DID YOU KNOW?

You can also work out the standard deviation using a function '=STDEV(A1:A10)' if the data is in cells A1 to A10 (not required for the exam).

Investigate

Set up a spreadsheet for one of the events and use an appropriate formula to work out the mean. Did you get the same result?

You can work it out using a formula and an ordinary calculator. In this qualification you are expected to use the statistical functions on a scientific or graphical calculator. The next example shows you how to do this. You will need to find out exactly how your calculator does it as, although they work in the same general way, there are slight differences in notation and sequences of key presses.

DID YOU KNOW?

μ (mu) is the Greek letter m, and is used to represent the mean. σ (sigma) is the Greek letter s, and is used to represent the standard deviation of a population.

WORKED EXAMPLE

Here are the times (in seconds) of 12 semi-finalists in the women's 50 m freestyle competition:

| 24.93 | 25.05 | 25.06 | 25.09 | 25.19 | 25.21 | 25.22 | 25.27 | 24.21 | 24.37 | 24.83 | 25.63 |

Calculate the mean and standard deviation of the results.

SOLUTION

The mean $= \dfrac{24.93 + 25.05 + 25.06 + 25.09 + 25.19 + 25.21 + 25.22 + 25.27 + 24.21 + 24.37 + 24.83 + 25.63}{12}$

$= 25.01$ seconds

To find the standard deviation, use your calculator and enter the statistics mode. Usually this is found by pressing Mode and selecting SD.

Then clear the statistics memory on your calculator – all calculators vary, but this is often done by selecting CLR and choosing the correct option.

We enter the data into the calculator by typing each value and pressing M+ after each one.

Once all the data is entered, locate the S-Var button on your calculator (you may need to use the shift key first) and select the option for x_{σ_n} or $x_{\sigma_{n-1}}$ and press the = button.

The '$n - 1$' version is recommended but both can be credited.

So the standard deviation of the data above is 0.389300071... = 0.39 seconds to 2 decimal places.

Investigate

> Find some data for a comparable men's swimming race. Calculate the mean and standard deviation and compare it with the women's race.

> What do you notice?

> Are there any differences between them?

> Which do you think was a more exciting race?

The frequency table shows the average number of visitors to a website each day during a two-week experiment.

You have seen, in Chapter 2.5, how to calculate an estimate of the mean number of visitors over the fortnight.

Number of visitors	Frequency
$0 \leqslant x < 1000$	3
$1000 \leqslant x < 2000$	5
$2000 \leqslant x < 6000$	4
$6000 \leqslant x < 10000$	2

What assumptions would you need to make? What are the limitations of your estimate?

How could you calculate an estimate of the standard deviation of the data? What are the limitations of your estimate?

Investigate

> Choose five single-digit numbers. For example, {1, 2, 6, 6, 9} or {1, 5, 5, 5, 8}. Calculate the standard deviation of the five numbers you chose. By choosing different sets of five numbers, what are the largest and smallest standard deviations you can make?

QUESTIONS

1 a Calculate the mean and standard deviation of the following car insurance quotes:

£424.00	£396.50	£516.87	£414.20
£489	£786	£439.75	£418.99

b Which quotes lie outside one standard deviation from the mean?

2 Here are the exchange rates between the pound and the euro over a fortnight. The rates show the number of euros that can be bought for £1.

1.29	1.23	1.24	1.30	1.31	1.29
1.28	1.25	1.21	1.22	1.22	1.25

a Calculate the mean exchange rate.

b Calculate the standard deviation of the data.

Another foreign exchange desk has a mean of 1.26 and a standard deviation of 0.03 over the same time period.

c Compare the exchange rates of the two desks.

3 A biologist has collected the pulse rates of people before and after exercise, measured in beats per minute.

The results are shown in the stem-and-leaf diagram below.

Before exercise								After exercise				
						9	5					
				9	8	8	6	8				
8	6	6	4	1	1		7					
			8	7	6	3	8	7	8	9	9	9
					5	1	9	2	3	5	5	
						4	10	1	1	3	6	
						0	11	0	2	5		
							12	0	1			

Key 12 | 0 means 120

a Calculate the mean and standard deviation of the pulse rate before exercise and of the pulse rate after exercise.

b Hence describe what the data shows about the impact of exercise on pulse rate.

E 4 The mean government spending per person in the UK as a whole in 2014/2015 was £8913.

The table below shows the mean government spending in £ per person across the different regions of the UK in the 2010/11 tax year and then in the 2014/15 tax year.

Region	2010/2011	2014/2015
England	9134	8638
Scotland	10662	10374
Wales	10304	9904
Northern Ireland	11355	11106

Compare the government spend per person by region and year, commenting on any trends. Compare the mean UK government spend of £8913 with the data in the table.

5 The table shows the number of emergency NHS England admissions each month in 2011 and 2015.

	2011	2015
January	442 004	463 401
February	401 206	424 635
March	446 846	474 941
April	419 243	450 253
May	427 277	468 416
June	413 321	458 790
July	425 798	473 914
August	410 554	455 432
September	410 761	464 195
October	435 080	479 987
November	424 887	475 564
December	441 655	487 798

a Calculate the mean number of admissions per month in 2011 and 2015.

b Calculate the standard deviation of the number of admissions per month in 2011 and 2015.

c Hence compare the admissions per month in 2011 and 2015.

PAYING FOR UNIVERSITY

> ## How can I afford to go to university?

I don't want to get into debt!

How else can I get a really good job?

Will it be worth it in the end?

WHAT YOU NEED TO KNOW ALREADY

→ Calculating with percentages

MATHS HELP

p352 Chapter 3.13 Formulae and calculation
p359 Chapter 3.15 Percentages
p364 Chapter 3.16 Interest

PROCESS SKILLS

→ Comparing financial options
→ Calculating repayments
→ Interpreting financial information

Many years ago students used to get a grant to support them through university. It was means tested, which meant that how much you got depended on your parents' income. With ever increasing numbers of young people going to university that became very expensive and so gradually students have had to pay more of the cost themselves. With most universities charging tuition fees of around £9000 per year it is necessary for most students to take out loans. The intention is that their increased earning power will allow them to repay the loan over their working life.

Many students choose to take out loans from the Student Loans Company (SLC), a government-owned organisation, to support them through university education. Student loans are available in two forms: tuition fee loans and maintenance loans. Tuition fee loans are paid to the university to cover the cost of studies whereas maintenance loans are paid directly to the student to cover their living costs. While studying, interest is added to student loans, although this tends to be at a lower rate than traditional loans since it is a loan from the Government. After graduating, and once earning a salary, students are required to pay back their loans when their gross income is over a particular threshold. In this chapter we will explore the repayment plans for student loans and how interest is calculated for student loans.

Investigate

> What are scholarships and bursaries and how can they support you through university education?

> What are the maximum amounts that you can borrow through a tuition fee or maintenance loan?

> How long do you have to repay a student loan?

> What are the two types of loans available to students?

> Interest is added to Bismark's student loan at a rate of 3.25% a year. Currently Bismark owes £34 895. How much interest will have been added to Bismark's loan in two years' time?

Repaying the loan

There are currently two types of repayment plan for student loans.

If you started your course before 1 September 2012 you will be on repayment plan 1.

If you started your course on or after 1 September 2012 you will be on repayment plan 2.

Repayment plan 1

You only start making repayments when your gross income is over the threshold of £17 495 a year (figure correct for the 2016/2017 tax year).

You pay 9% of anything you earn over the threshold.

Repayment plan 2

You only start making repayments when your gross income is over the threshold of £21 000 a year.

You pay 9% of anything you earn over the threshold.

For either plan you can also make additional voluntary repayments to the SLC at any time, which will reduce your balance earlier.

WORKED EXAMPLE

Louis started a three-year course at university in September 2012.

Louis' first job has an annual salary of £43 000, paid monthly.

He begins repaying his student loan at the end of the first month after he starts his job.

Work out his first repayment.

SOLUTION

Since Louis started his university course in September 2012 he will follow repayment plan 2.

The threshold for repayments is therefore £21 000.

£43 000 − £21 000 = £22 000

Louis will repay 9% of £22 000.

0.09 × 22 000 = 1980

Louis will repay £1980 annually.

Therefore his first repayment will be

1980 ÷ 12 = 165

Louis' first repayment will be £165.

HINT

Read the question carefully to work out which repayment plan to select.

Discuss

This may seem a considerable commitment. Many students now have part-time jobs to support themselves while at university. Others choose to live at home and go to a local university to reduce the cost. Some are lucky enough to have parents who can afford to pay their fees and tuition costs. Some are aiming for high earning careers and so will easily repay the loan.

> What is your situation?

> What is your likely salary after you graduate?

> Could you study from home or get a part-time job?

WORKED EXAMPLE

Carrie started a three-year course in September 2012.

At the start of the course she received a loan from the SLC to cover her tuition fees of £9000 per year and living costs of £4800 per year.

At the end of each year the SLC added 5.5% interest to the amount she owed.

Calculate the amount that Carrie owed the SLC at the end of her course.

SOLUTION

Year 1: £9000 + £4800 = £13 800
　　　　£13 800 × 1.055 = £14 559
Year 2: £14 559 + £9000 + £4800 = £28 359
　　　　£28 359 × 1.055 = £29 918.75
Year 3: £29 918.75 + £9000 + £4800 = £43 718.75
　　　　£43 718.75 × 1.055 = £46 123.28
Carrie owes the SLC £46 123.28 at the end of her course.

> **HINT**
> Use the answer key on your calculator to maintain accuracy in your answer.

Investigate

〉 Research the student loans available and use the current interest rates and thresholds for payments. Assume a starting salary that will allow you to start paying back the loan.

〉 Set up a spreadsheet to calculate how the amount owed varies over the first few years.

QUESTIONS

1 Saskia started her university course in September 2011 and graduated in 2014. Saskia then obtained a job earning £15 500 per year. She predicts that her wage will increase by 2.5% per year.

  a After how many years of work will Saskia have to start paying back her student loan?

  b How much will Saskia repay towards her student loan in the first year?

2 Alex finished university with a student loan of £14 000. He started work with a salary of £19 000 per year. After one year, he had a pay rise of £1500. At the end of each full year of work:

– 9% of his earnings above £17 495 (2016/2017 tax year) go towards paying off his loan.
– Interest of 1.5% of the outstanding amount is added to his loan.

How much will Alex still owe on his student loan after 2 full years of work?

 E 3 Tom is going to university. He has been granted a maintenance loan of £6000 per year. Tom's degree course is four years. His loan accrues interest at a rate of 3.9% APR while he is studying. Tom would like to repay his loan in a single repayment after he graduates. Calculate Tom's repayment.

CHAPTER
> 2.12

MAKING YOUR MONEY WORK FOR YOU

> ## What would you do with £1000?

Shall I spend it on a holiday?

Should I save it for when I have my own place?

Should I use it to buy a car?

WHAT YOU NEED TO KNOW ALREADY

→ Calculating percentages
→ Rounding to 2 decimal places

MATHS HELP

p356 Chapter 3.14 Approximation
p359 Chapter 3.15 Percentages
p364 Chapter 3.16 Interest
p368 Chapter 3.17 Graphs with a financial context

PROCESS SKILLS

→ Evaluating modelling assumptions
→ Comparing representations
→ Justifying decisions
→ Interpreting representations

Putting money aside 'for a rainy day' is a sensible thing to do and some ways of doing it are better than others. Deciding what to do with surplus earnings or a windfall or legacy needs knowledge of what is available. The advantages and disadvantages of different ways of saving or investing large and small sums of money depend on whether you are looking for safety or are willing to take a risk. It may also matter how quickly you may want to gain access to the money. In this chapter, we look at some aspects of saving and investing money and show you how to calculate the interest you might receive on those savings.

Discuss

> What's the difference between savings and investments?

> Can you find examples of each?

This diagram shows what could happen to your £1000 if you put it into an **ISA (Individual Savings Account)** and left it there for 10 years.

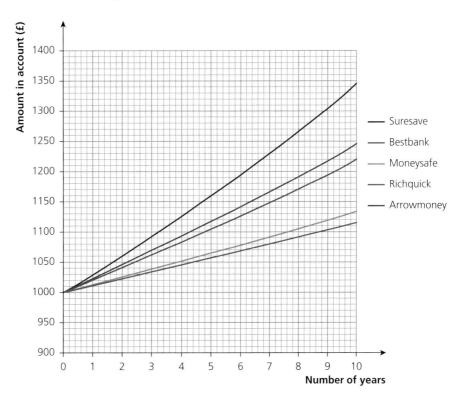

Discuss

> What assumptions have been made to draw these graphs?

> Which account would you choose?

> What else should you consider before deciding?

All the ISA graphs have positive gradients, indicating that after 10 years you have your original £1000 plus interest.

Notice that the graphs are not quite straight lines; this is compound interest, meaning that at the end of the year the interest is added on to the money in the account so next year more interest is earned because the amount is bigger. With compound interest, you get interest on the interest you've already gained.

Therefore, the gradients of the graphs become steeper year by year.

Simple interest is not added to what is already in the account, so the interest earned in the next year does not increase. The amount in the account stays the same and the interest is 'paid away' into a different account or given as a payment to the saver.

DID YOU KNOW?

Low interest rates mean the graphs are close to straight lines. High interest rates give graphs where the curve is more noticeable.

WORKED EXAMPLE

a What is the simple interest on £1000 at 1.5% after 10 years?

b What is the compound interest on £1000 at 1.5% after 10 years?

SOLUTION

a For a simple interest rate of 1.5% on £1000 saved for a year,

Interest $= £1000 \times \frac{1.5}{100}$

$= £1000 \times 0.015$

$= £15$

So interest earned over 10 years

$= £15 \times 10$

$= £150$

This can be done in one step:

Total interest $= £1000 \times 0.015 \times 10 = £150$

b To calculate the compound interest, we can work year by year.

For an interest rate of 1.5%,

Amount at end of first year $= £1000 \times 1.015$

Amount at end of second year $= £1000 \times 1.015 \times 1.015$

$= £1000 \times 0.015^2$

Amount at end of third year $= £1000 \times 1.015^3$

So, amount at end of 10th year $= £1000 \times 1.015^{10}$

$= £1160.54$ to the nearest penny

So the interest earned is £160.54

DID YOU KNOW?

The big advantage of ISAs used to be that you paid no tax on the interest, but at the time of writing, in 2017, no tax is paid on interest on any savings until the interest is more than £1000, which is more than the vast majority of savers would receive.

So after 10 years you have made £160.54 on your £1000 – not very impressive, because interest rates at the time of writing are generally very low. In 10 years, things may be very different.

› What if the interest rate was 8% per annum? How much would £1000 be worth after 10 years?

Investigate

› How much would you have to have in savings to gain £1000 interest each year if the interest rate was 2%?

Investigate

› In the diagram, the Suresave ISA has an interest rate of 3%. Calculate the amount in the account after 10 years and check that your answer agrees with the graph. Estimate the interest rates of the other ISAs.

Savings are generally a safe way of looking after your money – you don't lose what you put in, but are unlikely to gain much. Investments are more risky, because you have the chance of losing some or all of your money – but your investment could increase considerably.

Discuss

> This graph shows the actual performance of an investment fund over the three years to April 2016. Is this a better or worse way of looking after £1000 than the ISAs shown in the first diagram? Why?

Growth of £1000 over 3 years
Total return performance of the fund rebased to 1000. Your actual return would be reduced by the cost of buying and selling the fund, and inflation

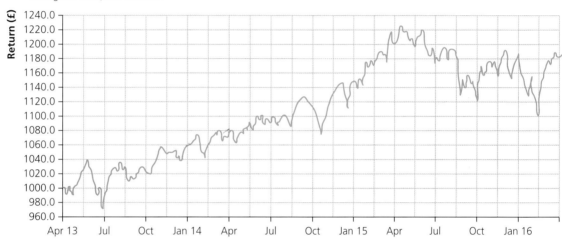

AER

The worked examples above show interest being calculated once each year, but in fact it can be calculated half-yearly, quarterly, monthly or even daily. If the interest is compounded, this results in (a bit!) more money in the account, as the example below shows.

WORKED EXAMPLE

£100 is put into a savings account with an annual interest rate of 2.4%. Interest is compounded quarterly. How much will be in the account at the end of the year?

SOLUTION

2.4% annual interest is 0.6% interest per quarter.

At the end of the year, the amount will be £100 × 1.006^4 ≈ £102.42

If the interest were compounded only at the end of the year, the amount would be £100 × 1.024 = £102.40

So the quarterly compounding results in a gain of 2 pence!

The interest of £2.42 is 2.42% of £100 and this is called the **AER (Annual Equivalent Rate)**. All savings accounts are required to quote the AER so that potential customers can compare like with like. This is the standard requirement for savings; for borrowing the corresponding term is **APR (Annual Percentage Rate)**; you will study this in the next chapter.

> What if the interest rate was 8.4%? What would be the effect of compounding the interest quarterly?

Investigate

You can work out the AER as shown in the example above. There is also a formula – see if you can use the method above to derive the formula:

$$r = \left(1 + \frac{i}{n}\right)^n - 1$$

where r is the AER, i is the annual rate (both expressed as decimals) and n is the number of times per year the interest is compounded.

> Use the formula to obtain the AER for the example above.

Where do all the pennies go?

Amounts of money are rounded to the nearest penny or truncated (rounded down) to appear on bank and building society statements.

WORKED EXAMPLE

£20 000 is invested at 2% interest for 12 years with the interest being compounded annually.

What difference does it make if the bank records the total to, say, 6 decimal places, or rounds it to the nearest penny, or truncates it?

Express these differences using error intervals.

SOLUTION

The table gives the amounts in £ at the end of each year using each method.

Years	Amount when storing 6 decimal places	Amount when rounded	Amount when truncated
1	20 400.00	20 400.00	20 400.00
2	20 808.00	20 808.00	20 808.00
3	21 224.16	21 224.16	21 224.16
4	21 648.64	21 648.64	21 648.64
5	22 081.62	22 081.61	22 081.61
6	22 523.25	22 523.24	22 523.24

Although amounts in the second column are stored using 6 dp the table shows them rounded to the nearest penny.

Years	Amount when storing 6 decimal places	Amount when rounded	Amount when truncated
7	22 973.71	22 973.70	22 973.70
8	23 433.19	23 433.17	23 433.17
9	23 901.85	23 901.83	23 901.83
10	24 379.89	24 379.87	24 379.86
11	24 867.49	24 867.47	24 867.45
12	25 364.84	25 364.82	25 364.79

The amount is greatest, in this case, when the more accurate figure is recorded. Truncating gives 5 pence less after 12 years, rounding gives 2 pence less. This is not much for an individual, but for a bank with 50 000 accounts of this size truncating would mean an extra £2500 in income.

The error interval in this case is [£25 364.82, £25 364.84] when the amount is rounded, and [£25 364.79, £25 364.84] when it is truncated.

Discuss

> Will both rounding and truncating give smaller amounts over time?

Investigate

> Investigate the effect of different interest rates on what you could receive after rounding or truncating and compare it with the amount calculated with more decimal places.

> Could it ever make a significant difference?

Use a spreadsheet to help you do the calculations.

> What is the effect of truncation if interest is compounded monthly?

Discuss

> Can you explain the difference between savings and investments?

> Use a spreadsheet to calculate the simple and compound interest each year for an investment at a rate of 2.5% per annum over 10 years.

> Adapt the spreadsheet to compare the return on an investment using both simple and compound interest for different rates, again over 10 years.

> Investigate the effect of the number of times interest is compounded each year for various annual rates. How does the AER vary?

You may wish to present your findings using a graph.

QUESTIONS

1 Calculate the simple interest earned on £800 saved for 6 years at an AER of 1.45%.

2 How much more is earned by compound interest than by simple interest when £8500 is invested at 2.5% per annum for 5 years?

3 Draw a graph to show how £10 000 put into a savings plan for 5 years at an AER of 5% for the first 2 years and 3% for the remaining 3 years grows. (You could draw this by hand or set up a spreadsheet to do the calculations and produce the graph.)

4 Calculate the AER for savings invested at an annual rate of 5% compounded twice a year (biannually).

5 Look at the graph of the investment fund on page 101.

a Calculate the percentage growth in the fund in the first year.

b Did the fund grow by this amount over the full three years? (Remember to compound the growth!)

c How accurate are your results, considering they are based on readings from the graph? Can you put upper and lower limits on the growth rate?

HINT

Decide on the range of accuracy you can get from the graph. Work through the problem using the lowest and then the highest values in that range.

LOOKING FOR A LOAN

Can I buy without a loan?

> Which loan should I get?

Will I qualify?

Will I be able to repay it?

How much will it cost?

Most people at some time in their lives decide to take out a loan. It may be to pay for a car, or furniture or to help them pay for an unexpected expense. You can choose between different times for repaying the loan, but the longer you have the money for, the more interest you have to pay. The monthly payments are lower if you choose a longer repayment term, which makes it more affordable in the short term, but you will be repaying it for a longer time.

WHAT YOU NEED TO KNOW ALREADY

→ Substituting values into formulae
→ Rearranging equations
→ Calculating with percentages

MATHS HELP

p352 Chapter 3.13 Formulae and calculation
p359 Chapter 3.15 Percentages
p364 Chapter 3.16 Interest

PROCESS SKILLS

→ Comparing options
→ Calculating using formulae
→ Deducing financial information

For most loans, which are often obtained from banks or independent sources, there are many ways of borrowing and paying back a sum of money. This may involve administration fees, regular instalments, insurance payments, or other fees which a borrower is required to pay. The Annual Percentage Rate (APR) is a way of comparing the costs of different schemes. It is, by law, given in all advertisements for borrowing money.

Investigate

There are things that you may come across when you are looking for a loan:

> What is a credit rating and how can this affect your ability to get a loan?

> What is *representative* APR and how does this affect the borrower?

> What are pay day loans and what are the rules relating to these types of loans?

> Why has the popularity of pay day loans increased?

This chapter is about short-term loans that you usually have some choice about taking out. You have seen how student loans work, and in the next chapter you will learn about mortgages. Both of these are loans that you repay over a long period of time.

Discuss

> What do you think are the advantages of taking out a loan for something rather than saving up for it?

> Why is APR important?

> Alice takes out a loan for £2000 at 4.9% APR. What does this mean?

> What is gross income?

How it works

The APR can be found using the following formula:

$$C = \frac{A_1}{(1+i)^{t_1}} + \frac{A_2}{(1+i)^{t_2}} + \ldots + \frac{A_m}{(1+i)^{t_m}}$$

where

i is the APR expressed as a decimal,

C is a loan of £C,

A_m is the amount in £ for instalment m,

t_m is the interval in years between the payment of instalment m and the start of the loan.

WORKED EXAMPLE

A loan of £5000 is repaid in three equal annual instalments of £2000.

The APR is quoted as 9.7%.

Is this correct?

SOLUTION

For three annual instalments the formula becomes

$$C = \frac{A_1}{(1+i)^{t_1}} + \frac{A_2}{(1+i)^{t_2}} + \frac{A_3}{(1+i)^{t_3}}$$

APR is quoted as 9.7% so $i = 0.097$.

Each instalment is equal and annual, so $A_1 = A_2 = A_3 = 2000$ and $t_1 = 1, t_2 = 2, t_3 = 3$.

Substituting these values into the formula gives

$$C = \frac{2000}{(1+0.097)^1} + \frac{2000}{(1+0.097)^2} + \frac{2000}{(1+0.097)^3}$$

$$= 5000.09$$

This is very close to the expected £5000 and therefore confirms that the APR quoted as 9.7% is correct.

WORKED EXAMPLE

A debt of £4000 to pay for a car is repaid in two equal annual instalments.

The APR is quoted as 10%.

Find the approximate value of the repayments correct to two decimal places.

SOLUTION

For two annual instalments the formula becomes

$$C = \frac{A_1}{(1+i)^{t_1}} + \frac{A_2}{(1+i)^{t_2}}$$

The loan is for £4000, so $C = 4000$.

APR is quoted as 10%, so $i = 0.1$.

Each instalment is equal and annual, so $A_1 = A_2 = A$ and $t_1 = 1, t_2 = 2$.

Substituting these values into the formula gives

$$4000 = \frac{A}{(1+0.1)^1} + \frac{A}{(1+0.1)^2}$$

The denominators can be written as decimals as follows:

$$\frac{A}{(1+0.1)^1} = \frac{A}{(1.1)^1} = \frac{A}{1.1} = \frac{1}{1.1}A = 0.90909A$$

Now solve the following equation to find the value of A:

$4000 = 0.90909A + 0.82645A$

$4000 = 1.73554A$

$A = 2304.76$

The value of each instalment is approximately £2304.76.

> ### HINT
>
> The formula used for working out APR is
>
> $$C = \sum_{k=1}^{m}\left(\frac{A_k}{(1+i)^{t_k}}\right)$$
>
> The mathematical symbol sigma Σ means *sum*.
>
> The formula may look complicated but just substitute in the given values and work out the required value.

Investigate

> ❯ Look for examples of deals that offer finance for a car, or double glazing or laptops.
>
> ❯ Do they quote an APR?

Discuss

> ❯ Is taking out a loan a better deal than going to a bank?
>
> ❯ If you have savings is it better to use those than to take out a loan?

WORKED EXAMPLE

Emily has just graduated and has recently started her first full-time job. She decides to borrow some money to buy a laptop. She will repay the loan over three years.

A lender agrees to give her a loan. They tell Emily that her repayments will be £206.40 a year and that her APR will be 11.5%.

Use the APR formula to find, to the nearest £100, how much Emily's loan was for.

➡

SOLUTION

$A_1 = A_2 = A_3 = 206.4$

$t_1 = 1, t_2 = 2, t_3 = 3$

$i = 0.115$

$$C = \frac{206.4}{(1.115)^1} + \frac{206.4}{(1.115)^2} + \frac{206.4}{(1.115)^3}$$

$$= 500.03$$

Emily's loan was for £500.

QUESTIONS

**Dreaming of a new car?
Get the car of your dreams with
this simple, great-rate car loan.**

4.9% APR representative
on loans of
£7,500–£15,000
over 2–5 years.
(Your rate may be different)

**Loans subject to status and early settlement fees
apply.**

1 a Matt has been granted the advertised 4.9%
APR on a car loan of £10 000 borrowed
over three years. Work out what Matt will
pay to settle the loan in one payment at the
end of three years.

b Melanie considers borrowing £7500 to buy
a new car. She would have to repay the loan
in a single repayment of £9000 made two
years after taking out the loan. Melanie was
not granted the advertised APR. Work out
the APR for Melanie's loan.

2 Tom is going to university. He has been
granted a maintenance loan of £6000. Tom's
degree course is four years. His loan earns
interest at a rate of 3.9% APR while he is
studying. Tom would like to repay his loan in a
single repayment after he graduates. Calculate
Tom's repayment.

3 Charlotte needs a loan to buy some furniture
for her first home. Charlotte wants to borrow
£3000. She has two options:

Option 1: Repay the loan in a single repayment
of £4500 at the end of two years.

Option 2: Repay the loan in two equal
instalments at 20% APR.

a Calculate the APR for repayment option 1.

b Calculate the size of each of the instalments
for repayment option 2.

4 Ruby has bought a camera costing £750. She
has taken out a loan to spread the cost over two
years. She will pay £450 after one year and
a further £450 after another year. The APR
advertised is 13.1%. Use calculations to confirm
that this is a good approximation.

5 Asif borrows £5000 to buy a car. The APR
quoted is 3.5%. He repays it in three equal
instalments. Work out the amount of each
instalment.

CHAPTER

> 2.14

CAN I AFFORD TO BUY A HOUSE?

> ## Mortgages

I'll never be able to buy a house!

What do I need to earn to get a mortgage?

Is a mortgage expensive?

home

SALE

WHAT YOU NEED TO KNOW ALREADY

→ Using spreadsheet formulae
→ Drawing and interpreting line graphs
→ Calculating simple and compound interest
→ Substituting into formulae
→ Solving equations

You often hear people say how much harder it is now to buy a house than it used to be. Getting a mortgage can also seem quite scary. Knowing a few key facts and understanding the terminology can help give you a more balanced view, and give you more confidence to approach a bank or building society to see what they can offer. This chapter will help you understand what is involved in getting and paying off a mortgage.

MATHS HELP

p356 Chapter 3.14 Approximation
p359 Chapter 3.15 Percentages
p364 Chapter 3.16 Interest
p368 Chapter 3.17 Graphs with a financial context

PROCESS SKILLS

→ Modelling financial problems
→ Justifying choices
→ Interpreting results of calculations
→ Deducing results
→ Estimating costs
→ Representing graphically and algebraically

Discuss

The table below shows three different mortgage offers. Answer the questions below to help you understand the table and basic mortgage terminology.

First Time Buyer Repayment mortgages payable over 25 years

Bank	Loan	Monthly repayment at initial rate	Initial interest rate	Type of mortgage	Max LTV	Product fees	Overall cost for comparison
Argent	£185 000	£851.53	2.73% then 4.49%	Fixed for 5 years	95%	Yes	3.9% APRC representative
Better	£185 000	£935.11	3.59% then 3.99%	Fixed for 2 years	95%	Yes	4.1% APRC representative
Choco	£175 000	£874.22	3.48% for term	Tracker for term	90%	No	3.5% APRC representative

HINT

The mortgage offered by Better Bank is fixed for two years. This means interest is charged on the loan at the initial rate for the first two years (24 months), giving a monthly repayment of £935.11 each month over this period. After two years, the rate will fluctuate depending on the markets, but is currently expected to be 3.99%. Because the rate has increased, the monthly repayment will increase.

Discuss

> What are the advantages of buying a place to live over renting?
> What are the disadvantages?
> Most mortgages are 'repayment mortgages'. What does that mean?
> How much is being borrowed under each of these mortgages?
> What is the difference between a fixed-rate mortgage and a tracker mortgage?
> What does LTV stand for?
> What does the term 'product fees' mean?
> What does APRC (or sometimes referred to as just APR) stand for?
> Which mortgage costs the least each month to start with? How do you know?
> Which mortgage costs the least each month after five years? How do you know? Can you be sure?
> Which mortgage costs the least overall? How do you know? Can you be sure?

How much can you borrow?

Iqbal and Gemma want to buy a house together. Between them they earn £50 000 per year. As a rough guide most banks and building societies will lend up to 4 times your total salary. (Sometimes more might be possible.) So Iqbal and Gemma are looking at properties around £200 000.

No matter which mortgage company they use, they will be expected

to pay a cash deposit towards their house. The minimum deposit they will be expected to pay is 5%, but some mortgage plans require more. The LTV (Loan to Value ratio) provides this information.

$$\frac{\text{Amount of loan}}{\text{Value of property}} \times 100\%$$

WORKED EXAMPLE

Iqbal and Gemma have saved a deposit of £15 000 and have found a property that is on the market for £200 000.

Will they be able to get a mortgage from Argent Bank, Better Bank or Choco Bank?

SOLUTION

They need a loan of £200 000 − £15 000 = £185 000

Hence LTV $= \frac{185\,000}{200\,000} \times 100\% = 92.5\%$

Iqbal and Gemma are therefore able to get a mortgage from Argent Bank or Better Bank but are not eligible for the mortgage from Choco Bank, which requires a maximum LTV of 90%.

Investigate

Iqbal also has an inheritance of £15 000 that he can use towards the deposit. However, when the property they want to buy is valued, it is only valued at £190 000.

> What is the new LTV?

Investigate

Iqbal and Gemma find another property that is valued at £200 000.

> How much of Iqbal's inheritance do they need to use to achieve a LTV of 90%?

DID YOU KNOW?

If Iqbal or Gemma had any other loans (for example, student loans) or financial commitments (for example, childcare) these would be taken into account and would reduce the mortgage loan available to them.

What other costs do you need to pay?

The main costs that need to be met when buying a property are:

- valuation fee (payable to the mortgage lender for valuing the property)

- conveyancing fees (payable to the solicitor for administering the purchase)

- stamp duty (a tax dependent on the property value that is payable to the Government)

- search fees (payable to the local authority to ensure there are no issues that could affect the purchase)

- product fees (bank costs associated with taking out the mortgage)
- survey fees (payable to a surveyor for checking that the house is structurally sound).

All these need to be taken into account when you decide whether you have enough money to buy a property.

How much does the mortgage cost?

WORKED EXAMPLE

Argent Bank require Iqbal and Gemma to pay 62 monthly payments of £851.53 and 238 monthly payments of £993.50 if they take up its mortgage offer. How much interest will they pay?

SOLUTION

This means they will pay a total of

$62 \times £851.53 = £52\,794.86$

$238 \times £993.50 = £236\,453.00$
_____
$£289\,247.86$

The original loan was for £185 000 so the total amount of interest paid is £289 247.86 − £185 000 = £104 247.86

DID YOU KNOW?

Sometimes due to internal bank processes, schedules are not always an exact number of years.

Mortgages are very expensive loans because they are taken out over such a long time. However, the majority of people simply wouldn't be able to buy a house without one.

Investigate

Better Bank require Iqbal and Gemma to pay 26 monthly payments of £935.11 and 274 monthly payments of £972.50.

> What is the total amount of interest they will pay Better Bank if they take up its offer?

How is the loan repaid?

Each monthly payment that Iqbal and Gemma make will pay off some of their loan and some of their interest. The split depends on how far they are into their repayment plan. At the start of the plan they owe the most capital, so the interest will be more and their loan will reduce slowly. Towards the end of the plan, the capital is small and so less interest is paid and the loan reduces quicker. The reduction of the loan in this way is called **amortisation.** Calculation of the amortisation is closely linked to compound interest.

The spreadsheet below shows the first year of the amortisation schedule for the Argent Bank mortgage offer.

	A	B	C	D	E	F
1	Month	Outstanding loan	Monthly repayment	Monthly interest rate	Interest	Capital
2	1	£185,000	£852	0.002275	£421	£431
3	2	£184,569	£852	0.002275	£420	£432
4	3	£184,138	£852	0.002275	£419	£433
5	4	£183,705	£852	0.002275	£418	£434
6	5	£183,271	£852	0.002275	£417	£435
7	6	£182,837	£852	0.002275	£416	£436
8	7	£182,401	£852	0.002275	£415	£437
9	8	£181,965	£852	0.002275	£414	£438
10	9	£181,527	£852	0.002275	£413	£439
11	10	£181,089	£852	0.002275	£412	£440
12	11	£180,649	£852	0.002275	£411	£441
13	12	£180,209	£852	0.002275	£410	£442

Discuss

> The monthly interest rate is calculated as 2.73 ÷ (100 × 12). Explain this calculation.

> What formula is used to calculate E3?

> What formula is used to calculate F3?

> What formula is used to calculate B4?

Another way of finding the **outstanding balance** on the loan at any time is by using the formula:

$$B = A(1 + r)^n - \frac{P}{r}((1 + r)^n - 1)$$

where

B = outstanding balance

A = original loan amount

P = monthly repayment amount

r = monthly interest rate

n = number of months

WORKED EXAMPLE

What is the outstanding balance on the Argent Bank mortgage offer at the end of month 10?

SOLUTION

For month 10,

$A = £185\,000$

$P = £852$

$r = 0.002275$

$n = 10$

So $B = 185\,000(1 + 0.002275)^{10} - \dfrac{852\left((1 + 0.002275)^{10} - 1\right)}{0.002275}$

$\quad = 189\,252.1 - 8607.75 = 180\,644$

So at the end of month 10 (which is the start of month 11) the outstanding balance is $£180\,644$.

Discuss

❯ Is this the same value as shown in the spreadsheet? Explain why there might be a difference.

Investigate

❯ Find the outstanding balance at the end of month 18.

The amortisation of the loan can be clearly seen by plotting a graph of the outstanding balance at the beginning of each year.

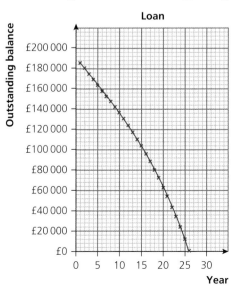

Discuss

❯ The graph is steeper at the end than at the beginning. What does that show?

Investigate

Use the graph to estimate:

❯ the outstanding balance at the end of year 17

❯ how much the balance reduces in year 20.

Working with APR (or APRC)

The APR for the Argent Bank mortgage deal is 3.9%. This means a mortgage with a consistent rate of 3.9% will pay the same amount of interest as the Argent mortgage which has a rate of 2.73% for 5 years followed by a rate of 4.49% for 20 years.

It is possible to calculate the monthly repayment amount that would be needed to give this APR by using the formula

$$A = \frac{P}{r}(1 - (1+r)^{-N})$$

where A = original loan amount

P = monthly repayment amount

r = monthly interest rate

N = total number of repayments

WORKED EXAMPLE

What is the monthly payment on a £185 000 mortgage whose interest rate is 3.9% each year of a 25-year term?

SOLUTION

For Argent's APR,

$A = £185\,000$

$r = \frac{3.9}{100 \times 12} = 0.00325$

$N = 25 \times 12 = 300$

Substituting into the formula above

$$185\,000 = \frac{P(1 - (1 + 0.00325)^{-300})}{0.00325}$$

$$= 191.45\,P$$

$$P = \frac{185\,000}{191.45}$$

$$= £966$$

The monthly repayment would be £966.

Investigate

❯ What would the monthly repayment be on a mortgage with an interest rate the same as the Better Bank mortgage APR over all 25 years?

QUESTIONS

E

1 A building society offers Biyu a mortgage on a new house worth £250 000. It is a fixed-rate repayment mortgage over 30 years. The initial rate of 2.79% is fixed for 3 years, rising to an expected rate of 2.99% thereafter. Biyu has a deposit of £30 000. She will have to pay 39 monthly payments of £902.80 and 321 monthly payments of £924.14.

    a Work out how much interest she will pay altogether.

    b Work out how much she will owe after the first three years.

2 Sam earns £35 000 per year. He wants to buy a small flat. He has saved £15 000 that he could put towards a deposit.

    a What size of mortgage is Sam likely to get?

    b He has found a flat he likes that is on the market for £150 000.

      i What size of mortgage will he need for this flat?

      ii What LTV will he have?

    c What other costs will Sam need to cover when he buys a house?

3 The table shows the annual amortisation of Abioye's mortgage.

    a How much was Abioye's mortgage for?

    b What are the **monthly** repayments?

    c How much interest does Abioye pay in total?

d How much of the loan does Abioye pay off in the first ten years?

e How much of the loan does Abioye pay off in the last ten years?

f Plot a graph to show the amortisation of the loan.

Year	Loan at start of year	Interest	Capital	Loan at end of year
1	£225 000.00	£6666.08	£6137.68	£218 862.36
2	£218 862.36	£6479.39	£6324.37	£212 538.03
3	£212 538.03	£6287.03	£6516.73	£206 021.35
4	£206 021.35	£6088.79	£6714.97	£199 306.46
5	£199 306.46	£5884.57	£6919.19	£192 387.32
6	£192 387.32	£5674.12	£7129.64	£185 257.73
7	£185 257.73	£5457.25	£7346.51	£177 911.29
8	£177 911.29	£5233.81	£7569.95	£170 341.40
9	£170 341.40	£5003.56	£7800.20	£162 541.27
10	£162 541.27	£4766.32	£8037.44	£154 503.88
11	£154 503.88	£4521.86	£8281.90	£146 222.04
12	£146 222.04	£4269.94	£8533.82	£137 688.29
13	£137 688.29	£4010.40	£8793.36	£128 894.98
14	£128 894.98	£3742.95	£9060.81	£119 834.21
15	£119 834.21	£3467.36	£9336.40	£110 497.85
16	£110 497.85	£3183.37	£9620.39	£100 877.51
17	£100 877.51	£2890.77	£9912.99	£90 964.57
18	£90 964.57	£2589.23	£10 214.53	£80 750.11
19	£80 750.11	£2278.56	£10 525.20	£70 224.97
20	£70 224.97	£1958.42	£10 845.34	£59 379.70
21	£59 379.70	£1628.55	£11 175.21	£48 204.56
22	£48 204.56	£1288.67	£11 515.09	£36 689.51
23	£36 689.51	£938.42	£11 865.34	£24 824.23
24	£24 824.23	£577.52	£12 226.24	£12 598.05
25	£12 598.05	£205.66	£12 598.10	£0.00

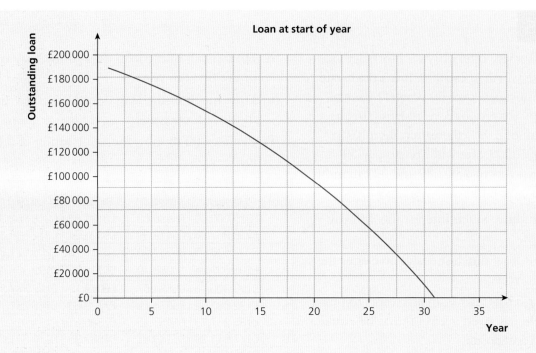

Loan at start of year

4 The graph shows the amortisation of Leanne's mortgage.

   a Estimate:

   i how much her mortgage was for

   ii what her outstanding loan will be at the end of the first 5 years

   iii how much of her loan she will pay off in the last 5 years.

   b Explain why the graph is steeper at the end than at the beginning.

   c Sketch a curve to illustrate the amortisation of the loan if it had been over 25 years.

5 Niamh has a mortgage with an APR of 3.7% and a 25-year term. Her repayments are £828.49 every month.

   a How much was her original loan for?

   b How much interest will she pay altogether?

WHAT HAPPENS TO YOUR PAY?

> ## Why don't I get all the money that I earn?

I work hard.

Why do I only get two-thirds of my pay?

Where does the rest of my pay go?

It would be good to think that if you earned £40 000 a year you would actually receive that amount of money – but of course if that were the case the Government would not have enough money for education, healthcare, social security, defence and all the other services that people expect to be provided. Depending on how much you earn, various deductions (such as income tax, National Insurance, student loan repayments and so on) are taken from your wage packet before you receive it or, if you are self-employed, paid following a self-assessment of your income and expenditure. It's a complicated area but this chapter introduces you to some of the basics.

WHAT YOU NEED TO KNOW ALREADY

→ Calculating percentages of an amount

MATHS HELP

p316 Chapter 3.6 Personal finance
p352 Chapter 3.13 Formulae and calculation
p359 Chapter 3.15 Percentages
p370 Chapter 3.18 Tax

PROCESS SKILLS

→ Checking answers
→ Estimating answers
→ Representing using algebraic expressions and graphs

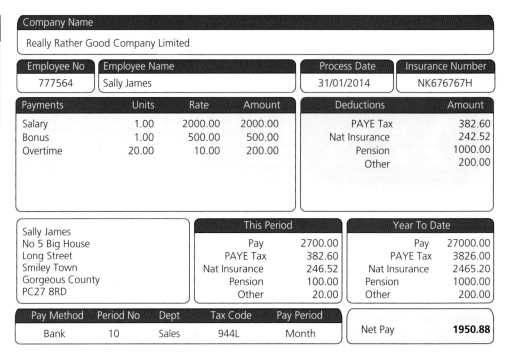

Company Name				
Really Rather Good Company Limited				

Employee No	Employee Name		Process Date	Insurance Number
777564	Sally James		31/01/2014	NK676767H

Payments	Units	Rate	Amount	Deductions	Amount
Salary	1.00	2000.00	2000.00	PAYE Tax	382.60
Bonus	1.00	500.00	500.00	Nat Insurance	242.52
Overtime	20.00	10.00	200.00	Pension	1000.00
				Other	200.00

Sally James No 5 Big House Long Street Smiley Town Gorgeous County PC27 8RD	This Period		Year To Date	
	Pay	2700.00	Pay	27000.00
	PAYE Tax	382.60	PAYE Tax	3826.00
	Nat Insurance	246.52	Nat Insurance	2465.20
	Pension	100.00	Pension	1000.00
	Other	20.00	Other	200.00

Pay Method	Period No	Dept	Tax Code	Pay Period	
Bank	10	Sales	944L	Month	Net Pay **1950.88**

Sally James earned £2700 in January 2014.

❯ Why did she only get £1950?

Discuss

❯ Why are deductions taken from people's earnings?

Income tax

The major deduction from most earnings is income tax. If your income exceeds a certain threshold, called the personal allowance (£11 000 a year in 2017 for most people), you have to pay income tax at the basic rate (20% in 2017) on the excess. Your **tax code** determines exactly how much of your income is taxable.

WORKED EXAMPLE

How much income tax will a person earning £18 000 a year pay?

SOLUTION

The tax-free allowance is £11 000, so the taxable amount is

£18 000 − £11 000 = £7000

The tax paid is 20% of £7000 = 0.2 × £7000 = £1400

The graph shows gross income and income after tax. Could you set up a spreadsheet to draw a graph like this?

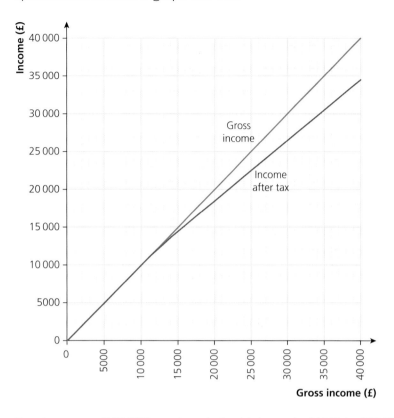

Earnings over £43 000 are taxed at a higher rate (40% in 2017).

So, someone with a salary of £60 000 a year will pay no tax on the first £11 000, tax at 20% on the next £32 000 and tax at 40% on the remaining £17 000.

So total tax to be paid = 0.2 × £32 000 + 0.4 × £17 000

$$= £6400 + £6800$$

$$= £13 200$$

For earnings over £150 000, there is an additional rate of tax (45% in 2017) and those earning over £122 000 do not get a tax-free personal allowance.

> How much more than £122 000 do you need to earn to make it worth it?

The table summarises the tax bands and rates in 2017:

Band	Taxable income	Tax rate
Personal allowance	Up to £11 000	0%
Basic rate	£11 001 to £43 000	20%
Higher rate	£43 001 to £150 000	40%
Additional rate	over £150 000	45%

Investigate

Are the tax bands the same now? Look them up to see if they have changed for the better.

National Insurance

Another major deduction from most people's pay is National Insurance. This is designed to raise money for State Pensions, Jobseeker's Allowance, Statutory Maternity Pay, the NHS and so on.

Anyone who is 16 or over, and whose gross pay is more than £155 a week (in 2017), or a self-employed person making a profit of more than £5965 a year, is required to pay National Insurance.

Your pay	Class 1 National Insurance rate
£155 to £827 a week (£672 to £3583 a month)	12%
Over £827 a week (£3583 a month)	2%

WORKED EXAMPLE

How much National Insurance (NI) will someone who earns £50191.44 per annum pay?

SOLUTION

£50191.44 per annum is £50191.44 ÷ 52 per week = £965.22 per week.

The first £155 per week is exempt.

The next £672 per week is charged NI of £672 × 0.12 = £80.64

The remaining £138.22 is charged NI of £138.22 × 0.02 = £2.76

So the total amount of NI paid is £80.64 + £2.76 = £83.40

Of course, this person's income is way above the national average – most people will pay far less National Insurance than this.

Other deductions from pay can include repayment of loans, occupational pension payments, private health insurance, savings, childcare schemes and so on.

Discuss

> It is said that roughly one third of your gross pay is taken in taxes. Is this true? (National Insurance is essentially a tax.)

QUESTIONS

1 What income per month will a person earning £23 000 a year have after income tax is deducted?

2 Estimate what percentage of her income a teacher earning £33 000 pays in income tax.

3 Which of these formulae will correctly calculate the tax paid (*TP*) by someone earning between £11 000 and £43 000 per annum?

- $TP = (GI - 11\,000) \times 0.2$
- $TP = GI - 11\,000 \times 0.2$

- $TP = 0.2\,GI - 2200$
- $TP = \dfrac{GI - 11\,000}{5}$

 where £*GI* is the gross income.

4 Jake pays basic rate tax of £1250 per annum. What is his gross income?

E 5 Use the information about income tax and National Insurance from the chapter.

 Aisha's gross pay is £19 300 a year. How much will she receive once income tax and National Insurance have been deducted?

OUT OF ONE HUNDRED

Is that a lot?

> Total monthly active users of Facebook increased by 15% in the last year to 1.65 billion

Can it continue to grow at that rate?

How many more people is that each second?

WHAT YOU NEED TO KNOW ALREADY

→ Recognising that 100% represents the whole
→ Calculating percentages of amounts
→ Converting percentages to decimals

MATHS HELP

p316 Chapter 3.6 Personal finance
p352 Chapter 3.13 Formulae and calculation
p359 Chapter 3.15 Percentages
p370 Chapter 3.18 Tax

PROCESS SKILLS

→ Estimating answers
→ Comparing methods
→ Checking answers
→ Evaluating methods

Percentages are used in all areas of daily life; for example, news bulletins regularly quote percentages, with headlines such as 'Total active users of Facebook increased 15% from a year earlier to 1.65 billion'. The idea of using 100 as a 'whole' originated in Roman times, when money lenders charged interest of 2, say, out of 100. We are still using the same idea 2000 years later. Despite the use of a place value system which enables us to express the same thing using decimals, we persist in using percentages. Tax rates are expressed as percentages of income (as in the previous chapter) and of expenditure, for example VAT, which is the focus of this chapter.

The increase in Facebook users is not 15% of 1.65 billion. It is 15% of the number of users a year earlier.

That means 1.65 billion is 115% of the previous year's number of users.

WORKED EXAMPLE

What was the previous year's number of active users of Facebook?

SOLUTION

There are two ways to work it out.

Method 1: 1.65 billion is 115%

Divide by 115 to get 1%

$\qquad$ 0.0143… billion is 1%

Multiply by 100 to get 100%

$\qquad$ 1.43 billion is 100%

The previous year's total is **1.43 billion**

Method 2: The previous year's total was multiplied by 1.15 (115%) to get 1.65 billion

Divide 1.65 billion by 1.15 = **1.43 billion**

You have used the decimal version of a percentage to calculate increased or decreased amounts in previous chapters. The decimal version used to calculate the answer in one step is called a **multiplier** because it multiplies! Multipliers are a particularly efficient and quick way to work out original value problems, such as in the example above. They are also very useful with repeated percentage change, such as when you work out compound interest. You used that in Chapter 2.12.

WORKED EXAMPLE

What is the multiplier for:

a   an increase of 40%

b   a decrease of 7%

c   a decrease of 2.5%.

SOLUTION

a   An increase of 40% takes the amount to 140%.

Converting to a decimal gives 1.4, which is the multiplier.

b A decrease of 7% takes the amount to 93%.

Converting to a decimal gives 0.93, which is the multiplier.

c A decrease of 2.5% takes the amount to 97.5%.

Converting to a decimal gives 0.975, which is the multiplier.

VAT

When we buy certain goods and services the Government adds a tax to the cost of the goods or services. The tax is called Value Added Tax (VAT). The price before VAT is called the net price; when VAT is added it is called the gross price.

Investigate

Different rates of tax apply to different goods and services. Look at a receipt from a weekly shop to see how some items have VAT and others do not.

The rates of VAT at the time of writing are 20%, 5% and 0%. Are they the same now? The rate was 10% for everything when VAT was introduced in 1973 and varied for several years before settling down to 15%, and then 17.5% with a lower rate for some things.

WORKED EXAMPLE

Samantha is buying a buttonhole to wear at a friend's wedding. The price is £4.00 plus VAT at the standard rate. Calculate the price of the buttonhole.

SOLUTION

The standard rate of VAT is 20%.

Work out 20% of £4

$$\frac{20}{100} \times 4 = \frac{80}{100}$$
$$= 0.8$$

Add it to £4 to get the full, or gross, price

Price = £4.80

There are quicker ways to work out the VAT.

20% converts to one fifth, which is a simple fraction, so the £4 is divided by 5 to get 80p.

When the rate was 15% people worked out 10% then halved it and added the two together.

> Why does this work?

> Can you think of a similar method for 17.5%?

A method that works for any percentage is to use multipliers.

WORKED EXAMPLE

Work out the price including VAT for Samantha's buttonhole using a multiplier.

SOLUTION

We are adding 20% so we want 120%.

120% is equivalent to the decimal 1.2.

1.2 is called the multiplier because it multiplies the original amount.

So we work out $1.2 \times 4 = 4.8$.

Because the context is money we write this as £4.80.

Multipliers are much quicker to use than the other methods when you have a calculator; they are also quicker for mental arithmetic when you are working with simple numbers.

> What is the multiplier for working out the amount including VAT at 5%?

> What was the multiplier for working out the amount including VAT at 15%? And at 17.5%?

What if you only know the gross price?

When a company buys raw materials to manufacture a product they pay VAT to their suppliers but they can claim it back to offset the VAT they add to the price of the product. Working out the amount to claim is not as straightforward as calculating 20% of the gross price.

> For the example above, work out 20% of the gross price of £4.80.

> Is it 80p? Why is it different?

WORKED EXAMPLE

Dai the welder makes a sculpture that he sells for £5028 including VAT at the standard rate. He uses £420 worth (including VAT) of gas in making the sculpture. Calculate the VAT Dai needs to pay the Government.

SOLUTION

VAT is charged at the standard rate of 20% on the sculpture

£5028 = original cost + 20% of original cost

Therefore £5028 = 120% of original cost

Original cost $= £\frac{5028}{120} \times 100$

$\qquad = £4190$

VAT is 20% of £4190 $= £4190 \times \frac{20}{100}$

$\qquad = £838$

VAT is charged at reduced rate of 5% on gas

£420 = original cost + 5%, so £420 = 105% of original cost

Value of 1% is $£ \frac{420}{105} = £4.00$

Therefore 5% (VAT) = 5 × £4.00

$\qquad = £20$

Dai must pay the Government the VAT he receives less the VAT he pays

£838 − £20 = £ 818

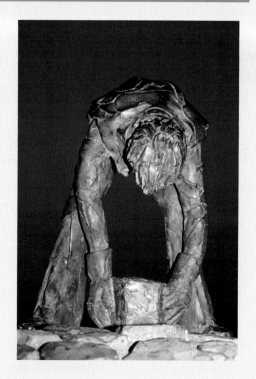

The first step is to identify what percentage is equivalent to the gross price.

In the first example above, 120% is equivalent to the gross price.

Then 100% was calculated.

In the second example 105% is equivalent to the gross price.

Then 1% was calculated.

You can choose your own route to the amount you need to work out.

WORKED EXAMPLE

Sadiq books a driving test costing £62. The price includes VAT at 20%.

Calculate the amount of VAT included in the test price.

➡

SOLUTION

£62 is equivalent to 120%.

120% is equivalent to 1.2.

The net price has been multiplied by 1.2 to get £62.

So divide £62 by 1.2 to get the net price.

Net price = £62 ÷ 1.2

$\qquad$ = £51.6666…

$\qquad$ = £51.67

So the VAT = £62 − £51.67

$\qquad$ = £10.33

Discuss

When a price is given that includes standard rate VAT, the amount of VAT can be calculated by dividing the price by 6. Can you think why this works?

> Alex is celebrating his friend's nineteenth birthday and buys a pack of cider costing £15 that includes VAT at the standard rate. Work out the amount of VAT he pays.

More uses of percentages

Percentages are used in many of the financial aspects of life. Discounts and price rises and things growing or depreciating in value are just two examples.

> Can you think of other examples where percentages are used in daily financial contexts?

Investigate

You have seen several ways to work out the VAT problems.

Can you apply them to other percentage problems?

> All staff in a company are given a salary increase of 2%. Calculate the increased salary if the previous salary was £15 250.

The VAT problems all involved an increase from net price to gross price.

> How can you use multipliers when there is a reduction in price?

The same percentage decrease is repeated for four years in the example below, and so the calculation can be written using powers to make it even quicker. You saw this with compound interest in Chapter 2.12.

WORKED EXAMPLE

A car depreciates by 20% a year. Calculate the value after four years of a car that cost £15 000 originally.

SOLUTION

After one year, the value is 80% of the original price, so you need to use a multiplier of 0.8.

Value of car after one year $= £15\,000 \times 0.8$

Value of car after two years $= (£15\,000 \times 0.8) \times 0.8$

Value of car after four years $= £15\,000 \times 0.8^4$

$$= £6144$$

Investigate

Working out the original value works in the same way as for VAT. Can you apply your understanding from those questions to this one?

> The annual pace of house price growth slowed to 4.9% in April 2016. Calculate the price of a house in April 2015 if the price in April 2016 was £262 250.

QUESTIONS

1 A discount store gives the cost of goods without VAT added. It advertises a television at £210 plus VAT. The price of the same television online is £249.99. Explain which television is better value.

2
Price Bonanza
28% off electricals priced £500 and over
18% off electricals priced under £500

a Calculate the total price of a television priced at £700 and a laptop at £450.

b The final total price includes standard rate VAT (20%). Calculate the amount of VAT.

3 A web designer charges her customers £15 000 + VAT for three months' work. In the same three months she buys two computers each costing £1200 inclusive of VAT. At the end of the three months she has to pay the Government the amount of VAT she charged less the amount she paid. Calculate the VAT she pays the Government.

4 A driving instructor gives 140 lessons in a month. VAT at the standard charge of 20% is applied to all lessons, and in the month the amount of VAT charged is £980. Calculate the price charged (including VAT) for a single lesson.

E 5 In 2011 the standard rate VAT was increased from 17.5% to 20.%. Before the increase in VAT a computer was priced at £645. Calculate the price after the increase in VAT.

6 A car depreciates by 35% in the first year and 27% each year after that. Calculate, to the nearest pound, the original price of a car valued at £5023 after three years.

MONITORING INFLATION

> The Government says the inflation rate has gone down this month.

> ## What is inflation?

> Every week my shopping seems to cost more.

> What is the difference between CPI and RPI?

THE DAILY NEWS

Inflation rate increases for the sixth consecutive month

Every month the UK Government monitors prices of goods and services. When prices go up, consumers buy less. When this happens, the Government might want to protect the economy by altering interest rates and tax allowances, so it is important that it monitors prices accurately and regularly. Wages and salaries also aim to keep up with the 'cost of living', which means it is important to monitor this closely.

WHAT YOU NEED TO KNOW ALREADY

→ Finding one quantity as a percentage of another
→ Drawing and interpreting line graphs
→ Finding the mean

MATHS HELP

p301 Chapter 3.1 Straight line graphs
p316 Chapter 3.6 Personal finance
p352 Chapter 3.13 Formulae and calculation
p359 Chapter 3.15 Percentages
p373 Chapter 3.19 Indices and currency

PROCESS SKILLS

→ Interpreting tables
→ Estimating financial measures
→ Representing rates of change
→ Comparing quantities

Inflation is when prices are generally increasing. Governments measure this using an **inflation rate** that is based on an index that represents the cost of a 'basket' of goods and services.

Consumer Price Index

Each month, the UK Government Office for National Statistics publishes the latest Consumer Price Index (CPI).

The CPI is an important indicator of how a country's economy is performing and shows the impact of inflation on family budgets. In most countries, the CPI is, along with the population census, one of the most closely watched national economic statistics.

The CPI is used by a government to adjust interest rates, tax allowances, wages, state benefits, pensions and many other payments. It is also used to help regulate prices.

The table shows the CPI for the period January 2010 to July 2016.

The CPI is calculated using a very large 'shopping basket' of around 700 goods and services on which people typically spend their money – from bread to ready-made meals and from the cost of a cinema seat to the cost of a bicycle. The content of the basket is

Month	CPI	Month	CPI	Month	CPI	Month	CPI
2010 Jan	87.8	2011 Sep	94.4	2013 May	98.5	2015 Jan	99.3
2010 Feb	88.2	2011 Oct	94.5	2013 Jun	98.3	2015 Feb	99.5
2010 Mar	88.7	2011 Nov	94.6	2013 Jul	98.3	2015 Mar	99.7
2010 Apr	89.2	2011 Dec	95.1	2013 Aug	98.7	2015 Apr	99.9
2010 May	89.4	2012 Jan	94.6	2013 Sep	99.1	2015 May	100.1
2010 Jun	89.5	2012 Feb	95.1	2013 Oct	99.1	2015 Jun	100.2
2010 Jul	89.3	2012 Mar	95.4	2013 Nov	99.2	2015 Jul	100.0
2010 Aug	89.8	2012 Apr	96.0	2013 Dec	99.6	2015 Aug	100.3
2010 Sep	89.8	2012 May	95.9	2014 Jan	99.0	2015 Sep	100.2
2010 Oct	90.0	2012 Jun	95.5	2014 Feb	99.5	2015 Oct	100.3
2010 Nov	90.3	2012 Jul	95.6	2014 Mar	99.7	2015 Nov	100.3
2010 Dec	91.2	2012 Aug	96.1	2014 Apr	100.1	2015 Dec	100.3
2011 Jan	91.3	2012 Sep	96.5	2014 May	100.0	2016 Jan	99.5
2011 Feb	92.0	2012 Oct	97.0	2014 Jun	100.2	2016 Feb	99.8
2011 Mar	92.2	2012 Nov	97.2	2014 Jul	99.9	2016 Mar	100.2
2011 Apr	93.2	2012 Dec	97.6	2014 Aug	100.2	2016 Apr	100.2
2011 May	93.4	2013 Jan	97.1	2014 Sep	100.3	2016 May	100.4
2011 Jun	93.3	2013 Feb	97.8	2014 Oct	100.4	2016 Jun	100.6
2011 Jul	93.3	2013 Mar	98.1	2014 Nov	100.1	2016 Jul	100.6
2011 Aug	93.8	2013 Apr	98.3	2014 Dec	100.1		

Source: Office for National Statistics

fixed. As the prices of individual products vary, so does the total cost of the basket. The quantities or 'weightings' of the various items in the basket are chosen to reflect their importance in the typical household budget. The CPI basket and weightings follow a standard international classification.

The CPI measures price changes, not price levels. It is therefore expressed in terms of the comparison of prices relative to a base year, when the index was given a value of 100.

DID YOU KNOW?

The index for January 2014 was 99, indicating that £99 would buy the same amount of goods and services as £100 would have in July 2015, the base month for this data.

Discuss

For the years covered in the table:

> In which month is the cost of the shopping basket the highest?

> How much money would buy the same amount of goods and services in August 2015 that £100 would buy in July 2015?

> Over which periods of at least three consecutive months does the CPI continually decrease?

Inflation rates

The annual rate of inflation is the percentage change in the CPI compared with the value recorded twelve months previously.

$$\text{CPI (\% change) or inflation rate} = \frac{\text{current index } - \text{ index twelve months previously}}{\text{index twelve months previously}} \times 100$$

WORKED EXAMPLE

What is the percentage change in the CPI between January 2013 and January 2014?

SOLUTION

$$\text{CPI (\% change) for January 2014} = \frac{\text{Jan 2014 index } - \text{ Jan 2013 index}}{\text{Jan 2013 index}} \times 100$$

$$= \frac{99 - 97.1}{97.1} \times 100$$

$$= 1.96\%$$

DID YOU KNOW

A month with a high CPI does not necessarily have a high inflation rate. If the CPI 12 months earlier was also high, then the inflation rate will be low.

Investigate

> Calculate the CPI (% change) for October 2011.

> Find out what is in the basket. Are there any surprises?

The graph below shows the inflation rates (CPI (% change)) between January 2006 and May 2016.

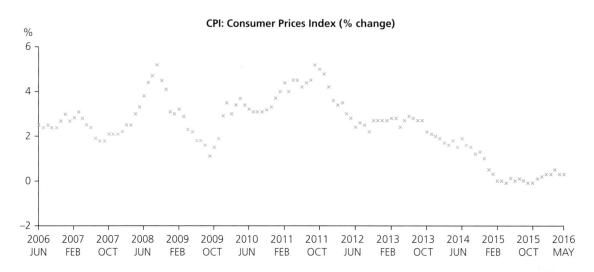

CPI: Consumer Prices Index (% change)

DID YOU KNOW

The line graph is drawn using a dotted line because only the values shown by the crosses exist. You cannot read off values in between. For example, there is a value for February 2015 and then for March 2015, but a value between the two makes no sense. There is no measurement taken of the prices between those points.

Investigate

Use the graph to estimate when during these years:

> the inflation rate was at its highest

> the inflation rate was at its lowest

> the inflation rate was about 4%

> the inflation rate was below 1%

> the inflation rate rose continuously over a six-month period or longer.

Retail Price Index

The Retail Price Index (RPI) is very similar to the CPI but it is based on a different 'basket' of goods and services, which is weighted according to its own unique system. Compared with the CPI, the RPI has a much longer history. The series started in 1947, and prices are expressed relative to January 1987, when the index has a value of 100.

The RPI has been assessed against the Code of Practice for Official Statistics and found not to meet the required standard for designation as a National Statistic. The RPI is therefore not so important today, but it is still used by the Government as a base for various purposes, such as the amounts payable on index-linked securities and social housing rent increases. Many employers also use it as a starting point in wage negotiation.

Retail Price Index from January 2010 to July 2016

Month	RPI	Month	RPI	Month	RPI	Month	RPI
2010 Jan	217.9	2011 Sep	237.9	2013 May	250.0	2015 Jan	255.4
2010 Feb	219.2	2011 Oct	238.0	2013 Jun	249.7	2015 Feb	256.7
2010 Mar	220.7	2011 Nov	238.5	2013 Jul	249.7	2015 Mar	257.1
2010 Apr	222.8	2011 Dec	239.4	2013 Aug	251.0	2015 Apr	258.0
2010 May	223.6	2012 Jan	238.0	2013 Sep	251.9	2015 May	258.5
2010 Jun	224.1	2012 Feb	239.9	2013 Oct	251.9	2015 Jun	258.9
2010 Jul	223.6	2012 Mar	240.8	2013 Nov	252.1	2015 Jul	258.6
2010 Aug	224.5	2012 Apr	242.5	2013 Dec	253.4	2015 Aug	259.8
2010 Sep	225.3	2012 May	242.4	2014 Jan	252.6	2015 Sep	259.6
2010 Oct	225.8	2012 Jun	241.8	2014 Feb	254.2	2015 Oct	259.5
2010 Nov	226.8	2012 Jul	242.1	2014 Mar	254.8	2015 Nov	259.8
2010 Dec	228.4	2012 Aug	243.0	2014 Apr	255.7	2015 Dec	260.6
2011 Jan	229.0	2012 Sep	244.2	2014 May	255.9	2016 Jan	258.8
2011 Feb	231.3	2012 Oct	245.6	2014 Jun	256.3	2016 Feb	260.0
2011 Mar	232.5	2012 Nov	245.6	2014 Jul	256.0	2016 Mar	261.1
2011 Apr	234.4	2012 Dec	246.8	2014 Aug	257.0	2016 Apr	261.4
2011 May	235.2	2013 Jan	245.8	2014 Sep	257.6	2016 May	262.1
2011 Jun	235.2	2013 Feb	247.6	2014 Oct	257.7	2016 Jun	263.1
2011 Jul	234.7	2013 Mar	248.7	2014 Nov	257.1	2016 Jul	263.4
2011 Aug	236.1	2013 Apr	249.5	2014 Dec	257.5		

Source: Office for National Statistics

Discuss

For the years covered in the table above:

> In which month is the cost of the shopping basket the highest?

> How much money would you need to buy the same amount of goods and services in August 2015 that you could buy for £258.60 in July 2015?

> Over which periods of at least three consecutive months does the RPI continually decrease?

> Calculate the RPI (% change) for October 2011.

> Comment on the differences you have found between RPI and CPI.

Investigate

> Find out the latest CPI, RPI and inflation rates.

> Describe the trends in CPI, RPI and inflation rates over the last 12 months.

QUESTIONS

1 Use the RPI table on page 135 to calculate the RPI (% change) for January, April, July and October in both 2014 and 2015. Draw a line graph showing these values. Describe the percentage change over this period.

2 Here is a table of possible values for CPI and inflation rates. Complete the table.

Month	CPI in year 1	CPI in year 2	Inflation rate in year 2
A	125	135	
B	130		0%
C	100		5%
D	135	125	

3 a Use the CPI table on page 132 to plot the CPI for every January between 2010 and 2016 inclusive. Describe the trends the graph shows over this six-year period.

b Use the CPI table to plot the CPI for every July between 2010 and 2016 inclusive. Describe the trends the graph shows over this six-year period.

c How do you think it is best to compare the CPI year by year?

E 4 The table shows the CPI for the first quarters of 1999 and 2000.

1999 Jan	71.4
1999 Feb	71.5
1999 Mar	71.9
2000 Jan	71.9
2000 Feb	72.2
2000 Mar	72.3

a Calculate the inflation rate for each month in the first quarter of 2000.

b Is inflation increasing over the first quarter of the year 2000?
You should show details of your calculations to support your answer.

HOLIDAY MONEY

What would you do?

Which way gives best value for money?

Which way gives most flexibility?

> ## What's the best way to get currency for foreign holidays?

Should you buy your holiday cash now?

Currency brokers bombarded with orders for euros as value of the pound plummets

- Families worried costs will soar before Easter and summer holidays
- Post Office has exchanged 43% more euros in the past week year-on-year
- A pound will now buy you just $1.35, compared with $1.51 a year ago

WHAT YOU NEED TO KNOW ALREADY

→ Using conversion graphs
→ Calculating percentages

MATHS HELP

p301 Chapter 3.1 Straight line graphs
p359 Chapter 3.15 Percentages
p373 Chapter 3.19 Indices and currency

PROCESS SKILLS

→ Checking by recalculating
→ Estimating
→ Representing mathematically
→ Comparing options

Most people will holiday abroad at some time and so need to exchange money.

It is difficult to know exactly how much you will need but thinking ahead can be helpful.

Increasingly, many people travel abroad as part of their job. Their employer may have arrangements for providing the foreign currency but someone will have had to work out the best way to exchange the money.

Companies who trade abroad, whether selling their goods or buying raw materials, need to know exactly what the costs involved in changing currencies are. They can determine appropriate prices for buying and selling once they have that information. They may not be getting notes and coins, but they are still switching between currencies using the exchange rate. They will also be subject to other charges.

You may hear news reports about exchange rates falling, which is good for exports but not for imports.

WORKED EXAMPLE

How many euros can you get for £200 holiday money?

SOLUTION

The exchange rate varies from day to day. A typical rate is €1 ≈ £0.79

If £0.79 ≈ €1

$$£1 ≈ €\frac{1}{0.79}$$

So $£200 ≈ €\frac{200}{0.79} ≈ €253.16$

Look up today's rate.

> What is £200 in euros at that rate?

When changing many different amounts of pounds to euros or vice versa, it's useful to use a spreadsheet, which can then be used to draw a conversion graph.

The formula in cell B2 (copied down the column) is A2/0.79

> Explain why that works.

Here is the conversion graph for pounds and euros.

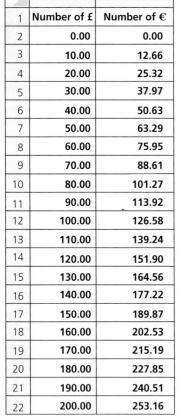

	A	B
	Number of £	**Number of €**
1		
2	0.00	0.00
3	10.00	12.66
4	20.00	25.32
5	30.00	37.97
6	40.00	50.63
7	50.00	63.29
8	60.00	75.95
9	70.00	88.61
10	80.00	101.27
11	90.00	113.92
12	100.00	126.58
13	110.00	139.24
14	120.00	151.90
15	130.00	164.56
16	140.00	177.22
17	150.00	189.87
18	160.00	202.53
19	170.00	215.19
20	180.00	227.85
21	190.00	240.51
22	200.00	253.16

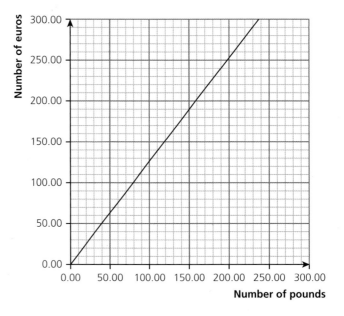

> Explain how to use it to convert pounds to euros and vice versa.

> What is the equation of the graph?

> How does this relate to the spreadsheet formula?

> What does the gradient of the line represent?

Discuss

> Why is the graph a straight line through the origin? (This indicates that the number of euros is directly proportional to the number of pounds.)

> Are all conversion graphs straight line graphs through the origin?

> Can you think of a measure where the conversion graph isn't a straight line through the origin?

Changing money before you set off

You can get foreign currency for use abroad before you start your holiday. For example, you could choose to order money from the Post Office online.

> What do you notice about these exchange rates?

Spend	£400+	£500+	£1000+	£1500+	£2000+
Euro	1.21	1.225	1.226	1.227	1.229

> Why does the Post Office offer more advantageous rates the more money you change?

> How do these exchange rates compare with the rate shown in the conversion graph above?

> What explains the differences?

> If you pay by credit card, an extra 1.5% is charged. How much would it cost altogether if you changed £800 to euros online at the Post Office using a credit card?

On the same day as the Post Office rates above were on offer, one of the major banks was offering an exchange rate of £1 = €1.1984.

> Is this better or worse than the Post Office rates?

> How do you know?

Both the bank and the Post Office claim not to charge **commission**, but the differences in the rates indicate that there are charges for the transactions. The service cannot be provided for nothing.

This extract from the *Guardian* shows how the charges affect the amount you get when you change money.

Converting £400 into euros would have produced €424 (0.99% below the true rate) at FairFX last weekend; €415.08 (3.16% below) at Tesco online; and €411.40 (4.08% below) at HSBC. The dollar rates show much the same story.

At the Eurostar terminal St Pancras, in central London, £400 bought only €378.52 (13.12% below the true rate) while Luton airport users would have received only £374.64 (14.3% below).

Most travellers would do better waiting until they arrive in Brussels, Lille or Paris, where they can use cash machines with UK debit cards; the rate is typically 2.75% below the true level.

Source: http://www.theguardian.com/money/2009/apr/04/foreign-currency-commission

WORKED EXAMPLE

The extract states that at Luton Airport, the exchange rate offered was '14.3% below the true rate'. What was the true rate?

SOLUTION

Here's how to work it out.

The airport changed £400 for €374.64

So, for each pound you got $€\frac{374.64}{400} \doteq €0.9366$

Less than one euro per pound!

This is 14.3% below the true rate, so 85.7% of the true rate is €0.9366 per pound

So, the true rate is $€\frac{0.9366}{0.857} \approx €1.0929$ per pound.

Quite a difference, so airports are not good places to get your holiday money as you will usually get better rates if you change your money beforehand.

The extract from the *Guardian* is from 2009.

> Are things any different now?

So, you can change money well before you set off at the Post Office, a bank or a travel agent. You can choose to change money at the airport, station or ferry port just before you leave the country or just after you arrive. That is convenient but likely to offer a poor exchange rate.

You can also get foreign currency from a cash machine in the country in which you are spending your holiday. In 2016, a cash machine charged £76.15 for €100. The exchange rate quoted for that day was £1 = €1.38889 and the bank charged a 'foreign cash fee' of £2 and a 'transfer fee' of £2.16.

> Work through the calculation to check that you can see how the total charge of £77.85 was arrived at.

Perhaps you don't need that much cash. You can choose to pay for most things using a credit or debit card. An example is a meal costing €46.03 paid with a debit card, which appears on the bank statement as a charge of £34.68.

> Does this seem like a good deal?

(The rate of exchange quoted for that day was £1 = €1.3666.)

> If you bring back unspent foreign currency from your holiday, how much does it cost you to change it back to pounds?

QUESTIONS

1 Draw a graph to convert pounds to Australian dollars, using the exchange rate of £1 = 0.54 Australian dollars.

2 You buy £500 worth of euros from the Post Office online and are charged 1.5% for using a credit card. How much will you pay altogether?

See page 139 for the exchange rates.

3 A bank offers an exchange rate of €1.20 to the pound.

 a Work out what one euro is worth at this exchange rate.

 b The cell A1, of a spreadsheet, has the amount of euros per pound. Write down the formula for B1 so that it contains the equivalent amount of pounds per euro.

4 A bill of €60.70 at Carrefour supermarket in France comes through at £45.63 on the bank statement. The exchange rate that day was £1= €1.3668. What percentage of the transaction did the bank take in charges?

5 Look at the *Guardian* extract on page 139 and find the number of euros offered for £400 at St Pancras. Use those figures to show that the true rate is approximately €1.0892 per pound.

6 Approximately how many euros would you get for £400 at a cash machine, where the exchange rate would be about 2.75% below the true level?

CAN I AFFORD TO LEAVE HOME?

> Room for rent £425 pcm

It looks lovely.

Can I afford it?

Is it really just one room?

Will I have to share stuff?

Moving into a place of your own is a big step. Your new-found independence comes with responsibilities, many of them financial. Knowing that you can meet those responsibilities is vital. It is all too easy to let your spending get out of hand as the 'coffee to go' and take-outs start to mount up. Budgeting and monitoring your bank account can help prevent that happening. In this chapter we will look at aspects of budgeting and understanding your bank account.

WHAT YOU NEED TO KNOW ALREADY

→ Substituting numerical values into algebraic expressions and formulae

→ Entering formulae into a spreadsheet

→ Recognising that 100% represents the whole

MATHS HELP

p304 Chapter 3.2 Spreadsheet formulae
p316 Chapter 3.6 Personal finance
p352 Chapter 3.13 Formulae and calculation

PROCESS SKILLS

→ Estimating amounts

→ Justifying decisions

Discuss

A room in a house is available to rent at £75 per week.

> What other bills are there that you will have to pay? Council tax? Electricity? Water?

> What other outgoings will you have? Travel? Socialising?

Investigate

> What are the likely costs of all of those items for you?

> How much will you need to earn to cover those expenses?

> Are there jobs that you can get which pay that much?

Renting a room in a shared house is one option. You will probably have to share the kitchen and bathroom facilities and share responsibility for other common areas like the hallway. How will you cope with that? One of the advantages is sharing the bills.

> An electricity bill of £112.50 is shared equally between 9 people. How much does that cost for each person?

> What do you think the difficulties might be with shared bills?

Another option is to rent a flat by yourself. You can control how much electricity and water are used and keep the flat as tidy as you find comfortable! However, you have to pay the bills yourself.

The skill in budgeting lies in judging your likely expenses and making a sensible estimate of them. The mathematics required to do the calculations is quite simple.

Working out if you can afford a holiday, for example, only needs some simple calculations as you will see in the following example.

WORKED EXAMPLE

Evie wants to go on holiday in six months' time. The cost of the holiday is £950 and she would like £200 spending money. Currently her monthly income is £875 and her monthly expenditure on rent, services, tax and food is £300, £72, £110 and £200 respectively. Can Evie save enough in the next six months to go on holiday?

SOLUTION

Cost of holiday is £950 + £200 = £1150

Monthly expenditure is £300 + 72 + 110 + 200 = £682

Monthly income − expenditure = £875 − 682 = £193

Therefore in six months Evie can save 6 × £193 = £1158; she has enough to go on holiday.

Discuss

> Is Evie wise in thinking she will have enough money?

> What do you think could go wrong with her plan?

When you rent a flat you need to find a deposit (usually one month's rent) and the rental agreement may be for six months or even a year. It is no good if you realise you can't afford it after one month. Careful planning in advance can help avoid a costly mistake.

WORKED EXAMPLE

Jeremy wants to move into a new flat that has a monthly rental charge of £425. He estimates his monthly costs for water, electricity and gas as £110. He also has to pay council tax of £140 per month. Jeremy also wants to have £500 a month for general living expenses. Calculate the annual amount of income that Jeremy needs to meet his expectations.

SOLUTION

Expenditure per month = £(425 + 110 + 140 + 500)
$$= £1175$$

Expenditure in a year = 12 × £1175 = £14 100.

Therefore Jeremy needs an annual income of £14 100 after deductions (see Chapter 2.15) to meet his expectations.

Bank accounts

When you are ready to move into your own place, you may already have a bank account. Your pay goes in each month on roughly the same date. You will have to give your bank details to your landlord and the rent will also come out on the same date each month. It is better if this is after your pay goes in!

It is very easy to keep track of your balance, but you also need to be aware of direct debits that you may have set up to pay electricity bills, council tax, broadband fees and so on. Your balance is not equal to the amount you can spend as you please each month. There will be rather less left over than you might wish.

DID YOU KNOW?

Amounts paid into your account are called credits and amounts paid out are debits.

Look carefully at the amounts and make sure you can see how they are worked out.

Understanding your bank statement

Date	Description	Type	In (£)	Out (£)	Balance (£)
	Brought forward				1,500.00
1st June 16	rent	DD		450	1,050.00
	insurance	DD		21	1,029.00
4th June 16	supermarket	DEB		14.78	1,014.22
6th June 16	electricity	DD		28	986.22
7th June 16	cheque	CHQ	15		1001.22
9th June 16	takeout	DEB		15.99	985.23
12th June 16	council tax	DD		40	945.23
	water	DD		32	913.23
14th June 16	cash	CPT		50	863.23
15th June 16	refund	BGC	23		886.23
	supermarket	DEB		26.78	859.45
16th June 16	cash	CPT		75	784.45
18th June 16	broadband	DD		30	754.45
	phone	DD		35	719.45

This is an example of what you may see on a bank statement.

Discuss

> What do you think CPT means?

> What does DEB mean?

> What does CHQ mean?

> Which item is the money for a birthday present?

> What does DD mean?

> If this is from a spreadsheet, what formula needs to go into the cells in the balance column?

> Why does the amount usually decrease in the balance column?

QUESTIONS

1 Kanu is planning to leave home and has worked out that he needs £780 a month to cover his living expenses. His current income is £750 so he is going to ask for a pay rise. Calculate the percentage increase that he needs to be able to afford to leave home.

2 James wants to start paying into a pension and has been told he should put 7% of his income into it. Currently, he earns £1200 a month and has car and flat expenses of £1140. Can James afford to put 7% into the pension?

You must show your working.

3 Emmanuel has an income of £1300 a month. Rent is 45% of his income, council tax is £129 per month, services cost £1308 a year and 5% goes into savings. Calculate the money remaining for other monthly expenses.

4 This spreadsheet contains the expenditure in the first three months of the year. What formula should be entered into cell E3 to find the total? Calculate the total expenditure for these three months.

	A	B	C	D	E
1					
2		Jan	Feb	Mar	Total Jan-Mar
3	Spending	450	515	490	
4					

5 Gail is constructing a spreadsheet to estimate the amount of money she has available after paying her rent, services and council tax. Write a suitable formula to enter into cell E3.

	A	B	C	D	E
1					
2	Income	Rent	Services	Council tax	Spending
3					

PRACTICE QUESTIONS: PAPER 1

1 Alan uses his car to give his three friends a lift to college each day. They each pay him £10 a week towards petrol. Alan has looked on the internet and found that his car does 40 miles to the gallon, and the round trip each day is 22 miles. Petrol costs £1.20 per litre. He decides it needs to cover a further 9p per mile for 'wear and tear'. Has Alan charged the right amount? There are 4.5 litres in one gallon.

You **must** show your working. **[4 marks]**

2 The graph shows how the blood collected from donors in the UK is used.

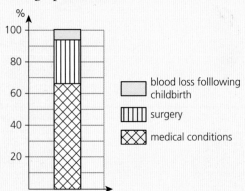

Data from https://www.blood.co.uk/why-give-blood/how-blood-is-used/

Around 4% of the population between ages 17 and 65 regularly donate blood.

The population of the UK is approximately 65 million.

Estimate how many donors' blood is used to treat blood loss following childbirth.

Show details of your assumptions and calculations. **[4 marks]**

3   A school buys a set of textbooks that are priced at £15.50.

A publisher gives 13% discount for buying 150 textbooks.

a   Calculate the price the school pays for buying 150 textbooks and how much
they saved compared with the regular price. **[3 marks]**

The books are also available as e books, costing on average £3.20 per year, per
student, and this includes VAT at 20%.

The school may reclaim the VAT on the e book.

b   Work out how much the school would save by buying the e books instead
of the paper books over a period of four years. **[3 marks]**

There are other costs involved with both of the options including replacing books and printing
pages.

c   Estimate the cost of each option over four years. Are e books still the better option?

Show details of your assumptions and calculations. **[4 marks]**

4   A university wants to find out how many hours its students work per week on average.

Here are the results of the university's initial survey with a pilot group:

12	10	12	8	6	12	9	15	18	10
10	12	6	0	11.5	14	8	13.5	0	4

a   What type of data is collected by the university?

Choose all that apply from:

discrete   continuous   primary   secondary   qualitative   quantitative **[2 marks]**

The university then conducts a more extensive survey across a larger number of students.

Here are the results:

Number of hours worked, x	Frequency
$0 \leqslant x < 4$ 2	23
$4 \leqslant x < 10$ 7	39
$10 \leqslant x < 15$ 12·5	62
$15 \leqslant x < 20$ 17·5	21
$20 \leqslant x < 30$ 25	6

The mean number of hours worked from the extensive survey was 10.7 hours and the standard
deviation was 5.6 hours.

b Compare the results from the pilot group and the extensive survey. **[6 marks]**

The pilot group were 20 students who were in the canteen of the main university building one lunchtime.

c Give two reasons why this is not a good sample to take. **[2 marks]**

d Describe a better sampling method that the university could use. **[2 marks]**

5

Income tax rates and bands 2016/2017

Band	Taxable income (per year)	Tax rate
Personal allowance	Up to £11 000	0%
Basic rate	£11 001 to £43 000	20%
Higher rate	£43 001 to £150 000	40%
Additional rate	over £150 000	45%

National Insurance Rates 2016/2017

Your pay	Class 1 National Insurance rate
£155 to £827 a week (£672 to £3,583 a month)	12%
Over £827 a week (£3,583 a month)	2%

Student loan repayment (2016/2017 tax year)

Start date of course	Threshold for repayment	Percentage of earnings above threshold
Before 1st September 2012	£17 495	9%
On or after 1st September 2012	£21 000	9%

Marc started a three-year course at university in September 2012.

He had a student loan of £28 000 when he started work.

He started a job in April 2016 with an annual salary of £23 000.

Marc says

'I shall have £1600 per month to live on'

Is he correct?

You **must** show your working. **[9 marks]**

→

6 A geographer is measuring the maximum width of pebbles (in mm) on a beach under erosion.

   Here are the results:

19	23	37	30	9	21	42
31	32	29	20	17	34	5
16	35	23	24	41	38	39

   a Draw a stem-and-leaf diagram to represent the data.   **[3 marks]**

   The geographer collected similar data one year ago.

   Here are the results:

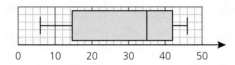

   b Compare the pebbles on the beach now with one year ago.   **[4 marks]**

   c Is there evidence to suggest that the pebbles are eroding?   **[3 marks]**

      You **must** show working to support your decision.

7 Estimate how many sandwiches are eaten in the UK each day.

   Show details of your assumptions and calculations.   **[5 marks]**

8 Jackie decides to invest £10 000 in a scheme that earns compound interest at an
   annual rate of 'between 3% and 5%'.

   a Draw a graph to show the potential performance of the scheme over five years. **[4 marks]**

   b Write down an inequality to show the possible outcome of her investment
      at the end of five years.   **[2 marks]**

ALCOHOL - FRIEND OR FOE?

> ## Don't drink and drive

You put others in danger.

Who would be so stupid?

I drive better after a drink.

WHAT YOU NEED TO KNOW ALREADY

→ Calculating with index numbers
→ Interpreting graphs

MATHS HELP

p304 Chapter 3.2 Spreadsheet formulae
p316 Chapter 3.6 Personal finance
p322 Chapter 3.8 Representing and analysing data

PROCESS SKILLS

→ Interpreting data presented in tables and graphs
→ Evaluating arguments
→ Criticising presentation and claims

We are bombarded by information throughout our waking hours, some of it true, some of it originating from something true but presented in a misleading way and much of it false. So many people have a vested interest in persuading us of the merits or otherwise of products, ideas and other people that it becomes extraordinarily difficult to tell fact from falsehood. It is vital to be able to analyse these claims and headlines critically and this chapter will show how mathematics can help to do that.

Alcoholic drinks have been around for several thousand years in a variety of civilisations. They were often safer to drink than water and had a much longer 'shelf life'. Now we frequently see reports of the damaging effects of alcohol interspersed with stories of the benefits of (moderate) drinking.

Drinking and driving

There are campaigns each Christmas to raise awareness of the dangers of drinking alcohol and then driving.

The number of fatalities from drink drive accidents in 2014 is estimated as 240. This number has remained fairly stable for several years now.

Discuss

What are the effects of drinking alcohol on peoples' driving? Why is this dangerous? Is 240 acceptable?

Comparing the 2014 data on drink drive casualties with baselines

Compared with 2005–2009 average			Compared with 2010–2014 average		
Killed	↓	48%	Killed	↑	1%
Serious	↓	40%	Serious	↓	9%
KSI	↓	42%	KSI	↓	7%
All casualties	↓	40%	All casualties	↓	11%
Accidents	↓	38%	Accidents	↓	10%

Investigate

> What does KSI mean?

> Which figures would you use in a road safety campaign? Why?

> Who might prefer to use the 2005–2009 figures? Why?

> Is this a good way to present the data? Explain why you think that.

> Use the graph in the example below and other data you might find to check the information above.

WORKED EXAMPLE

Afzal, Beth and Cal look at the graph and make these claims:

Afzal: There were only about 60 casualties in drink drive accidents in 2014.

Beth: Fatalities from drink drive accidents have decreased by 84% since 1979.

Cal: Since 2009 the same number of people have died as had serious injuries in drink drive accidents.

Comment on the validity of the claims.

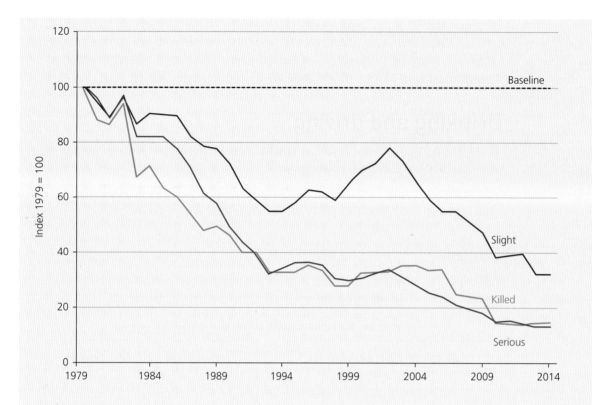

SOLUTION

Afzal has read numbers from the vertical axis of the graph:

Slight ≈ 32, serious ≈ 15, killed ≈ 13 and added them to get 60.

The vertical scale is an index number referring to 1979 as the base year which represents 100. It is not numbers of casualties so you can't add the numbers. Afzal is not correct.

Beth has read off 16 as the index number of fatalities in 2014. That is 84 points down from 1979. Since 1979 is the base year and represents 100, and she is comparing like with like, she is correct in saying the drop is 84%.

Cal has noticed that the lines for 'serious' and 'killed' are almost coinciding for 2009 to 2014 and so assumes they represent the same number of people. The reading for 'serious' is about 14 which is compared with the 100 for serious accidents in 1979 so approximately one seventh of the number. However, the reading for 'killed' is compared with the number killed in 1979 which may not be the same as the number with serious injuries in 1979. The drop to one seventh of 1979 numbers is about the same in both cases but we do not know what the actual numbers are. Cal is not justified in his claim as he needs more information.

Notice the solution explains where the numbers mentioned in the claims come from as well as identifying whether there is an error in the interpretation. There is also a decision as to whether the claim is justified.

Negative effects of alcohol

Alcohol is also a factor in many violent crimes, days off work and health problems. This costs the economy billions each year.

Discuss

How do the effects of alcohol result in such a huge cost to the economy?

Investigate

The information below is taken from a report published in March 2016 by Alcohol Concern.

More than 9 million people in England drink more than the recommended daily limits
In the UK, in 2014 there were 8697 alcohol-related deaths
Alcohol related harm costs England around £21 billion per year, with £3.5 billion to the NHS, £11 billion tackling alcohol-related crime and £7.3 billion from lost work days and productivity costs
Alcohol was 61% more affordable in 2013 than it was in 1980
Alcohol now costs the NHS £3.5 billion per year; equal to £120 for every tax payer
The alcohol-related mortality rate of men in the most disadvantaged socio-economic class is 3.5 times higher than for men in the least disadvantaged class, while for women the figure is 5.7 times higher
Alcohol-related crime in the UK is estimated to cost between £8 billion and £13 billion per year
A fifth (29%) of all violent incidents in 2013–14 took place in or around a pub or club. This rises to 42% for stranger violence. Over two thirds (68%) of violent offences occur in the evening or at night
193 males and 121 females between 15 and 34 years of age died from alcohol-related causes in 2011 in the UK
The number of alcohol-related hospital admissions of 15- to 24-year-old male patients increased by 57%, from 18 265 to 28 747 from 2002 to 2010
The number of hospital admissions of 15- to 24-year-old female patients increased at a faster rate (76%), from 15 233 in 2002 to 26 908 in 2010
In a sample of over 2000 15–16-year olds from the UK, 11% had had sex under the influence of alcohol and regretted it

Source: Statistics on Alcohol, March 2016 published by Alcohol Concern

> What does 'more affordable' mean?

> Why do you think so many more violent offences occur in the evening or at night?

> How many of the sample of 15–16-year olds regretted having sex under the influence of alcohol? How many would this be for all 15–16-year olds in the UK?

> How could you improve the presentation of this information?

> A headline in an online article states 'over two-thirds of violent crime is caused by alcohol'. Why is this not supported by the facts in the table?

> Make up some questions about this data.

Interpreting statistics, often presented as percentages, relating to the risks of developing serious medical conditions is important for clinicians and the general public alike. Policy-makers and those in charge of health budgets also need a good understanding of the significance of statistics. A study may report that a particular behaviour results in a 50% increase in the risk of premature death. This sounds serious but it could accurately describe a risk increasing from 2 in a million to 3 in a million which is not very great.

WORKED EXAMPLE

1 in 8 women develop breast cancer during their lifetime. Drinking more than about 3 drinks a day multiplies this risk by a factor of 1.5. This represents an increase of 50% in the risk of getting cancer.

Adapted from https://www.cancer.gov/about-cancer/causes-prevention/risk/alcohol/alcohol-fact-sheet#r7

Critically analyse this statement.

SOLUTION

1 in 8 represents a percentage of 12.5% of women.

Multiplying this by 1.5 = $1.5 \times 12.5 = 18.75\%$

This is an increased risk of just over 6% for any woman, not 50%.

50% is the extra women developing cancer, who wouldn't have done otherwise, because of the amount they drink.

They would have to drink at least three drinks a day, every day, to have this increased risk.

The statement is not false, just easy to misinterpret.

How your body is affected by alcohol

The quantity of alcohol in drinks is measured in units. The recommended limit for drinking is 14 units per week with at least 2 alcohol-free days to allow the liver to recover. One unit is 10 ml or 8 g of pure alcohol.

WORKED EXAMPLE

	A	B	C	D	E	F
1	Drink	ABV	Size	Alcohol units	Number drunk	Units consumed
2	wine	14%	125ml glass	1.8		
3	wine	14%	175ml glass	2.5		
4	wine	14%	250ml glass	3.5		
5	wine	14%	750ml bottle	10.5		
6	beer	2.80%	pint	1.6		
7	strong beer	4.80%	pint	2.7		
8	vodka	40%	25ml shot	1		
9	vodka	40%	50ml double	2		
10	flavoured cider	4%	330ml bottle	1.3		
11						

From UK Chief Medical Officer's guidelines

a Critically analyse the presentation of the information in the table.

b Ashley decides to keep a record of her alcohol intake. She fills in how many of each drink she has each week and writes a formula in column F of the spreadsheet. In cell F2 she writes = C2 ⋆ E2. What is the error in her formula?

c How many bottles of wine could you drink in a week without going over the recommended limit?

d Some people think that surveys of alcohol use regularly underestimate the amount drunk. Why might they think that?

SOLUTION

a The first four lines of the table give the same information.

The number of units for a 250 ml glass is not double the number for a 125 ml glass. This could be due to rounding.

Row 9 is unnecessary because it is the same information as row 8.

There are many drinks not included in the table.

There is only one strength of wine included in the table.

b The formula should be = D2 ⋆ E2.

c The number of bottles of wine is less than one and a half.

d People may not be honest in their responses.

The information can be compared with alcohol sales.

Discuss

Alcohol Concern recommends:

A minimum unit price is one of the most effective strategies of reducing alcohol-related harm. Selling alcohol for no less than 50p a unit would tackle health inequalities, reduce alcohol related crime, hospital admissions, lost productivity days and save lives.

› Do you think this would make a difference?

The effect of a unit of alcohol varies depending on many factors including gender, body weight and whether you have recently eaten. It takes approximately an hour for the body to absorb the alcohol that has been drunk and it then eliminates it at a rate of approximately one unit per hour. The speed of absorption and of elimination varies between individuals and circumstances but these figures give a rough guide.

Investigate

This graph shows one scenario of absorption and elimination of alcohol following an evening of beer-drinking.

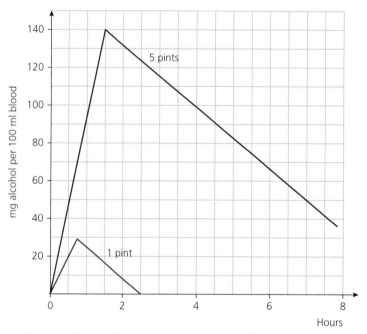

› How well does this graph model how the level of alcohol varies?

› Is this consistent with losing one unit per hour?

› 80 mg of alcohol per 100 ml of blood is the legal limit for driving. How soon can someone drive legally after drinking five pints of beer? What if they had drunk eight pints of beer?

› Make up a question about this graph.

The benefits of alcohol

Many people enjoy a glass of wine with a meal. Most people report that a drink helps them relax after work or when they are socialising. Celebrating occasions usually involves 'raising a glass' to make a toast and the glass usually contains an alcoholic drink.

Discuss

'Red wine boosts good cholesterol.'

New research shows that a daily glass of red wine for four weeks increases HDL or good cholesterol by up to 16%, and reduces the amount of the clotting compound fibrinogen by up to 15% (*from http://www.dailymail.co.uk/health/article-336208/Red-wine-boosts-good-cholesterol.html*).

> What do you think?

> The words 'up to' are used in both statements – how does that affect your opinion?

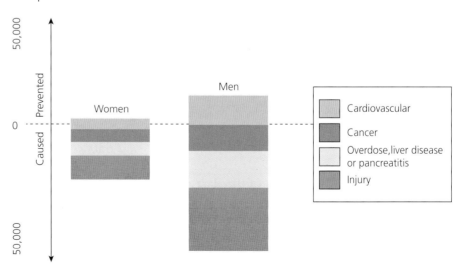

Investigate

The graph on page 157 shows the effect of alcohol on health.

› What measure is used in the graph to indicate the effect of alcohol?

› What benefits are suggested by the graph?

› What is the biggest risk from alcohol, according to the graph?

› Write a brief report on whether drinking alcohol should be recommended.

› Make up some questions about this data.

QUESTIONS

1 Look at the graphs and answer the questions following it.

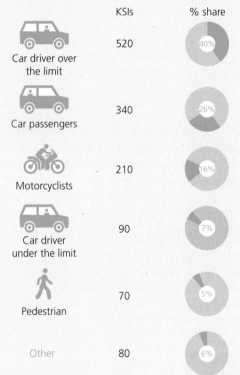

	KSIs	% share
Car driver over the limit	520	40%
Car passengers	340	26%
Motorcyclists	210	16%
Car driver under the limit	90	7%
Pedestrian	70	5%
Other	80	6%

a Suggest some road user types that would be classed as 'Other'.

b Comment on the presentation of the data.

c How does the data suggest that you shouldn't accept a lift from someone who has been drinking?

d 'Drink drivers killed more than 410 innocent people in 2014'. Critically analyse this claim.

2 A college summarised the plans for the students in the first year in a spreadsheet.

	A	B	C	D	E
1		M	F		
2	Uni	48	62	110	44%
3	not	76	50	126	60%
4		124	112	472	

a What error has been made in presenting this data?

b Suggest two further improvements that could be made to the presentation of this data.

c What does the figure of 44% in cell E2 represent?

d What proportion of the females plan to go to university?

3 Jed invests £500 in a savings account that offers an interest rate of 3% per annum. He does the following calculation to see how much it will be worth after three years.

£500 × 1.3 = £650

£650 × 1.3 = £845

£845 × 1.3 = £1098.50

a Explain and correct the error Jed has made.

b Suggest a more efficient way for him to work it out.

4 Use the diagram to answer the questions.

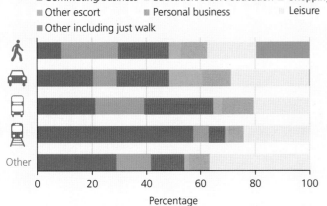

- Commuting/business ■ Education/escort education ■ Shopping
- Other escort ■ Personal business Leisure
- Other including just walk

Percentage

a For what purpose are people most likely to travel by bus?

b What percentage of people who use the train do so for work purposes?

c Don says 'Trains are the most popular method for travelling to work'.

Terrie says 'This chart must be wrong, more than 100% of people travel for work.' Critically analyse their claims.

Show working to support your comments.

5 Use the data to decide whether to recommend travelling by car, foot or motorbike.

| Road Type | Car occupant | Pedestrians | Motor cyclists | Pedal cyclist | Other |

Urban

Rural

Motorways

On **Urban roads** there were 616 fatalities.

On **Rural roads** there were 1,063 fatalities.

On **Motorways** there were 96 fatalities.

Give reasons for your decision.

MISLEADING CLAIMS

> ## Claims of the 2016 EU referendum campaign

We pay far too much money to the EU

It's better to be part of a larger group

Everyone is flooding into our country

Lots of businesses will move out of the UK

MATHS HELP

p322 Chapter 3.8 Representing and analysing data

PROCESS SKILLS

→ Criticising claims and representations
→ Analysing charts and images
→ Checking facts
→ Evaluating arguments

It is human nature to believe things that are said to us or written, whether on paper or electronically. Too many people rely on this and say things that are, to say the least, grossly misleading and often simply untrue. Apologies and retractions do not make the headlines, so people often go on believing these false claims. It is vital for every citizen to criticise these claims that are made, particularly when someone has a vested interest in making them, and to pay attention to what is said by more neutral observers.

In June 2016 there was a referendum in which the UK voted to leave the EU. During the political campaign before the referendum several claims were made by the competing sides –'Vote Leave' and 'Britain Stronger in Europe' – regarding the impact that a 'Brexit' (exit from the EU) would have on the UK.

Investigate

The most highly disputed claims of the referendum related to the financial cost to the UK of being a member of the EU and the financial impact that leaving the EU would have on average households.

> Research these claims. What were the arguments of each side?

> Do the claims seem realistic?

> What assumptions do you think these claims were based on?

> Can you find arguments for and against each of these claims?

WORKED EXAMPLE

These claims were reported during the EU referendum campaign:

- A 'Britain Stronger in Europe' member claimed that more cars were made in the north-east of England than in the whole of Italy.
- A 'Vote Leave' member claimed that EU law forbade shops from selling bananas in a bunch of more than three.

Use the information below to critically analyse these claims.

Nissan Motor Manufacturing Ltd had a turnover of £5.3 billion in 2013/14
DB Arriva Public Transport had a turnover of £3.5 billion in 2014
The Go-Ahead Group PLC Public Transport had a turnover of £3.2 billion in 2014/15
Nissan manufactured 476 000 cars in 2015
Jaguar Land Rover produced 489 000 vehicles in 2015
Nissan is based in the North-East of England
Jaguar Land Rover has plants in the Midlands and Merseyside
Almost 1.6 million cars were produced in the UK in 2015, 80% of which were exported
1 014 223 vehicles were manufactured in Italy in 2015
697 864 vehicles were manufactured in Italy in 2014
Commission regulation 2257/94 decreed that bananas in general should be 'free from malformation or abnormal curvature'. Those sold as 'extra class' must be perfect, 'class 1' can have 'slight defects of shape' and 'class 2' can have full-scale 'defects of shape'
Nothing is banned under the regulation, which sets grading rules requested by industry to make sure importers – including UK wholesalers and supermarkets – know exactly what they will be getting when they order a box of bananas
There is no regulation about bunches of bananas having to be a particular size

→

SOLUTION

— A 'Britain Stronger in Europe' member claimed that more cars were made in the north-east of England than in the whole of Italy.

Nissan is based in the north-east and manufactured 476 000 cars in 2015 which is less than the 1 million vehicles produced in Italy that year. Jaguar Land Rover is not in the north-east. 2015 is the most recent year that figures would have been available so that does not support the claim. Vehicles are not necessarily all cars and so the claim may be justified if the majority of the vehicles made in Italy are not cars.

There were nearly 1.6 million cars produced in the UK in 2015 which is more than Italy.

Many components for cars made in a given country are imported from elsewhere so you cannot really say that a car is manufactured in a particular country. It is manufactured in many countries and assembled in one.

There are many firms in the motor industry in the north-east but Nissan is the only one making a significant number of cars. The others mentioned are dealerships, successful but not manufacturers.

— A 'Vote Leave' member claimed that EU law forbade shops from selling bananas in a bunch of more than three.

The directives about bananas are to define the quality of different classes of produce, as named on the pack. This is to stop bruised and damaged bananas being sold to shops. There is no restriction on what can be sold, as long as they are described accurately. There is no information about the size of a bunch. Supermarkets sell bananas in bunches of more than three all the time. This particular claim is false.

HINT

You may have to make some assumptions as well.

Discuss

❯ Can you think of any examples of misleading claims or data in the media?

❯ Can you find some examples of headlines that need checking?

❯ How are advertisements regulated?

❯ Are newspaper headlines regulated?

QUESTIONS

1 Following the 2014/15 Budget announcement, the *Daily Mail* produced this three-dimensional pie chart about Government income.

 a What criticisms could you make about this pie chart?

 b How could the chart be improved?

 c What point do you think the article was trying to make?

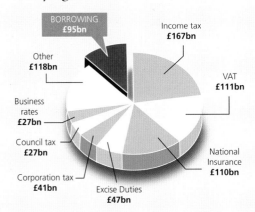

2 *The Sun* newspaper used this image to support an article about the rising costs of renewable energy to customers.

 a What criticisms would you make of this chart?

 b Why might the image have been presented in this way?

 c How could this image be improved?

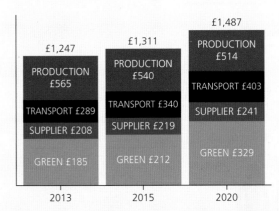

3 This graph, released by the Treasury in 2014, displayed the level of Government investment in various types of major infrastructure projects. The bar chart has a 'logarithmic' scale, where the gaps between £1 million, £10 million,

£100 million, £1 billion, £10 billion and £100 billion were each represented by increments of the same size.

 a What is meant by the term 'logarithmic' scale?

 b What criticisms would you make of the chart produced by the Treasury?

 c Why might the bar chart have been presented in this way?

 d How could this bar chart be improved?

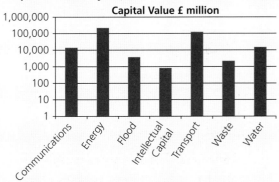

4 Dave wants to know how much he should be paying in tax. The first £11 000 of his salary is tax-free and he pays 20% on the next £32 000. After that he pays 40% tax.

Dave's annual salary is £38 000

He works out his tax using this calculation:

Tax = 0.2 × 11 000 + 0.2 × 32 000 = £6620

Critically analyse Dave's calculation, correcting it where necessary.

E 5 Jo is updating her monthly budget. She says, 'I usually allow £650 for rent, fuel, water and council tax but my rent has gone up 5%, fuel has increased by 6%, the water charges have stayed the same and council tax has gone up by 2%. That means I need another 13% so that is £84.50 per month.'

Explain why Jo is wrong.

Show working to justify your comments.

PRACTICE QUESTIONS: PAPER 2 COMPULSORY

1 Tom is using a spreadsheet to work out how profitable his business projects are.

	A	B	C	D
1	Date	cost	income	percentage profit
2	24/01/2016	£284	£410	14.43661972
3	31/01/2016	£120	£156	23
4	38/01/2016	£45	£68	51.11111111
5				

a Analyse Tom's spreadsheet, identifying any errors. **[3 marks]**

b Suggest some improvements he could make. **[3 marks]**

c Tom looks at the results and says:

'The third one is the most profitable. 51% profit is very good. I shall quit my job and spend my time doing more projects like that.'

Explain why Tom may be unwise to do that. **[2 marks]**

2 Figure 2.3 Evolution of penicillin resistance in *Staphylococcus aureus*: a continuing story

1928	Penicillin discovered
1942	Penicillin introduced
1945	Fleming warns of possible resistance
1946	14% hospital strains resistant
1950	59% hospital strains resistant
1960s–70s	Resistance spreads in communities
1980s–90s	Resistance exceeds 80% in communities, 95% in most hospitals

When this information was published, many people said that doctors should stop using penicillin.

Critically analyse the suggestion. **[3 marks]**

3 Hospital admissions with a primary diagnosis of drug-related mental and behavioural disorders

Hospital Episode Statistics (HES) 2014/15

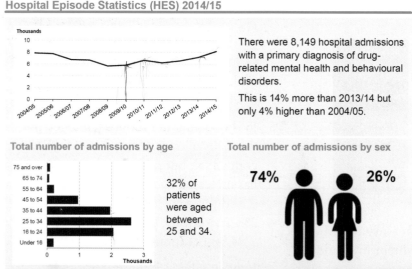

There were 8,149 hospital admissions with a primary diagnosis of drug-related mental health and behavioural disorders.

This is 14% more than 2013/14 but only 4% higher than 2004/05.

32% of patients were aged between 25 and 34.

For more information: Tables 1 and 2 of Statistics on drug misuse, England, 2016

Teri says, 'Only a fifth of admissions to hospital were female.'

Dave says, 'Most of the people admitted to hospital were under 30.'

Tony says, 'Drug related hospital admissions have soared since the election in 2010.'

Does the data support their claims?

Show working to justify your comments. **[9 marks]**

CONSUMER PROTECTION FOR PACKAGED GOODS

A dreamy bed of organic
oat flower, lavender
& limeflower

Wt 20g ℮ (0.71oz)   20 herbal tea sachets

> When I opened the jar it looked half empty. How do I know I've got what I paid for?

› What is the e-mark?

> When I weighed the contents, there was less than it said on the packet.

> What does 390 g followed by an e mean?

Many goods we buy these days, such as coffee, baked beans, cooking oil and peanuts, are pre-packaged in pre-determined quantities. The packages are filled through an automated mechanical process. A degree of variation in the content of the packages is inherent in all such processes. This cannot be helped, but we need to be sure we are not being short-changed by the packaging process. The Weights and Measures (Packaged Goods) Regulations 2006 provides consumer protection against this through a regulatory framework for the automatic filling of packages. The e-mark on a package indicates that it complies with these regulations.

WHAT YOU NEED TO KNOW ALREADY

→ Calculating mean and standard deviation
→ Working with percentages
→ Working with total probability
→ Drawing and interpreting histograms

MATHS HELP

p320 Chapter 3.7 Statistical terms
p333 Chapter 3.10 Calculating statistics
p378 Chapter 3.21 Normal distribution

PROCESS SKILLS

→ Modelling situations
→ Interpreting graphs
→ Deducing probabilities
→ Inferring properties of data
→ Estimating statistics
→ Representing data

The e-mark shows that a pre-packaged product is regulated by the 2006 regulations. It tells the consumer that the content weight shown on the package is an average weight and that the actual weight lies within defined acceptable tolerances. Thus in some packages the contents will be below the stated quantity.

This average system applies to most goods that are pre-packed into pre-determined quantities by weight or volume, including most foodstuffs. The regulations apply to all packages intended for sale in constant nominal quantities, which are between 5 g and 25 kg (or 5 ml and 25 l) inclusive.

A tin of chickpeas has an average weight of 400 grams. If you weighed the content of all the tins of chickpeas that were produced one morning and plotted the histogram, you would get something like this:

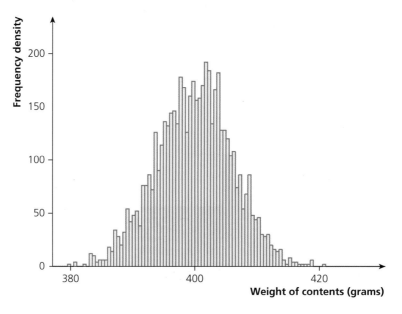

The larger the sample size, the closer the histogram will be to the **normal distribution curve**, shown in blue below.

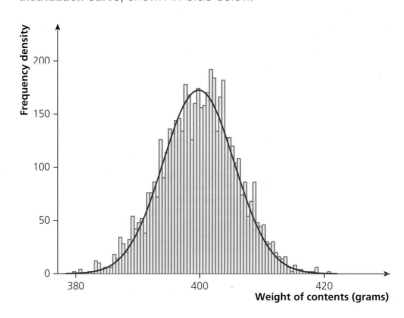

Data distributed normally shows two key characteristics:

> it is a bell shape

> it is symmetrical.

One interesting piece of information about every normal distribution is that about $\frac{2}{3}$ of the data falls within one standard deviation of the mean. In other words, the probability that a piece of data selected at random will lie within one standard deviation of the mean is $\frac{2}{3}$.

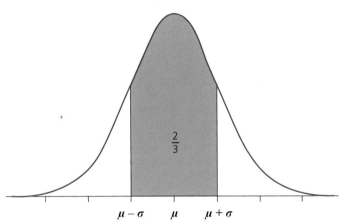

The peak of the curve is at the centre and represents the mean, median and mode – which are all identical for the normal distribution.

The spread of the distribution is described by the standard deviation.

Discuss

> What fraction of the data will fall above the mean plus one standard deviation $(\mu + \sigma)$?

> What percentage of the data will fall below the mean minus one standard deviation $(\mu - \sigma)$?

Another really useful piece of information about every normal distribution is that approximately 95% of the data falls within two standard deviations of the mean. That's pretty much all the data!

Discuss

> What percentage of the data will fall above the mean plus two standard deviations $(\mu + 2\sigma)$?

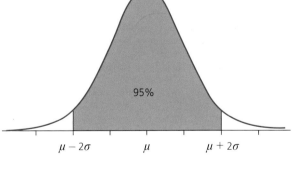

> What percentage of the data will fall below the mean minus two standard deviations $(\mu - 2\sigma)$?

> What is the probability that a piece of data selected at random will fall within two standard deviations of the mean?

WORKED EXAMPLE

The contents of a tin of chickpeas has a mean weight of 400 g with a standard deviation of 15 g. Above what weight will 97.5% of the tins be?

SOLUTION

95% of the data is within two standard deviations.

$400 + 2 \times 15 = 430$

$400 - 2 \times 15 = 370$

$95\% + 2.5\% = 97.5\%$ of the tins will be above 370 g.

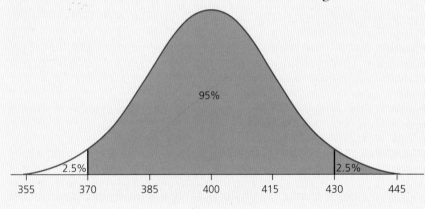

Investigate

> A bottle contains 250 ml of olive oil on average. The standard deviation is 9 ml. Below what volume will only 2.5% of bottles fall?

> What is the probability that the volume of the contents of a bottle selected at random will fall outside of two standard deviations from the mean?

The properties of the normal distribution form the basis of the criteria for the e-mark. The directive sets out that the proportion of packages that are short of the stated quantity by a defined amount (the 'tolerable negative error' or TNE) should be less than a specified level.

For our 400 g tin of chickpeas, the directive stipulates that the TNE is 3% of the nominal weight and that only 3 tins in every 50 should be below this.

3 ÷ 50 = 0.06
    × 100
= 6%.

3% of 400 = 12 g	Tolerable negative error
400 + 12 = 412 g	12 g above the mean
400 − 12 = 388 g	12 g below the mean
3 ÷ 50 = 6%	3 tins in every 50 fall below.

These criteria can be represented using the normal distribution curve shown below.

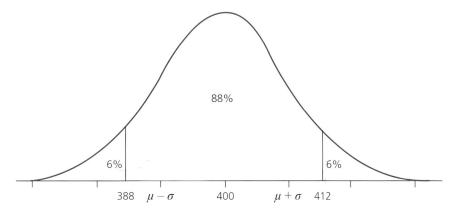

Discuss

› Estimate the standard deviation for this distribution, which meets the TNE directive.

› If the actual standard deviation was more than your estimate, would the contents meet the directive?

QUESTIONS

E **1** A bag of pre-packed potatoes has an average weight of 1 kg with a standard deviation of 40 g.

   a Write down the fraction of bags that have a weight below 960 g?

   b Write down the fraction of bags that have a weight below 920 g?

   c Sketch a normal distribution curve to show this information.

2 Here are two histograms that show the distribution of the heights of 1000 Italian women and 1000 Italian men.

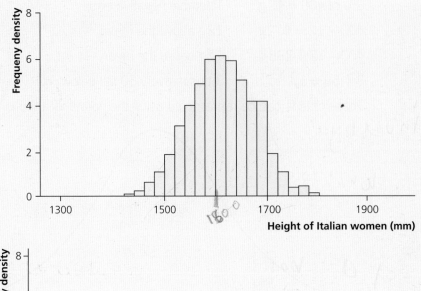

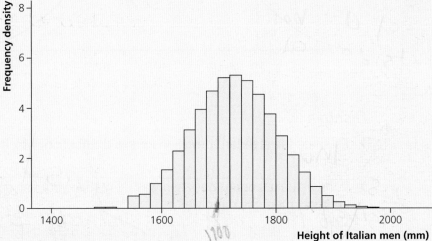

Estimate the mean and standard deviation of both distributions. Use your estimates to compare the heights of Italian women with Italian men.

3 The contents of a carton of tomatoes has a mean weight of 390 g with a standard deviation of 4 g. What content weight will only 2.5% of the cartons be below?

4 The speed of all cars passing a speed camera on a 30 mph stretch of road is recorded. The mean speed is found to be 34 mph with a standard deviation of 4 mph. What is the probability that a car chosen at random was breaking the speed limit?

5 A manufacturer wants to offer guarantees on the lifetime of its new light bulb. The bulbs have a mean lifetime of 1200 hours with a standard deviation of 85 hours. The manufacturer wants to give refunds on no more than 2.5% of sales. What maximum lifetime should they guarantee?

CHAPTER
> 2.23

AM I NORMAL?

> ## What is normal?

I'm really short.

I want to be different.

I want to be like everyone else.

This chapter continues your study of the important normal distribution, which models the distribution of much naturally occurring data such as height, weight and performance in tests of various kinds. You will learn how to match any normal distribution to a standardised one, and hence find the probability of data being in particular parts of the distribution – for example, finding the probability of a woman's height being more than 180 cm (6 feet).

WHAT YOU NEED TO KNOW ALREADY

→ Understanding probability
→ Calculating mean and standard deviation
→ Using a calculator
→ Interpreting graphs

MATHS HELP

p322 Chapter 3.8 Representing and analysing data
p333 Chapter 3.10 Calculating statistics
p378 Chapter 3.21 Normal distribution
p381 Chapter 3.22 Normal distribution – finding probabilities

PROCESS SKILLS

→ Modelling using normal distributions
→ Checking answers
→ Deducing from known facts
→ Representing data
→ Comparing models with real data

How is the normal distribution used?

You learnt in the last chapter about the normal distribution and how it is a good model for much familiar data, such as heights, weights, exam results, time for a battery to recharge, sizes of items produced in a manufacturing process, and so on.

The shape of the curve reflects the fact that for the types of data listed above, most results are somewhere in the middle, with a minority of results at the extremes.

This graph shows the standard normal distribution, with a mean of 0 and a standard deviation of 1. This is written as N(0, 1)

The total area under the curve is 1, and the symmetry of the curve means that the probability of a data item being greater than zero is exactly the same as the probability of a data item being less than zero, that is, one half.

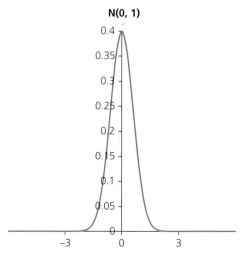

You know from the last chapter that about two-thirds of the data lies within one standard deviation of the mean (that is, between –1 and +1) and that about 95% of the data lies within two standard deviations of the mean (that is, between –2 and +2). We can do estimates of probabilities using these approximations.

In this chapter you will learn how to use **statistical tables** to find probabilities more accurately.

The diagram on the left is a sketch of the standardised normal distribution N(0, 1). The **normal distribution table** tells you the probability that a data item drawn from a standard normal distribution, N(0, 1), lies in the shaded region under the curve.

> **Note**
>
> You can find normal distribution tables on AQA's website
>
> http://www.aqa.org.uk/subjects/ mathematics/aqa-certificate/ mathematical-studies-1350/ statistical-tables

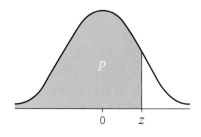

That is, it gives you the probability that the data item is less than or equal to z.

This is an extract from the table.

It shows that when z is 0.7, the shaded area in the diagram above is 0.75804.

This means that the probability that a random data item is less than or equal to 0.7 is 0.75804.

We can deduce further facts from this one piece of information:

The probability that a random data item is greater than 0.7 is $1 - 0.75804 = 0.24196$ (the unshaded area under the curve, as the shaded area and the unshaded area together make 1).

This is, by symmetry, equal to the probability that a random data item is less than –0.7.

Putting this together means that the probability of a random data item being between –0.7 and +0.7 is

$1 - 2 \times 0.24196 = 0.51608$

z	0.00
0.0	0.50000
0.1	0.53983
0.2	0.57926
0.3	0.61791
0.4	0.65542
0.5	0.69146
0.6	0.72575
0.7	0.75804
0.8	0.78814
0.9	0.81594
1.0	0.84134
1.1	0.86433
1.2	0.88493

Can you explain why the probability that a random data item is between 0 and 0.7 is 0.25804? *Find Probability at 0 = 0.5 so 0.75804 −*

0.5000
——————
0.025804.

z	0.00	0.01	0.02	0.03
0.0	0.50000	0.50399	0.50798	0.51197
0.1	0.53983	0.54380	0.54776	0.55172
0.2	0.57926	0.58317	0.58706	0.59095
0.3	0.61791	0.62172	0.62552	0.62930
0.4	0.65542	0.65910	0.66276	0.66640
0.5	0.69146	0.69497	0.69847	0.70194
0.6	0.72575	0.72907	0.73237	0.73565
0.7	0.75804	0.76115	0.76424	0.76730

Note that the columns further to the right of the probability table enable you to find probabilities for values of z that have two decimal places.

Discuss

Check that you can use the table to show that the probability of a random data item being between −0.3 and + 0.3 is 0.23582.

Not all normal distributions, of course, have a mean of 0 and a standard deviation of 1.

The diagram shows the graphs of the standard normal distribution N(0, 1) and a normal distribution with mean 2 and standard deviation 3. The variance is the square of the standard deviation, and the distribution is expressed in terms of its mean and variance, that is, N(2, 9).

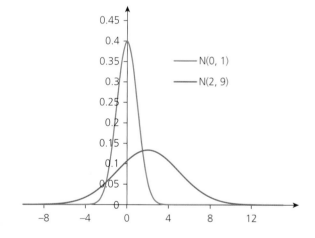

Discuss

❯ What is the same and what is different about the two graphs?

Investigate

The graph of any normal distribution N(μ, σ^2), where μ is the mean and σ is the standard deviation, can be transformed into the standard normal graph in two steps:

− a horizontal shift to make the graph symmetrical about the y-axis

− a horizontal shrink (or stretch) to make the spread of the graph match the spread of the standard normal graph.

This is achieved by replacing x by $\dfrac{x - \mu}{\sigma}$.

Note that the expression $\frac{x - \mu}{\sigma}$ is the number of standard deviations that x is from the mean. Make sure you understand why. The expression $\frac{x - \mu}{\sigma}$ is called the **standardised score**, or **z-score**.

So, the probability that a random data item from the distribution N(2, 9) is less than 4 is the same as the probability that a random data item from the distribution N(0, 1) is less than the z-score, so $\frac{4 - \mu}{\sigma} = \frac{4 - 2}{3} \approx 0.67$. From the table, the probability is 0.74857.

This diagram shows a normal distribution, N(166, 53.9) that closely approximates female heights in centimetres.

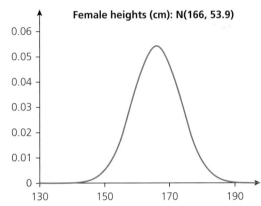

Female heights (cm): N(166, 53.9)

Discuss

> Is it possible for a population to match the normal distribution perfectly?

Investigate

> Your calculator can probably work with normal distributions and calculate probabilities so that you don't have to use tables. Find out what your calculator can do!

WORKED EXAMPLE

From the graph above, what percentage of females are shorter than 150 cm?

SOLUTION

The standard deviation of the distribution is $\sqrt{53.9} \approx 7.34$

The z-score for 150 is $\frac{150 - 166}{7.34} \approx -2.18$

From the normal distribution table, the probability that a random data item from the standard normal distribution is less than 2.18 is 0.98537.

HINT

The tables don't include negative values for z so use the symmetry of the curve to work them out.

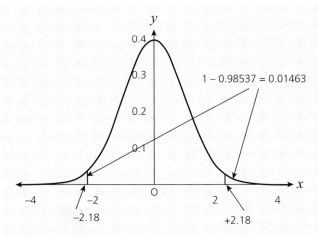

So the probability that it is greater than 2.18 is $1 - 0.98537 = 0.01463$, which is the same as the probability of it being less than -2.18.

So the percentage of females shorter than 150 cm is approximately 1.5%.

Check that this seems reasonable by looking at the graph.

Comparing the model with real data

The normal distribution is a model that attempts to describe the distribution of real data so that you can make predictions based on probabilities. A measure of how good the model is, is how closely it fits the data. Trying to get a closer fit often means the mathematics becomes too complex to work with so a compromise between closeness of fit and usability is inevitable.

WORKED EXAMPLE

The weight of female guinea pigs shows a normal distribution $N(800, 50^2)$.

The weights in grams of 10 female guinea pigs are:

691   717   735   756   786   789   794   809   814   826

How well does the model $N(800, 50^2)$ fit the distribution?

SOLUTION

The mean of the 10 weights is 771.7 g, which is nearly 30 g less than the mean in the model. The standard deviation is 45 g, which is also less than the standard deviation in the model.

We can also work out some probabilities:

$P(X < 800) = 0.5$ for the model, but 0.7 for the data

$P(X < 750) = 0.17$ for the model, but 0.3 for the data.

So the model is not a good fit for the data, which may be partly due to the small size of the sample. The guinea pigs may be underweight for some reason, perhaps through lack of proper food or illness, or they may be a smaller breed.

Investigate

Use the normal distribution of female heights described on page 174, or find one for adult heights in general, and compare it with the distribution of heights for a sample from your college. You could do this by working out the expected numbers of students with heights less than a given value, 160 cm say, and comparing it with the actual frequencies.

Can you get a better fit by using the mean and standard deviation of your sample for your normal distribution model?

Did your data fit either model well? Is there any reason why your sample may be biased?

Note

You can find normal distribution tables on AQA's website

http://www.aqa.org.uk/subjects/mathematics/aqa-certificate/mathematical-studies-1350/statistical-tables

QUESTIONS

1 Using the fact that 95% of the data drawn from the standard normal distribution $N(0, 1)$ lies within two standard deviations of the mean, find the probability that a data item is greater than 2.

2 Use tables to find the probability that a random data item from $N(0,1)$ is:

  a less than or equal to 0.2

  b between −0.2 and 0.

3 What percentage of the data in the standard normal distribution lies between −0.54 and +0.54?

4 For the normal distribution $N(4, 4)$:

  a Write down the mean and standard deviation.

  b Calculate how many standard deviations the number 3 is from the mean.

5 What is the probability that a random data item from the distribution $N(2, 9)$ is:

  a less than 1

  b between 0 and 1?

6 The distribution of male heights in metres can be modelled by $N(1.7, 0.04)$. What percentage of men are over 1.8 metres tall?

7 Male diastolic (measured between heartbeats) blood pressure is normally distributed, with mean 82 mm of mercury and standard deviation 10 mm of mercury. Diastolic blood pressure of 90 mm or more is a possible cause for concern. What percentage of men have blood pressure this high?

E 8 The weight of male guinea pigs are normally distributed with mean 1050 g and standard deviation 75 g.

The weights in grams of 10 male guinea pigs are:

970 979 1016 1035 1039 1044
1056 1070 1124 1139

How well does the model $N(1050, 75^2)$ fit the sample of guinea pigs?

★★

POKER HANDS

It's a 'once in a lifetime' hand.

> ## What is the 'population' of poker hands?

What are the chances of getting a hand like that?

Can statistics help me win at poker?

WHAT YOU NEED TO KNOW ALREADY

➜ Methods of sampling

MATHS HELP

p320 Chapter 3.7 Statistical terms
p322 Chapter 3.8 Representing and analysing data
p330 Chapter 3.9 Sampling methods

PROCESS SKILLS

➜ Justifying choices
➜ Critically analysing sampling methods

The term 'population' in statistics refers to the set of all the items about which information is sought. The word 'population' is used not just for human or animal population, but to represent the complete group under study. For example, all the adult voters in the UK, all the households in your town, all the whales in the Pacific Ocean, all the cars produced by a factory, all the possible outcomes when you roll a dice for an indefinite number of times or even all the possible poker hands.

As mentioned in Chapter 2.9, you can only be completely accurate about the number and characteristics of a population if you collect information from every member of the population. However, this can be both costly and time-consuming and in some cases is nearly impossible to accomplish.

We often use sampling as a way to make predictions about the desired population instead. Take a look at Chapter 2.9 to find out more about various sampling techniques that can be used.

> Within your school or college, can you give five examples of populations that could be sampled?

A census can be used to collect information from an entire population.

> Think of and explain three barriers to carrying out a census of the people in the UK.

> Think of and explain three barriers to carrying out a census of the people in other countries.

Investigate

> How can you estimate a human population?

> How can you estimate a wildlife population?

> What are the problems with making estimates of wildlife populations?

> What is opinion polling and what may make it inaccurate?

From Chapter 2.9 we have seen that random sampling (also known as simple random sampling) is where each member of a population has an equal chance of being chosen. In this way we aim to reduce the opportunity for systematic error, or **bias**, to ensure that the sample is representative of the population from which it has been taken.

The advantage of simple random sampling is that it is easy to apply to small populations. However, every person or item in the population has to be listed before the random sample can be taken. This can be very time-consuming to use for large populations.

DID YOU KNOW?

The list of the population is called a **sampling frame**.

WORKED EXAMPLE

Biologists are investigating the average length in millimetres of freshwater fish.

A population of 80 fish is being investigated. Each fish has been measured from the tip of its nose to the middle of its tail and the results recorded.

Each fish was tagged when it was measured.

Describe how random numbers could be used to select a sample of 25 fish.

SOLUTION

Each of the 80 fish within the population of the lake is assigned a number.

The random number generator on a calculator is used to generate two-digit integers between 01 and 80. Only numbers in the range 01–80 are accepted and any repeats are ignored to ensure that each fish has an equal chance of being selected.

This process is repeated until a random sample of 25 fish is generated.

> **HINT**
>
> **Ran#** and **RanInt** are common functions on calculators that can be used to generate random numbers.

Discuss

> ❯ What are the potential difficulties in collecting the data for this population?
>
> ❯ Why would this scenario never happen?

Investigate

> ❯ How can you use a table of values to generate a random sample?
>
> ❯ What are the random number functions on your calculator? How do they work?
>
> ❯ Can you generate a random integer on your calculator?
>
> ❯ Can you generate a random number between a specified start value and end value on your calculator?

> **HINT**
>
> Calculate the same statistics for the whole population (since you can).
>
> You may wish to collaborate to examine more random samples of size ten, and use a spreadsheet to do the calculations.

Investigate

> ❯ Find a list of the top 100 films. There are several lists to choose from, depending on the criteria chosen to select the films. This is the population that you will work with. Choose one of the properties of the films, such as year of release or the money taken so far.
>
> ❯ Select a random sample of ten films and work out some statistics for that sample. For example, you could work out the mean and the range, or the median and interquartile range.
>
> ❯ Now select a second random sample of ten films and work out the same statistics as before.
>
> ❯ What do you notice?
>
> ❯ Which sample is more representative of the population? How do you know?

Discuss

› How can you ensure a representative sample in your investigation?

› What if the samples were of size five? Or twenty?

It is notoriously difficult to ensure a representative sample. If you had access to the data for the whole population you would use that and not bother with a sample.

A larger size of sample tends to be more representative than a smaller one. Spending time eliminating sources of bias is time well spent, but it is hard to anticipate all of the factors that could bias the sample. There have been several high-profile failures to predict the results of elections by polling organisations because of bias that has crept into the sampling process.

Discuss

› Sometimes people complete online surveys to earn money. Do you think the results from these surveys are a true reflection of the views of the population under study? Why do you think that?

› Opinion polls are surveys of public opinion. They are usually designed to represent the opinions of a population. One method of polling is to take phone numbers from a directory and randomise the last digit. What are the advantages and disadvantages of using this method?

WORKED EXAMPLE

A dentist wants to ask a random sample of patients about their levels of satisfaction with the service they receive at the dental practice.

Describe how the dentist could select a random sample and what decisions or issues may be involved.

SOLUTION

The dentist will need to make decisions about the age categories that they want in their sample. In addition to this they will need to consider how they will contact the patients to obtain their opinions. For instance, the dentist may wish to contact individual patients or may prefer households as this may increase the coverage of the survey. Other issues include the reliability of the responses from the patients.

To generate the sample, each patient can be assigned a number and then a random number generator can be used to create a sample of the required size.

There are many ways of generating a random sample so many solutions are possible. They should all, however, be justified.

Another type of sampling is a **systematic** sampling. The population is listed using a sampling frame, and, for example, every fifth, or tenth, or seventeenth item is chosen for the sample. The starting point is chosen randomly and the gap depends on what proportion of the population you wish to sample. The dentist in the previous example could ask every tenth patient to get a 10% sample, but it would no longer be a random sample.

WORKED EXAMPLE

The principal of a college wants to know about the study habits of the students. There are 5000 students at the college and she wants a sample of 100.

Describe how a systematic sample of 100 could be selected.

SOLUTION

A sampling frame is needed. The college has an admission number for each student and the ordered list of the students, using these numbers, would be appropriate.

A sample of 100 is one fiftieth of the college so a systematic sample would select every fiftieth student. The first one could be selected using a random number between 1 and 50.

This is not a random sample but should be representative as it is unlikely to have a built-in bias towards a particular age or type of student.

QUESTIONS

1   A group of students attempted to generate a random sample. This was done by putting a number of pieces of string of varying lengths into a bag and having students pull out a 'random sample' of lengths of string.

   a   Criticise this method of generating a random sample.

   b   How could the method be improved?

2   Afzal investigates the different types of plants growing in a field near his house for a project in biology.

> **HINT**
>
> Think about the type of constraints that need to be in place to ensure that a sample is random.

   a   Describe the population for his investigation.

   b   Describe a method he could use to collect a random sample from that population.

3   For a coursework task, Natalie chooses to investigate the number of cars that pass her college between the hours of 2 p.m. and 3 p.m.

   a   Describe the population in Natalie's investigation.

   b   How could you collect a sample from this population?

4   There are 65 boys and 78 girls in Year 13 at a school.

   A sample of 25 girls is being selected for a survey.

   a   Describe how random numbers could be used to select a sample of 25 girls.

   b   Explain why a sample carried out using the same method might generate a different sample of 25 girls.

E 5   There is a by-election in a marginal constituency. The local paper wants to have regular polls of the electorate to monitor how the parties are faring. Describe how a systematic sample of 500 voters could be selected.

QUALITY CONTROL

〉 How do you check that a lightbulb lasts as long as it should?

The only way would be for customers to record when they turn it on and off.

They could test some of them in the factory.

It's impossible!

In Chapter 2.22 we saw that packaged quantities were regulated by the e-mark so that consumers could be sure the contents (on average) were as stated. But how is this checked? A can of soup could be weighed, but as well as the weight of the contents varying slightly so will the weight of the tin. So how can you check the exact weight of the contents? The tin could be opened and the contents weighed, but it is not possible to do this for all the tins in order to calculate a mean because there would be none left to sell! A lightbulb that is claimed to last for an average of 2000 hours cannot be tested without using the lightbulb up! This chapter explains one way of solving this problem.

WHAT YOU NEED TO KNOW ALREADY

→ Understanding the shape of the normal distribution

→ Using tables to find probabilities from any normal distribution

→ Calculating mean and standard deviation

MATHS HELP

p 378 Chapter 3.21 Normal distribution
p 381 Chapter 3.22 Normal distribution – finding probabilities
p 385 Chapter 3.23 Confidence intervals for the mean

PROCESS SKILLS

→ Modelling data

→ Justifying decisions

→ Representing graphically

→ Comparing distributions

The only sensible method of quality control for the majority of manufactured products, including lightbulbs, is for the manufacturer to test a sample. But how big should the sample be?

A lightbulb manufacturer claims that a particular type of bulb lasts for 2000 hours on average.

Investigate

> How many days is 2000 hours?

> Estimate how many hours the light would be on each day.

> Estimate how many days the bulb should actually last.

Point estimates

Suppose a manufacturer tests one randomly selected lightbulb and it lasts for 2100 hours. This is good for the manufacturer because it means they can sell all their lightbulbs except one and they can say they estimate the mean to be 2100 hours. This is called a **point estimate for the mean.**

Discuss

> Do you think this is a reliable estimate?

> What do you think the reaction of the manufacturer might have been if the bulb had lasted 1900 hours?

A better point estimate is obtained by increasing the sample size. Suppose four bulbs are tested and their lifetimes are found to be 1950, 2100, 2050 and 1975 hours. The sample mean is now

(1950 + 2100 + 2050 + 1975) / 4 = 2018.75 hours.

The larger the sample size, the closer you would expect the sample mean to be to the population mean, but you still only have a point estimate.

Intervals for the mean

Suppose instead you took several samples. If you took 100 samples then you would have 100 point estimates for the mean. By looking at the distribution of these means you could begin to estimate an interval within which the true mean should lie.

It is claimed a particular type of lightbulb is manufactured with a lifetime of 2000 hours and a standard deviation of 25 hours.

Here is the normal distribution curve for these bulbs.

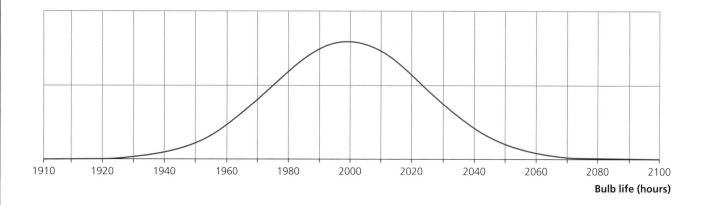

Investigate

> What percentage of bulbs have a life between 1950 and 2050 hours?

The blue dots in the graph below are the sample means for 100 samples of four lightbulbs.

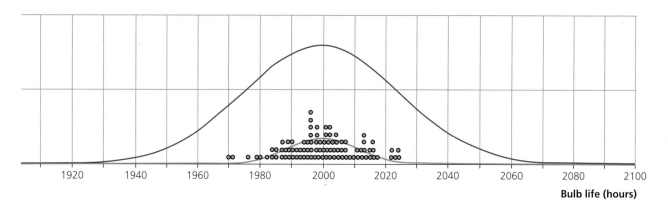

Discuss

> Estimate the mean of the blue dots (i.e. the mean of the sample means).

> Estimate the standard deviation of the blue dots (i.e. the standard deviation of the sample means).

The green curve shows the normal distribution with a mean of 2000 and a standard deviation of 12.5. Do you think this is a good fit for the sample means? What relationships are there between the distribution of the population of bulb lifetimes (red curve) and the distribution of the sample means (green curve)?

The blue dots on the graph below are the sample means for 100 samples of one hundred lightbulbs.

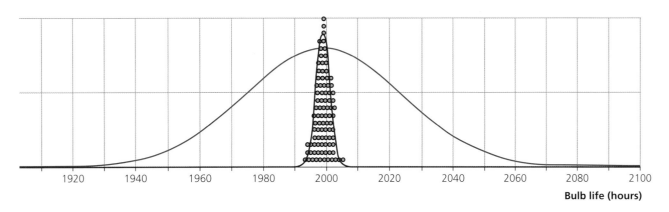

Bulb life (hours)

Discuss

> Estimate the mean of the blue dots (i.e. the mean of the sample means).

> Estimate the standard deviation of the blue dots (i.e. the standard deviation of the sample means).

The purple curve shows the normal distribution with a mean of 2000 and a standard deviation of 2.5. Do you think this is a good fit for the sample means?

In summary:

Distribution	Mean	Standard deviation	
Population	2000	25	
Sample mean – sample size 4	2000	12.5	The standard deviation is half that of the population
Sample mean – sample size 100	2000	2.5	The standard deviation is a tenth that of the population

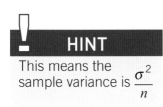

> **HINT**

This means the sample variance is $\dfrac{\sigma^2}{n}$

Discuss

> Can you see a relationship between the sample size and the standard deviation of the sample means?

Sample means are distributed normally with the same mean as the original data but with a standard deviation of $\dfrac{\sigma}{\sqrt{n}}$ (often called the **standard error**).

It is possible to use this information to find an interval estimate for the population mean when it is not known.

Confidence intervals

For the bulbs with mean life 2000 hours and standard deviation 25 hours, the probability that a value x from a normal distribution of mean μ and standard deviation σ will lie in the interval $\mu \pm 2\sigma$ is 0.95. But x is known and μ is not, so there is a probability of 0.95 that μ is in the interval $x \pm 2\sigma$.

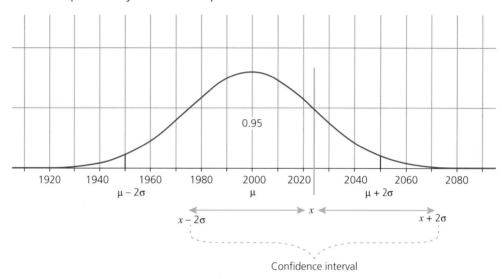

95% of the time, any such confidence interval will contain the mean.

However, if we use a sample mean from a sample of size 4, our confidence interval becomes smaller, providing us with a better estimate.

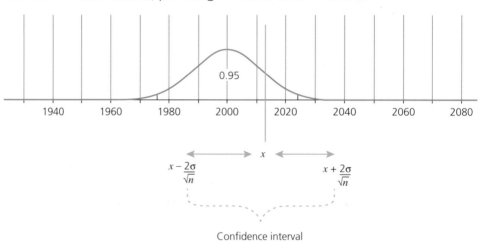

The larger the sample, the better the estimate.

DID YOU KNOW?

Look up $P(Z < z) = 0.975$ in the standard normal table and you get a z-value of 1.96. In Chapter 2.22 the normal distribution is described as having approximately 95% of the data within two standard deviations of the mean. We use 2 here as an approximation for 1.96.

WORKED EXAMPLE

The lifetimes of Superbulbs are normally distributed with a standard deviation of 35. A random sample of four Superbulbs were tested and the number of hours they lasted were recorded. They were:

2130   2543   2450   2654

Calculate a 95% confidence interval for the mean lifetime of Superbulbs.

SOLUTION

$2130 + 2543 + 2450 + 2654 = 9777$

$\frac{9777}{4} = 2444.25$

95% confidence interval for the mean is $2444.25 \pm \dfrac{2 \times 35}{\sqrt{4}}$ hours

That is, 2444.25 ± 35 hours

or 2409.25 to 2479.25 hours

Investigate

The lifetimes of Ecobulbs are normally distributed with a standard deviation of 60. A random sample of ten Ecobulbs were tested and the number of hours they lasted were recorded. They were:

1460   1583   1575   1674   1546   1489   1627   1477   1578   1532

> Calculate a 95% confidence interval for the mean lifetime of *Ecobulbs*.

Although a 95% confidence interval is a common one to use, any percentage is possible, depending on the accuracy needed. 90% and 99% are also commonly used intervals.

A 99% confidence interval

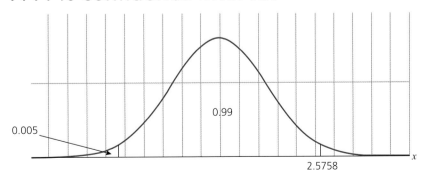

From normal tables the z-score for a probability of 0.995 is 2.5758.

So a 99% confidence interval for the Ecobulbs is

Sample mean $\pm \dfrac{2.5758\sigma}{\sqrt{n}}$

Note

You can find normal distribution tables on AQA's website

http://www.aqa.org.uk/subjects/ mathematics/aqa-certificate/ mathematical-studies-1350/ statistical-tables

Investigate

> What probability would you look up in the tables for a 90% confidence interval?

> What z-score would you get?

WORKED EXAMPLE

The lifetimes of Superbulbs are normally distributed with a standard deviation of 35. A random sample of five Superbulbs were tested and the number of hours they lasted were recorded. They were:

2130   2543   2450   2654   2481

Calculate a 90% confidence interval for the mean lifetime of Superbulbs.

SOLUTION

$2130 + 2543 + 2450 + 2654 + 2481 = 12\,258$

$\dfrac{12\,258}{5} = 2451.6$

A probability of 0.95 in the tables gives a z-score of 1.6449.

90% confidence interval for mean is $2451.6 \pm \dfrac{1.6449 \times 35}{\sqrt{5}}$ hours

That is, 2451.6 ± 25.7 hours or 2425.9 to 2477.3 hours

QUESTIONS

1   A new machine produces bags of nails with (E) a standard deviation of 1 g. Tim chooses 10 bags and weighs them, obtaining the following results (in grams):

11.7   12.2   12.5   11.4   12.0
14.0   11.5   12.6   13.3   11.2

Calculate a 99% confidence interval for the mean gross weight of a bag produced by the new machine.

2   A fishing boat weighs its latest catch of 132 cod. The catch weighs 967 kg. The standard deviation for the weight of a cod is 2.5 kg.

Calculate a 95% confidence interval for the mean weight of a cod in the waters used by the fishing boat.

3   Each of the eggs in a box of six medium-sized eggs was weighed, obtaining the following results:

54 g   62 g   57 g   59 g   63 g   62 g

Calculate a 90% confidence interval for the mean weight of medium-sized eggs. The standard deviation of the weights is 2.5 g.

E 4   Nathan produces pistons for a particular car engine. The diameters of the pistons have a standard deviation of 0.3 mm. He measures a random sample of five of them and finds their diameters are:

73.5 mm   73.5 mm   73.7 mm
73.8 mm   74.2 mm

Calculate an 85% symmetric confidence interval for the mean diameter of the pistons he produces.

★★

CHAPTER 〉 2.26

ARE BIRTHDAYS GOOD FOR YOUR HEALTH?

Does it make sense?

That sounds good.

Do you agree?

Birthdays are good for you!

Research shows that the more birthdays you have the longer you live!

WHAT YOU NEED TO KNOW ALREADY

→ Plotting points
→ Interpreting graphs
→ Thinking skills

MATHS HELP

p320 Chapter 3.7 Statistical terms
p322 Chapter 3.8 Representing and analysing data
p388 Chapter 3.24 Correlation

PROCESS SKILLS

→ Checking outliers
→ Presenting arguments
→ Interpreting real data
→ Representing data
→ Criticising interpretations

Significant research goes into seeing how things such as income and educational success, or health and amount of exercise taken, are connected. This is used to enable good policies to be made, and money to be spent effectively. If such measures appear to be linked either positively (where as one goes up so does the other) or negatively (one measure gets smaller as another gets bigger), they are said to be correlated. This chapter shows you some aspects of correlation and reminds you that things can appear connected even though there is no causal relationship.

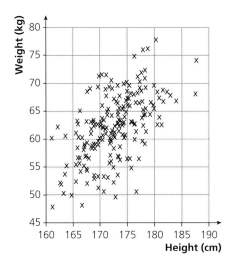

Correlation is associated with **bivariate data** – data sets with two pieces of information about each subject, such as the heights and weights of a set of people, the number of birthdays and the length of life of a set of people, the marks students get in two different tests, or parental income and educational success of their children.

The scatter diagram on the left shows a plot of height and weight for a group of people.

Discuss

> Where on the diagram are the tallest and thinnest people? The shortest and fattest? The shortest and thinnest? The tallest and fattest?

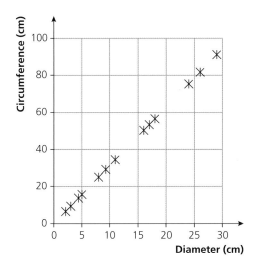

This graph shows a plot of the circumferences of a set of circles against their diameters.

There is **perfect positive correlation**. The points all lie on a straight line going diagonally upwards as there is a directly proportional relationship between the diameter of a circle and its circumference $(C = \pi d)$.

The graph can be used to find the circumference of a circle if you know its diameter and vice versa. Since this data has perfect positive correlation, you know that your reading from the chart is very likely to be correct.

DID YOU KNOW?

A straight line that passes through these points is called a **line of best fit**.

The diagram below shows bivariate data for every secondary school in England: the percentage of pupils getting good GCSE results and the percentage of pupils receiving free school meals.

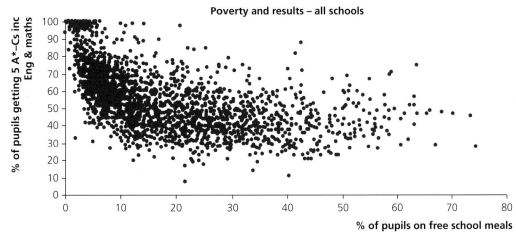

Poverty and results – all schools

191

HINT

The line of best fit here has a negative gradient although it could be argued that the correlation is not linear.

HINT

This graph can be converted to a scatter graph showing the bivariate data by using the year as the common aspect with the phone sales and deaths giving the two coordinates for plotting the points. Using the year like this often produces results that are nonsense.

The diagram shows **negative correlation**, which is particularly strong for percentages of pupils receiving free school meals from zero to 20%

Discuss

> What do you think of the suggestion that making everyone pay for school meals would improve GCSE results?

Correlation does not imply causation!

There are plenty of examples of data sets that correlate with one another but that have no connection at all – here is one:

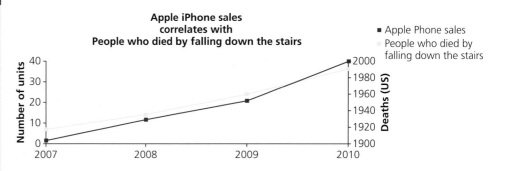

There are also some interesting examples of differing interpretations of correlation.

Discuss

> There is a negative correlation between a country's level of debt and its economic growth (debt above 90% of GDP is correlated with slow growth). In 2010, two economists argued that high debt causes slow growth. Another economist challenged this argument, arguing that it was slow growth that caused debt, not the other way round.

"Sleeping with one's shoes on is strongly correlated with waking up with a headache.

Therefore, sleeping with one's shoes on causes a headache."

The above example commits the correlation-implies-causation fallacy, as it prematurely concludes that sleeping with one's shoes on causes a headache. A more plausible explanation is that both are caused by a third factor, in this case going to bed drunk, which thereby gives rise to a correlation. So the conclusion is false.

Source: https://en.wikipedia.org/wiki/Correlation_does_not_imply_causation

HINT

Many people base arguments on the association between two measures when there is no causation involved. Often the common aspect is time, and many quantities change with time but have no relationship with each other.

Discuss

There are various possibilities when two data sets, X and Y, are correlated:

– X causes Y.

– Y causes X.

– X and Y partly cause each other.

– X and Y are both caused by something else.

– The correlation is just chance; there is no causal relationship.

> Discuss the two examples given above, and other examples you find, and decide which of the five possibilities fits the case.

Outliers

An outlier is a data item that is a long way from others, such as the final item in this list: 3, 4, 4, 5, 7, 8, 30.

The median is 7; if we omit the outlier, the median is 6.

The mean is 8.71; if we omit the outlier, the mean is 5.17.

You can see that the outlier affects the mean much more than the median, which is one reason for using the median rather than the mean in some circumstances.

> Can you explain why the outlier affects the mean more than the median?

You should consider whether outliers should be included in the data or ignored. If you have collected the data yourself (primary data), you should check whether you have made a mistake of some kind that has produced the outlier and, if so, correct the mistake.

If the data is secondary, you should consider how likely it is that the outlier is a mistake, and remove the outlier if you think that it is (unless you are able to correct it). If not, include it in your data but consider doing your calculations both with and without the outlier to see how significant the differences are.

An outlier can also affect the position of the line of best fit. One point can suggest correlation when, without it, there is no correlation. The line passes through the double mean point and so this also affects its position if outliers are removed or included.

HINT

Another source of false arguments is using **extrapolation**, where a value of the data is predicted beyond the range for which it is valid. The time for running a mile is an example of this – eventually someone will run the mile in zero time!

QUESTIONS

1 The diagrams show examples of bivariate data. For each diagram, say whether there is positive, negative or no correlation and if the correlation is perfect, strong or weak.

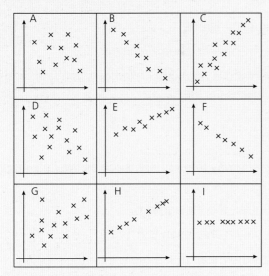

2 These two scatter graphs show body mass index (BMI) (weight in kilograms divided by square of height in metres). The first diagram shows BMI against height, and the second shows BMI against weight.

   a What type of correlation is shown in each and how strong is it?

   b Can you explain the difference between the two graphs?

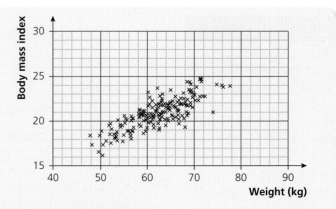

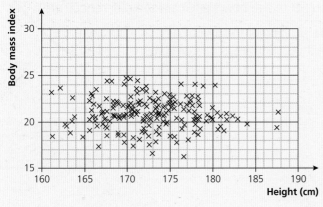

3 The data in the table is represented in the scatter graph.

Height (cm)	Weight (kg)
167.1	55.2
181.7	66.7
176.3	74.8
173.3	69.6
172.2	70.5
60.3	174.5
177.3	69.2
177.8	66.7
172.5	54.9
169.6	59.0

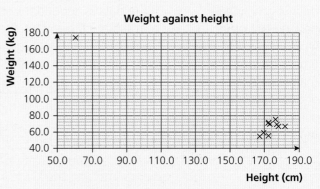

   a Identify the outlier in the data and explain how you would deal with it.

   Give a reason for your decision.

   b Describe the correlation, now that the outlier has been dealt with.

CHAPTER

> **2.27**

THE OLD WINE IS THE BEST

Is the flavour different?

Does it always get better?

How can you tell?

Is it worth the extra money?

> ## Does wine really get better with age?

There are many things that we would like to change in the world. Knowing what causes them to be as they are helps us to decide how to effect that change. Noticing factors that correlate with those things is part of the process. We can measure the strength of the correlation using mathematics, but we rely on common sense and other research to see if there is a cause and effect relationship.

WHAT YOU NEED TO KNOW ALREADY

→ Plotting a scatter graph
→ Recognising types of correlation
→ Identifying outliers

MATHS HELP

p320 Chapter 3.7 Statistical terms
p322 Chapter 3.8 Representing and analysing data
p388 Chapter 3.24 Correlation

PROCESS SKILLS

→ Deducing correlation
→ Calculating the pmcc from raw data and from scatter graphs
→ Interpreting relationships
→ Criticising the arguments of others

A wine specialist grades ten wines on a scale of 1 to 30 (30 being the highest) for their taste. The wine specialist then records their grading against their age.

Wine	A	B	C	D	E	F	G	H	I	J
Age	40	7	33	15	21	24	37	34	29	9
Grade	30	3	21	9	13	15	24	21	19	5

Investigate

We now look for evidence that the age and quality of the wine are linked.

> Plot a scatter graph to represent this information.

> Describe the correlation between the age of a wine and the grade given to it.

> Comment on whether you believe the taste of wine does improve with age.

Do we have enough information from the scatter graph based on one specialist's opinion to comment on the relationship between the age of wine and the taste of wine? There is a positive correlation between the age and the grade it receives; does this mean all old wines taste better? In Chapter 2.26 we learnt that correlation does not necessarily imply causation. In this chapter we will look in detail at correlation coefficients and how we use them to measure the strength of correlation between two variables. We will only consider linear correlation.

Techniques

DID YOU KNOW?

The pmcc is named after its inventor, Karl Pearson.

There are several techniques available within the field of statistics to measure the strength of correlation between two variables. **Pearson's product moment correlation coefficient (pmcc)** is one of the more frequently used techniques. The letter r is used to represent the pmcc, and it always lies within the range $-1 \leqslant r \leqslant 1$, where $r = 1$ represents perfect positive correlation and $r = -1$ represents perfect negative correlation.

You can see how these values relate to the scatter graphs.

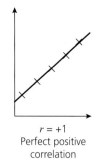

$r = +1$
Perfect positive correlation

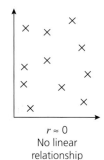

$r \approx 0$
No linear relationship

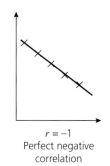

$r = -1$
Perfect negative correlation

HINT

r is used on calculators to stand for pmcc.

Investigate

> It is appropriate to use technology to do the number crunching involved in calculating the pmcc. Work out how you can use your calculator to find the pmcc.

> How can you use a spreadsheet to find the pmcc?

HINT

Use the formula PEARSON in spreadsheets to find the pmcc.

> Which other measures of correlation are used in statistics?

> For each of the following values of r, describe the correlation and draw a sketch of what you predict its corresponding scatter graph will look like.

$r = -0.7$ $r = 0.1$ $r = 0.9$

You can make up some points to check your predictions.

WORKED EXAMPLE

Car manufacturers test the braking distance of cars travelling at various speeds.

Speed (miles per hour)	Distance (metres)
20	6
30	14
40	24
50	38
60	55
70	75

a Calculate the pmcc between the braking distance of cars and the speed the car is travelling at.

b Use the pmcc to comment on the relationship between the braking distance of cars and the speed the car is travelling at.

SOLUTION

a Use your calculator to find the value of r.

b The braking distance of cars and the speed the car is travelling at are positively correlated.

0.986 is very close to 1 so there is a very strong association between braking distance and speed.

Investigate

> Can you find five coordinates on a scatter diagram that give a pmcc of exactly 0.99? Use a spreadsheet to check your answers. Is there more than one solution?

> Give one advantage of plotting data on a graph before calculating the value of a pmcc.

WORKED EXAMPLE

The scatter diagram gives the length and the wingspan of the 12 most common garden birds in the UK.

a Use statistical techniques to comment on the relationship between the wingspan and length of garden birds.

You may wish to use a copy of this table to support you.

Length l (cm)									
Wingspan w (cm)									

b The correlation coefficient between wingspan and length for seabirds is 0.744. Explain what this tells you about the relationship between wingspan and length of seabirds when compared with that of garden birds.

Scatter graph of wingspan against length of garden birds

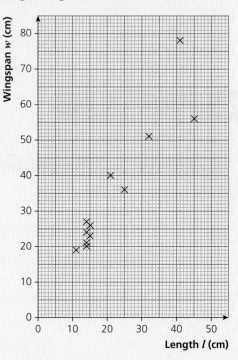

SOLUTION

a Use the pmcc to comment on the relationship between wingspan and length of garden birds.

Length l (cm)	11	14	14	14	14	15	15	21	25	32	41	45
Wingspan w (cm)	19	20	21	24	27	23	26	40	36	51	78	56

Use your calculator to find the value of r.

$r = 0.932$ (3 sig. fig.)

The pmcc indicates that the wingspan and length of garden birds are positively correlated.

This value of the pmcc indicates that they are strongly related and that as the length of a bird increases the wingspan increases.

b The correlation coefficient is lower for seabirds compared with garden birds.

Therefore there is a weaker relationship between the wingspan and length of seabirds compared with garden birds.

QUESTIONS

1 An examination has two papers: Paper A and Paper B.

The table and graph show results for 10 students.

Student	Paper A	Paper B
A	8	9
B	9	18
C	16	22
D	14	28
E	21	34
F	25	34
G	22	38
H	28	37
I	25	52
J	39	53

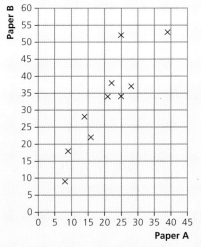

a Describe the correlation between the two papers using the pmcc and hence comment on the skills being tested by each paper.

b Find the mean scores for Paper A and Paper B. Add the mean point to a copy of the graph.

c Draw a line of best fit onto your diagram through the mean.

d One student has a medical certificate to show that his performance in Paper A had been affected by illness. Which candidate is this? Justify your answer and suggest how the overall score for this student could be adjusted.

2 Ten students sat tests in maths, English and science.

Michael was absent from the maths test.

Here are the test scores:

Student	Maths	English	Science
Connie	15	51	26
Alice	29	42	25
Michael	Absent	65	94
George	40	67	31
Leah	45	45	38
Anna	51	34	45
Max	58	83	39
Murray	67	50	60
Louis	78	44	60
Frances	61	74	75

A teacher wants to estimate Michael's maths score.

Should the teacher use the English or the science scores to help with their estimation.

Justify your decision with calculations to support your argument.

E 3 Two methods are used to predict the height that a child will reach as an adult.

Method A uses the heights of the child's parents.

Method B uses the height of the child at age 18 months.

The table shows the predicted and actual adult height for ten women using both method A and method B.

Method A results

Height prediction (cm)	157	158	159	162	162	169	175	176	177	178
Actual height (cm)	153	159	161	157	162	163	180	165	172	172

Method B results

Height prediction (cm)	159	160	161	161	161	169	165	167	167	169
Actual height (cm)	153	159	161	157	162	163	180	165	172	172

Use statistical methods and comment on the accuracy of each method.

★★★

CHAPTER ❯ 2.28

APPORTIONING BLAME IN TRAFFIC ACCIDENTS

❯ ## How fast were the vehicles travelling?

They must have been going too fast.

It all happened so quickly I'm not sure whose fault it was.

You can see when they started to brake from the skid marks.

WHAT YOU NEED TO KNOW ALREADY

→ Using scatter diagrams
→ Drawing lines of best fit by eye
→ The equation of a straight line
→ Calculating the mean
→ Substituting into equations

MATHS HELP

p301 Chapter 3.1 Straight line graphs
p352 Chapter 3.13 Formulae and calculation
p388 Chapter 3.24 Correlation
p391 Chapter 3.25 Regression

PROCESS SKILLS

→ Modelling real data
→ Justifying conclusions
→ Estimating speeds
→ Evaluating arguments
→ Summarising arguments
→ Predicting unknown quantities

In many vehicle accidents a key factor in apportioning blame is the speed of the vehicles involved. A crucial piece of information to help determine this is the skid marks (or 'tyre impressions') left by the vehicles. Skid marks are left when the driver applies the brakes so hard that the wheels lock, but the car continues to slide along the road. The investigating police officer will measure the skid marks and then an expert will determine the speed of the vehicle that left them. This chapter offers a simplified explanation of how this is done.

A car took part in 20 trials. The speed was recorded when the driver performed an emergency stop and the resulting skid marks were measured. The results are shown in the graph below.

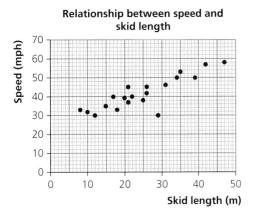

Discuss

> What is this type of graph called?

> What sort of correlation does it show?

> How would you use this graph to find the speed of a car that, under emergency stop conditions, left a skid mark 30 metres long?

> There is an outlier. Which point is it? What might have caused this outlier?

> Do you think the outlier should be excluded from the data?

Some oil had in fact been spilt on a patch of the road and this caused the outlier. As the oil spill did not affect any of the other skids, it was decided to remove the outlier from the data set.

Discuss

> The graphs below show possible lines of best fit. Which do you think is the 'best fit'? Why?

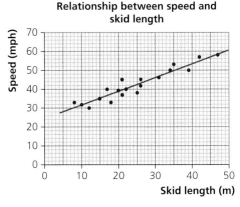

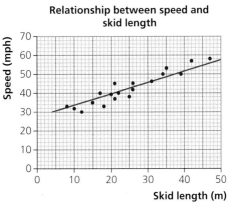

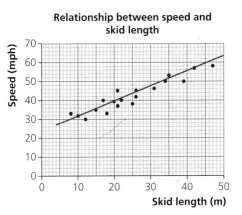

The 'best' line of best fit should go through the **mean point**.

Adding the 19 values for speed and the 19 values for skid length and then dividing by 19, gives a mean point of (24.7, 42.3). This point is shown in red on the graphs below.

																				Total	Mean
Skid length (metres)	12	22	34	42	8	21	31	39	26	47	15	17	35	21	10	18	26	25	20	469	24.7
Speed (mph)	30	40	50	57	33	37	46	50	45	58	35	40	53	45	32	33	42	38	39	803	42.3

Discuss

> The two graphs below show possible lines of best fit that go through the mean point. Which do you think is the best fit? Why?

> How would you find the equation of these lines?

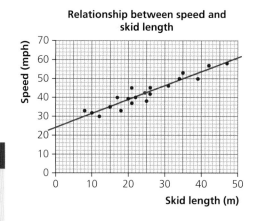

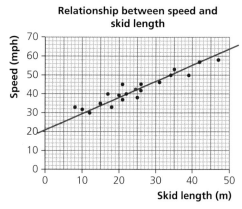

The best line of best fit through the mean point is called the **regression line of y on x.**

It can be found mathematically directly from a scientific or graphic calculator by inputting the pairs of coordinates for every point and using the linear regression option.

HINT

Either option 4 or 8 may be used on this calculator. Make sure you know which value is the gradient and which is the y-intercept.

As it is a straight line, the equation of the regression line is of the form

$y = ax + b$

where a is the gradient and b is the y-intercept.

The equation can then be used to predict a value of y from a given value of x.

HINT

Make sure you assign the two variables to the correct axes. The variable you want to predict must go on the y-axis.

Discuss

> For the data above:

  $a = 0.72928406$

  $b = 24.2613564$

  Check these values using your calculator.

> If you push the wrong key on your calculator, you end up with completely the wrong result. As well as being very careful, what methods could you use to check your work?

The values of *a* and *b* are usually rounded to 3 significant figures.

So $a = 0.729$ and $b = 24.3$.

For the skid mark data, the regression line of speed on skid length is therefore given by the equation

$y = 0.729x + 24.3$

Plotting the regression line

After you have keyed the pairs of coordinates into your calculator, you should also be able to read off the coordinates of the mean point.

Discuss

> Check you know how to find this on your calculator.

From the equation of the regression line, you also know the *y*-intercept.

So for the regression line $y = 0.729x + 24.3$ we know the points (24.7, 42.3) and (0, 24.3) on the line.

Plot these two points on your scatter diagram and join them to give you the regression line.

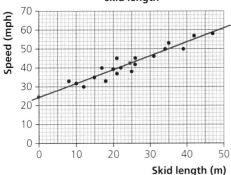

Does your regression line look like it's in the right place? This is a quick way to check you haven't made any silly mistakes.

Another useful check is to find a third point by substituting a different value of *x* into the regression line equation and calculating *y*.

The two points we have plotted are at the beginning of our *x* range and in the middle of our *x* range, so a good point to choose would be near the end of our *x* range. Say $x = 40$.

$y = 0.729x + 24.3$

$ = 0.729 \times 40 + 24.3$

$ = 53.46$

Investigate

> Check the point (40, 53.46) lies on the regression line in the diagram.

Using the regression line to make predictions

WORKED EXAMPLE

In an accident, one car came to a stop after applying its brakes and then a second car hit it. The first car produced skid marks 38 metres long. At what speed was the first car travelling when it applied its brakes?

SOLUTION

$y = 0.729x + 24.3$

When $x = 38$,

$y = 0.729 \times 38 + 24.3$

   $= 52$

We can predict that the first car was travelling at 52 mph.

Using the regression line to find values within the range of the original data is called **interpolation**. Because we have evidence that the relationship appears to be linear in that range, the prediction is fairly reliable.

Extrapolation is the process of predicting a value of y when x is outside the range of the given data, that is, if the skid mark was less than 8 m or more than 47 m). This is a very dubious thing to do. It is not possible to tell whether the linear relationship will continue for all values of x. Values obtained through extrapolation should not be relied upon. They are the source of many false arguments.

WORKED EXAMPLE

At the scene of a pile-up on the motorway, a car managed to stop just 0.5 m from the car in front of it. It produced skid marks 70 m long. What speed was the car doing when it put on its brakes?

SOLUTION

$y = 0.729x + 24.3$

When $x = 70$,

$y = 0.729 \times 70 + 24.3$

   $= 75.3$

The regression line predicts that the car was travelling at 75.3 mph. However, this is **not** a reliable estimate because the data has been extrapolated in order to obtain the estimate.

Investigate

> Predict the speed at which car A braked leaving a skid mark 5 m long at the scene of an accident.

> Predict the speed at which car B braked leaving a skid mark 11 m long at the scene of the same accident.

> The accident was in a 30 mph zone. Is it likely that either car was speeding?

Often when an accident occurs, a car applies its brakes but does not stop in time and hits another car or object. In these cases, the skid will only tell you the **loss in speed** from when the brakes are applied until the moment of impact. Experts can estimate impact speed from the damage to the objects involved. Skid length will therefore provide a prediction of a minimum speed for the car under these circumstances.

HINT

Use the equation of the regression line in the worked example. You may need to make some assumptions.

Investigate

> Car B braked to avoid a child running across the road in a 40 mph zone. Car A hit the back of car B. Car B produced skid marks of 24 m long and car A produced skid marks 15 m long. Do you think either driver was blameless? Explain your reasoning.

> You are a police accident investigator. Write an email to the insurance companies of both cars explaining the outcome of your preliminary findings.

QUESTIONS

1 The skid lengths of 14 cars under wet conditions were obtained.

Skid length (metres)	25	32	40	25	48	15	18	36	20	11	19	26	25	20
Speed (mph)	38	50	54	44	62	37	44	57	44	36	38	42	38	39

a Use the data to predict the speed of the following cars, which swerved to avoid a dog on a wet road whose speed limit is 40 mph.

　i Car A, which stopped without hitting anything and left skid marks of length 30 m.

　ii Car B, which hit a lamppost and left skid marks of length 32 m.

b Write a short report for the local paper explaining your preliminary findings.

2 A vocational exam consists of a practical and a written exam. Two students miss the written exam because of sickness. The exam board predicts the exam result for the missing students in several ways in order to arrive at a grade based on the practical only. First, they predict the exam results based on the results of the other students in their college.

Practical	70	84	63	77	76	95	83	80	73	77	79	85
Exam	48	68	43	56	A	A	61	69	46	63	65	72

a Plot the data on a scatter diagram for the students who attended the exam.

b Find the regression line and add it to your diagram.

c Predict exam results for the two students who were absent.

d When the exam board thinks the prediction is not very reliable, they use alternative methods. Explain how reliable you think each result is. Which, if either, would you recommend the exam board uses?

E

3 A food hamper company will need to take on extra packing staff for Christmas. On one day in September, the packing manager records the number of items in each hamper and the time it takes to pack each hamper.

Number of items	21	8	13	20	15	17	17	10	8	19	20	9
Time to pack (min)	33	23	29	35	14	25	28	23	20	33	33	23

a Draw a scatter diagram for this data and identify any possible outliers. Decide whether to include them in the data and give reasons for your decision(s).

b Use a regression line to predict how long it will take to pack a hamper with 18 items.

c The packing manager thinks that on typical day in December, they will get:

3 orders for hampers with 10 items

3 orders for hampers with 15 items

5 orders for hampers with 20 items

3 orders for hampers with 25 items.

Estimate the total packing time required each day. How reliable is your estimate? Explain your reasoning.

4 The average price of oil purchased by the major UK power producers and the average annual domestic electricity bill in the UK were recorded over a 15-year period.

a Calculate the pmcc for this data.

b Find the equation of the regression line for cost of electricity bill on price of oil based on this data.

c Use your equation to predict the average annual domestic electricity bill when the oil price is £300 per tonne.

d What does the correlation coefficient tell you about the reliability of your prediction?

Price of oil (£ per tonne)	Annual domestic electricity bill (£)
120.96	286
118.59	278
127.92	276
158.40	277
145.60	284
233.45	318
254.61	379
240.27	426
287.36	490
268.32	502
419.48	488
531.39	529
577.20	561
539.93	599
488.65	616
325.84	614

Published by Department of Energy & Climate Change on 30 June 2016

PRACTICE QUESTIONS: PAPER 2A

1 The heights of Chinese women are normally distributed with mean 1554.4 mm and standard deviation 57.41 mm.

The heights of Japanese women are normally distributed with mean 1557.9 mm and standard deviation 51.17 mm.

Work out the probability of a randomly chosen woman of each nationality being taller than 1.65 m. **[4 marks]**

Compare the heights of Chinese and Japanese women. **[2 marks]**

2 The length of great white sharks is normally distributed with mean μ and standard deviation of 76.5 cm. It is thought that their food sources within the ocean are being threatened by pollution. Following a spate of attacks on humans some sharks were caught and their lengths measured to see if they were smaller than the expected mean length of 4.677 m.

The lengths, in centimetres, of the sample were:

508	359
513	465
498	416
526	430
516	416

a Calculate a point estimate of the mean, μ of the population of great white sharks. **[1 mark]**

b Calculate a 90% symmetric confidence interval for μ and comment on the likelihood that the sharks are smaller than expected. **[6 marks]**

3 Body mass and brain size for some medium sized mammals are shown in the graph.

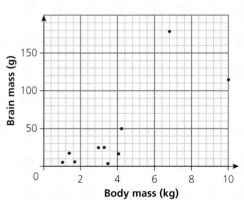

a Identify the outlier and give a reason for including it in the data set. **[2 marks]**

b Describe the correlation between body mass and brain mass for medium sized mammals. **[1 mark]**

Body mass and brain mass are shown in the table for some small mammals.

Small mammals (less than 1 kg)	Body mass (kg)	Brain mass (g)
Arctic Squirrel	0.92	5.7
Tenrec	0.9	2.6
Hedgehog (European)	0.785	3.5
Rock Hyrax (Hetero b)	0.75	12.3
Hedgehog (Desert)	0.55	2.4
Chinchilla	0.425	6.4
Rat	0.28	1.9
Galago	0.2	5
Mole Rat	0.122	3
Hamster (golden)	0.12	1

   c Calculate the equation of the regression line of body mass on brain mass.

      Interpret the equation of the regression line in the context of the data.   **[5 marks]**

   d The owl monkey has brain mass 15.5 g. Use your regression line to estimate its body mass.   **[2 marks]**

   e Draw a scatter graph and use it to compare the body mass predicted by the regression line with the actual body mass of 0.48kg.   **[3 marks]**

      Give a reason why larger brain mass might imply larger body mass, and a reason why it might not.   **[2 marks]**

3 The scores of the first ten athletes in the heptathlon in the 2012 Olympics in the High Jump and the Shot Put are shown in the table.

Athlete	High Jump	Shot Put
A	1054	813
B	1016	845
C	978	805
D	1132	1016
E	978	811
F	830	848
G	1016	845
H	1016	704
I	978	725
J	1016	720

Data from https://en.wikipedia.org/wiki/Athletics_at_the_2012_Summer_Olympics_%E2%80%93_Women%27s_heptathlon

Write a report about the events for an assignment.

Include your value for the pmcc to support your comments.   **[6 marks]**

5 Nicky is collecting data for a project on the time students at her college spend on sporting activities. She interviews a sample of students from her Psychology class.

   a Describe the population for her project.   **[2 marks]**

   b Give a reason why her sample may not be representative.   **[2 marks]**

   c Describe how she could collect a simple random sample for her project.   **[2 marks]**

★★★

CHAPTER

> 2.29

COOKING A ROAST DINNER

The beef needs to be cooked properly.

> ## How can I make sure everything is ready at th right time?

I don't want cold carrots.

I like my roast potatoes crispy.

Planning complex projects, such as building houses or roads, developing and marketing new products or managing big events can be made easier with the use of some mathematical tools. This chapter gives you the opportunity to learn about using some planning tools such as identifying a 'critical path' through a project in simple cases.

WHAT YOU NEED TO KNOW ALREADY

→ Working systematically
→ Checking your work

PROCESS SKILLS

→ Modelling
→ Interpreting results
→ Representing projects
→ Evaluating different representations

MATHS HELP

p396 Chapter 3.26 Critical path analysis

Cooking a roast dinner

The **critical path analysis** method of planning was developed in the late 1950s to control large defence projects, and has been used routinely since then. It helps to plan which tasks must be completed, and in what order, to complete a project successfully.

You probably would not formally use critical path analysis to help you to cook roast beef and Yorkshire pudding, but most cooks would have in their heads an idea of what has to be done before what, and approximately how long each activity would take, so it is a useful project to analyse.

First, here is a table of the activities that need to be completed:

	Activity	Time (minutes)	Preceding activities
A	Heat oven	10	
B	Put beef in oven and leave to roast	75	A
C	Remove beef from oven and let it rest	15	B
D	Make gravy and leave to simmer	10	B
E	Prepare potatoes, carrots and cabbage	15	
F	Parboil potatoes	15	E
G	Roast potatoes	60	F
H	Cook carrots	20	E
I	Cook cabbage	3	E
J	Put vegetables in serving dishes	2	G, H, I
K	Make batter for Yorkshire pudding	5	
L	Heat oil in baking tin in oven for Yorkshire pudding	7	
M	Cook Yorkshire pudding	25	K, L
N	Dish up and transfer everything to the dining table	5	C, D, J, M

The following diagram is an **activity network** drawn from the list of activities, showing the time needed for each activity in green.

Discuss

› Make sure that you understand how the activity network diagram relates to the list of activities.

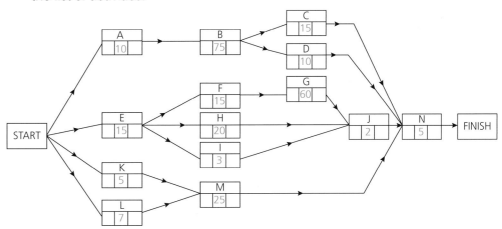

Investigate

> Before we go on, have a go at making an activity list and drawing an activity network for a small project such as making tea and toast or going on a trip to the shops.

Now we work through the activity network from left to right (a 'forward pass') showing the earliest possible starting times in red.

Activities A, E, K and L have no precedents and so their earliest starting time is 0.

Activity B follows A and so can only start when A is completed, after 10 minutes so its earliest starting time is 10.

Similarly activities F, H and I can only start when E is completed.

Activity M follows both K and L and can only start when the longer of those two, L, is completed and so its earliest starting time is 7.

Activities C and D follow the completion of B so their earliest starting time is 10 + 75 = 85. G follows F so has earliest start time of 15 + 15 = 30.

For J to start, G, H and I must be completed so choose the largest value from 30+60 (G), 15+20 (H) and 15+3 (I) to get 90.

The earliest starting time for N is 100 because that is the time for C (85 + 15) and longer than D (85+10), J (90+2) and M (7+25).

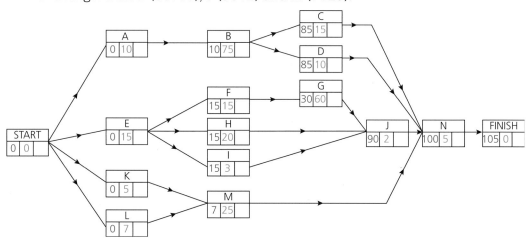

Now work through the activity network from right to left (a 'reverse pass') to show the latest possible finish times in blue.

The FINISH has earliest start time of 105 minutes so that becomes the latest finishing time for the project and for activity N.

Activities C, D, J and M precede N making their latest finishing times 105 – 5 = 100.

Activities G, H and I precede J and so their latest finishing times are 100 – 2 = 98

Activity B precedes C and D so its latest finishing time is the lower of 100 – 15 (C) and 100 – 10 (D) at 85 minutes.

Activity F precedes G so its latest finishing time is 98 – 60 = 38.

Activity E precedes F, H and I so its latest finishing time is the lowest of 38 – 15 (F), 98 – 20 (H) and 98 – 3 (I) at 23 minutes.

Activities K and L precede M so their latest finishing times are 75. Activity A precedes B so its latest finishing time is 10.

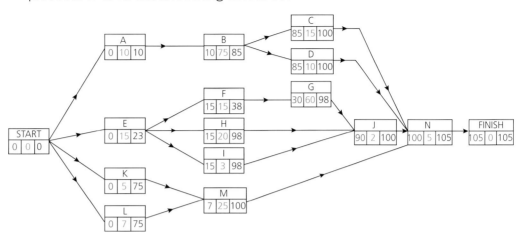

The resulting diagram allows us to identify the **critical path** for this project. The activities on the critical path are those for which the earliest starting times plus the time taken for the activity are the same as the latest finishing times – there is no slack in the system for those activities.

So the critical path is A, B, C, N, and shows that cooking and serving the dinner can be completed in 105 minutes.

Gantt charts

A Gantt chart is another way of presenting the information needed for planning a project. Gantt charts are widely used in business.

We will make a simple Gantt chart for the roast beef dinner; this shows the activities on the critical path.

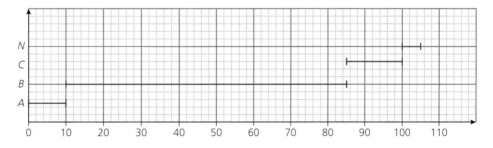

This shows which task follows which and how long the critical path is. The dinner cannot be completed in less than 105 minutes

The Gantt chart shows all the activities; the critical path is shown in red; there is no flexibility in the starting and finishing times for activities on the critical path.

This is similar to the format you are likely to see in the examination.

The other activities are shown in solid black and dotted black, the dotted black showing how long the activity will take and the latest finishing time and the solid black showing the amount of 'spare time' there is for that activity and the earliest starting times.

There is some choice about when these activities can be completed.

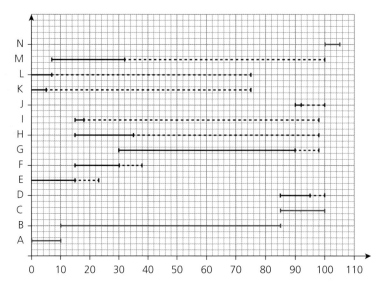

Look carefully at the chart to see how the original information is represented within it.

QUESTIONS

1 For the small project you investigated earlier (making tea and toast):

a work through the forward pass and reverse pass process and find the critical path

b draw a Gantt diagram.

2 Here is a table for a project involving five activities. Draw an activity diagram and find the minimum time necessary for completing the project.

Activity	Time (hours)	Preceding activities
A	2.5	
B	1.5	A
C	1	A, B
D	3	A
E	1	C, D

3 The activity network below shows the activities needed to design and manufacture a product, as described in the list of activities:

	Activity	Time (days)
A	Development of design	4
B	Detail product design	8
C	Specification and sourcing of materials	6
D	Testing	6
E	Manufacture of components	10
F	Sub-assembly	6
G	Main assembly	8
H	Finishing	6

a Complete the activity network and find the critical path.

b Draw a Gantt chart for the process.

c Write down what you see as the advantages and disadvantages of the two different representations.

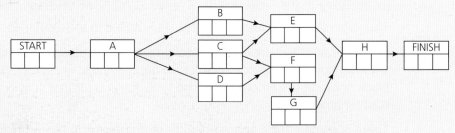

CHAPTER

> ## 2.30

WINNING COMBINATIONS

> ## What are my chances of winning?

Should I have a go?

Is it biased against me?

I've been watching – no one has won for a while – that gives me a better chance.

WHAT YOU NEED TO KNOW ALREADY

→ Listing exhaustive sets of outcomes for random events

→ Listing systematically

→ Working out the probability of a single event

MATHS HELP

p400 Chapter 3.27 Calculating probabilities
p404 Chapter 3.28 Venn diagrams
p409 Chapter 3.29 Using probability for expected amounts

PROCESS SKILLS

→ Calculating expected outcomes

→ Justifying calculations

→ Modelling random events

Understanding probability is useful in many situations. It is the theoretical basis for statistics, and thus for statistical tests and predictions. It is required in any assessment of risk, when the outcomes of a decision are uncertain. This is almost always the case as we cannot know what will happen in the future. It is particularly important to have an understanding of probability when investing money or gambling.

Discuss

Gaming machines are designed to give large payouts and also make a profit.

> How does the manufacturer use mathematics when designing the machine and deciding on what combinations pay out? How profitable are the machines? How many places have you seen slot machines near where you live? Are they popular? Why do you think people use them?

The mathematics of **probability** is used to calculate the likelihood of a particular **outcome** from the set of possible outcomes. The outcomes are random, or at least that is what you assume when you play the game! To work out how likely you are to win, you need to calculate the probability of the winning outcomes. To work out your expected winnings when you play a number of games, you need to use the probabilities to find the expected 'value' of a series of random events.

Investigate

Consider a biased tetrahedral die with each side having one of the numbers 1, 2, 3 or 4. The probability $P(N)$ of each number, N, being on the bottom when the die is rolled is:

$P(4) = 0.25$

$P(3) = 0.3$

$P(2) = 0.4$

$P(1) = 0.05$

> Why do the probabilities add to 1?

The probabilities have been estimated by rolling the die many times and recording the outcomes. For a fair tetrahedral die the probability of each number being on the bottom is 0.25, as each outcome is equally likely.

For the biased die, thrown once, work out:

> the probability of getting an even number

> the probability of getting a prime number.

Suppose the die is thrown 20 times.

The probability of throwing 4 is 0.25, so you expect to get a 4 on $\frac{1}{4}$ of those throws, that is 5 times.

If you score the value on the bottom of the die (4) each time, that is a total score of $5 \times 4 = 20$.

You expect to score 18 from throwing 3s.

> How was that worked out?

> What do you expect to score from throwing 2s?

You can work out the total score you expect by adding the expected score for each face of the die.

Total $= 20 \times 0.25 \times 4 + 20 \times 0.3 \times 3 + 20 \times 0.4 \times 2 + 20 \times 0.05 \times 1$

$\qquad = 20 \times (0.25 \times 4 + 0.3 \times 3 + 0.4 \times 2 + 0.05 \times 1)$

$\qquad = 55$

Investigate

> What is the expected score for 20 throws of the unbiased tetrahedral die?

A gaming machine needs to take at least £100 per jackpot if it is not to make a loss. If the cost to play the machine is £2.00, then on average a payout can occur every 50 attempts. So, if the probability of a winning line is $\frac{1}{50}$, we expect the machine to pay out every 50 attempts.

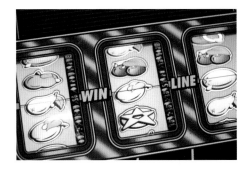

In 50 attempts the money paid in is 50 × £2.00 = £100, so if the jackpot is £100 the expected profit is £0.

If the profit for every payout is set at £50, the probability of winning needs to be $\frac{1}{75}$ because

payout + profit = £100 + £50

$\qquad\qquad\qquad$ = £150

This requires 75 attempts at £2.00. So 1 attempt in 75 can be a winner.

Discuss

> Is this a reasonable profit margin? How many times would a jackpot happen in one evening? Would you recommend a different model? There are usually other prizes apart from the jackpot – how does that affect your evaluation?

Throwing 10 heads in a row

> What are the chances of throwing 10 heads in 10 tosses of a coin?

We can start by looking at a simpler case such as throwing 2 heads in 2 tosses of a coin.

We assume the coin is fair, which means a head and a tail are equally likely, so the probability of each is $\frac{1}{2}$.

The possible outcomes are HH, HT, TH, TT.

So the probability of 2 heads in 2 throws is $\frac{1}{4}$.

Investigate

> Work out the probability of 3 heads when you toss a coin 3 times in the same way, then 4 heads when you toss a coin 4 times.

You should notice a pattern, which you can then extend to 10 heads when you toss the coin 10 times.

Now work out $(\frac{1}{2})^{10}$.

> What do you notice?

> How many sets of 10 throws of the coin do you need so that you can expect 10 heads once?

> What difference does it make if you throw 10 coins at once rather than tossing one coin 10 times?

> What difference does it make if you throw one coin repeatedly many times until you throw 10 heads consecutively?

These calculations are based on a theoretical model. You could throw 10 heads the first time you try, but it is unlikely.

How well does the model match what really happens?

Investigate

> You need to gather some data. It is easier to work with throwing 10 coins and counting the heads but you could throw one coin 10 times and do the same. Combine the data for the whole class and compare it with the model. How good is the match? You could use a spreadsheet to simulate the throws although this is not real data. Let 1 represent heads in the spreadsheet and see how many 1s come up each time. Does it give a better match?

There are two main approaches to solving probability problems involving one event. These are shown in the following example.

WORKED EXAMPLE

A raffle is held at a local fundraising event. There are 100 tickets numbered from 1 to 100.

The prizes are:

Number ends in a 7	Free drink at a local cafe
Number is a multiple of 25	2 tickets for the local cinema

Jim buys one ticket. What is the probability that it is not a winning ticket?

SOLUTION

There are two approaches to working this out.

Method 1 Using frequencies:

Number of ticket numbers ending in 7 = 10

Number of ticket numbers that are multiples of 25 = 4

No ticket will be both a multiple of 25 and end in 7.

Number of losing tickets = 100 − 10 − 4

$$= 86$$

So probability of a losing ticket $= 86 \div 100$

$$= 0.86$$

Method 2 Using probabilities:

Probability of a ticket number ending in 7 = 0.1

Probability of a ticket number that is a multiple of 25 = 0.04

Probability of a losing ticket $= 1 - 0.1 - 0.04$

$$= 0.86$$

Here we make the assumption that all outcomes are equally likely in order to calculate the probabilities. Do you think that is a valid assumption in this case?

We make the same assumption in the next example. Is it a valid assumption here?

WORKED EXAMPLE

An examiner is designing a multiple-choice examination paper and wants to make sure that a candidate who takes random guesses ends up with zero marks. The paper has 20 questions and each question has four possible options. A question answered correctly is to be given 6 marks. Work out how many marks must be deducted for a wrong answer so that zero marks will be awarded for a candidate who answers the questions randomly.

SOLUTION

Since there are 4 possible outcomes for each question and only 1 outcome is correct, P(correct) – the probability of being correct – is $\frac{1}{4}$ and P(wrong) is $\frac{3}{4}$.

Number of questions answered correctly $= 20 \times \frac{1}{4}$

$$= 5$$

Number of questions answered incorrectly $= 20 \times \frac{3}{4}$

$$= 15$$

Number of marks from correct answers $= 5 \times 6$

$$= 30$$

Therefore marks for each wrong answer $= -30 \div 15$

$$= -2$$

The examiner needs to award -2 marks for every wrong answer.

QUESTIONS

1 To raise £500 for a World Challenge trip, Evie creates a lucky dip. She fills an opaque box with 39 red balls, 10 white balls and 1 black ball. The prize is dependent on the colour of ball picked out as shown in table below:

Red ball prize nothing

White ball prize £5

Black ball prize £50

a What is the probability of picking each colour of ball?

b What is the probability of winning £50?

2 Three fair coins are thrown.

a What is the probability of at least one head?

b How many times would you expect no heads in 1000 throws of the three coins?

E 3 In a college of 300 students, 30 study three science subjects, 45 study two science subjects and 120 study one science subject. Nobody studies more than three science subjects.

What is the probability that a randomly selected student does not study any science subjects?

4 The rail network needs to close a road to raise the height of a bridge over the railway line. The work will take 56 days to complete, but work cannot occur if it rains on any day. The probability of it raining on any day is 0.15. Calculate the number of days the road is expected to be closed.

★★

FILM STARS

Do men get more lead roles?

What genre of films do women star in?

How fair is the film industry?

> ## Are women getting better roles in films?

WHAT YOU NEED TO KNOW ALREADY

➜ Using fractions to represent probabilities
➜ Constructing tree diagrams

MATHS HELP

p400 Chapter 3.27 Calculating probabilities
p404 Chapter 3.28 Venn diagrams
p409 Chapter 3.29 Using probability for expected amounts

PROCESS SKILLS

➜ Deducing probabilities
➜ Interpreting diagrams
➜ Representing problems

Venn diagrams are used in many contexts to provide a visual representation. Examples include classification of animals or plants, diagnosis of medical conditions and achievement in schools and colleges. In statistics, they are particularly helpful when calculating probabilities of dependent events. A Venn diagram makes it clear what the whole population is, and so finding the correct probability is straightforward. Tree diagrams, which you will be familiar with from GCSE, are another way to calculate probabilities when combining events. This chapter looks at some practical applications of both types of diagram.

HINT

The romance films are within the romance circle.

The comedy films are outside the romance circle.

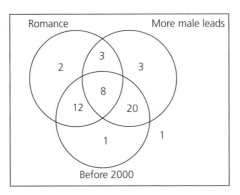

Female actors are sometimes quoted as saying that there are not enough good roles for women in films. The data in the Venn diagram compares the number of men and women in lead roles in 50 comedy and romance films before/since 2000. We assume for this investigation that a film is either a romance or a comedy but not both.

Discuss

> How many of the films represented in the diagram were made before the year 2000?

> How do you know that there is only one comedy film since 2000 with at least an equal number of female lead roles?

> Does the data suggest there are plenty of good roles for women?

The Venn diagram shows the number of films for each combination of the characteristics we have chosen. From this we can calculate the proportions, or probabilities, for these characteristics.

First some notation is needed for clarity.

R = event that a film is a romance.

R′ = event that a film is not a romance, which in this case means a comedy.

M = event that a film has more male leading roles than female ones.

M′ = event that a film has **not** got more male leading roles than female ones (it has the same number of, or more, leading roles for female actors).

You should be able to work out what the events B and B′ are now!

An event is represented by an upper case letter as you have seen above.

The probability of that event, A, is written P(A).

The probability of the complement, A′, is P(A′).

Since A and A′ are **exhaustive**, P(A) + P(A′) = 1.

Two events can be combined using the ideas of **union** and **intersection**.

DID YOU KNOW?

R′ is called the **complement** of event R as between them they cover all the possibilities. R and R′ are exhaustive and their probabilities add to 1.

A∪B is the **union** of A and B, meaning either A happens, or B happens or both happen.

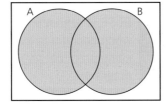

A∩B is the **intersection** of A and B, meaning both A and B happen.

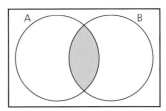

WORKED EXAMPLE

Use the information about films above to work out:

a   P(M)

b   P(R∪M)

c   P(M'∩B)

d   the probability that there are at least as many females as males in leading roles for films made before 2000

e   what R∩M∩B represents

f   the probability that a comedy made since 2000 has more males in leading roles than females.

SOLUTION

a $P(M) = \frac{3+3+8+20}{50} = \frac{34}{50}$

b $P(R \cup M) = \frac{34+2+12}{50} = \frac{48}{50}$

c $P(M' \cap B) = \frac{12+1}{50} = \frac{13}{50}$

d   For films made before 2000, the probability that there are at least as many females as males in leading roles $= \frac{12+1}{12+1+8+20} = \frac{13}{41}$

e   R∩M∩B represents romances made before 2000 with more male than female leads.

f   The probability that a comedy made since 2000 has more males in leading roles than females $= \frac{3}{3+1} = \frac{3}{4}$

Investigate

❯   Is it more likely that a comedy or romance has more leading roles for female actors? How did you decide?

❯   Find similar data for other genres of film.

❯   Do some genres offer more opportunities for female actors?

Independent and dependent events

In the example above, parts d and f put a condition on the probability.

The films considered for d had to be made before 2000. The denominator for the fraction was 41 as there were 41 films made before 2000.

The films considered for f had to be comedies made since 2000. The denominator for the fraction was four as there were only four such films.

Conditional probability is where an event is dependent on another event. This helps us investigate questions like the one above.

The probability of a film that has at least as many leading roles for female actors as male actors,

$$P(M') = \frac{16}{50}$$

It is interesting to compare this with the same probability but just for romances.

Probability for romances only $= \frac{14}{25}$, a much greater chance.

For comedies the probability $= \frac{2}{25}$, a much smaller chance.

Quite a difference between the three probabilities! They are dependent on what genre is under scrutiny.

Discuss

> Why do you think that is?

False positive paradox

Another way to represent probabilities with events that may be dependent is to use a tree diagram.

WORKED EXAMPLE

You are concerned that you may be suffering from a rare disease. It occurs in 1 in every 10 000 of the population. However, you can get tested for it and the test is accurate 99% of the time. You test positive. How worried should you be?

SOLUTION

D is the event that you have the disease.

P is the event that you test positive.

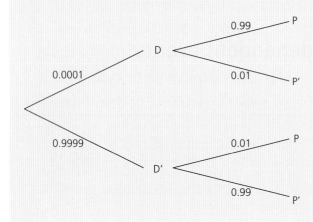

The probabilities at the second stage are dependent on the event at the first stage.

0.99 is the probability for testing positive if you have the disease but 0.99 is the probability for testing negative if you don't have the disease because those are the correct results in each case.

P(D then P) = 0.0001 × 0.99 = 0.000099

P(D′ then P) = 0.9999 × 0.01 = 0.009999

So the probability of testing positive = 0.000099 + 0.009999 = 0.010098

This is just over 1%.

You have tested positive; we need the probability that you have the disease:

$$\text{Probability} = \frac{\text{P(D then P)}}{\text{P(D then P)} + \text{P(D′ then P)}}$$

$$= \frac{0.000099}{0.010098}$$

$$= 0.0098\ldots$$

which is slightly less than 1%.

So you still have only less than a 1% chance of having the disease – save the worrying for another occasion!

The structure of a tree diagram is always to have an exhaustive set of outcomes coming from each point so that the probabilities for the first stage add up to 1. The probabilities for each set for the second stage also add up to 1. You can have more than two outcomes at each stage.

QUESTIONS

1 Ruth did a survey of 40 people at her college. She asked them whether they had a cat, a dog or both as a pet.

Her results are shown in the Venn diagram.

D is the event 'owns a dog'.

C is the event 'owns a cat'.

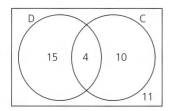

a Write down P(C∩D).

b Work out P(C∪D)′.

c Describe the event (C∪D)′.

2 Dave asks 30 of his friends about their favourite takeaway.

C is those who prefer curry.

P is those who prefer pizza.

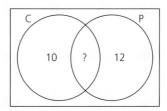

P(P) = 0.5

a Copy and complete the Venn diagram.

b Work out P(C′).

3 Deborah asks 40 of her friends about which forms of exercise they enjoy.

The Venn diagram shows the results.

S is those who enjoy swimming.

R is those who enjoy running.

T is those who enjoy playing a team sport.

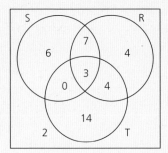

a Work out the probability of a randomly chosen friend not enjoying any of the three sports.

b Work out the probability that a randomly chosen friend who enjoys team sports also enjoys swimming.

c Describe the people represented by the region containing 14 on the diagram.

4 A survey is conducted with 40 young adults about which food they like when they go out.

24 liked Mexican food; 16 liked Italian food; 19 liked Japanese food; 6 liked all three; 4 liked Mexican and Japanese but not Italian; 2 liked Japanese and Italian but not Mexican; 11 liked both Mexican and Italian.

a Draw a Venn diagram showing this information.

b Work out the probability of a randomly chosen young adult liking none of the three types of food.

5 The probabilities of an award for passing a level in a computer game are shown in the table.

Award	Gold	Silver	Bronze
Probability	0.2	0.35	0.45

Reuben passes two levels in the game. Calculate the probability that he got the same level of award in both games.

6 Maria travels to college by bus. The probability that she gets up in time is 0.7, otherwise she is late. If she gets up in time, she catches a bus that will get her to college on time with probability 0.8. If she is late, the probability of the bus getting her there on time is 0.1.

Calculate the probability that Maria is late for college.

E 7 Sadiq is taking part in two events at the Olympic Games.

In the first event, the probability that he wins gold is 0.2, silver 0.25 and bronze 0.35.

In the second event, the probability that he wins gold is 0.1, silver 0.2 and bronze 0.4.

Calculate the probability that he returns with two medals of the same colour.

★★

PREMIUM BONDS

Line up everyone with £1,000 worth of premium bonds in order of their year's winnings, and the person halfway along would have won...not a penny! In fact, you'd need to walk past almost two-thirds of the line until you hit the first £25 winner.

Source: http://www.moneysavingexpert.com/ savings/premium-bonds

> ## Premium Bonds – are they worth it?

Would you buy Premium Bonds?

Why not buy a lottery ticket instead?

At least you get your money back if you don't win.

About £62 billion is currently invested in Premium Bonds in the UK, so they are clearly a popular way of keeping your money safe – and you could win a million pounds! This chapter explores the likelihood of doing so, and shows you how to calculate your expected reward in games of chance or your expected return on a business investment.

WHAT YOU NEED TO KNOW ALREADY

→ Using percentages
→ Calculating probability

MATHS HELP

p304 Chapter 3.2 Spreadsheet formulae
p352 Chapter 3.13 Formulae and calculation
p400 Chapter 3.27 Calculating probabilities
p409 Chapter 3.29 Using probability for expected amounts

PROCESS SKILLS

→ Modelling financial situations
→ Checking reasoning and calculations
→ Representing financial situations
→ Comparing returns on investments

Prize	Number per month
£1 000 000	2
£100 000	2
£50 000	5
£25 000	9
£10 000	24
£5 000	46
£1 000	1 257
£500	3 771
£100	64 198
£50	64 198
£25	1 879 199

Premium Bonds do not pay interest in the way other savings accounts do – instead, the interest is paid out according to a random choice of Premium Bonds done by a machine called Ernie (Electronic Random Number Indicating Equipment). So you might get nothing at all (by far the most likely outcome) or you might win a million pounds (extremely unlikely).

About £62 billion is currently invested in Premium Bonds.

The random numbers for the bonds that win prizes are drawn every month. The odds vary somewhat from month to month. The table shows typical monthly payouts.

So, each month, the owners of two Premium Bonds will receive one million pounds, and nearly two million Premium Bonds will earn their owners a £25 prize.

The total paid out each month is typically around £63 million – about £755 million a year. It sounds like a lot of money – but there are 62 billion bonds so the chances of winning anything are low.

The quote above from Money Saving Expert indicates that the median prize money per year is £0.

To calculate the mean, we divide the total annual payout by the total number of bonds:

Mean payout per bond per year $= \frac{£755 \times 10^6}{62 \times 10^9} \approx £0.0122$

Not very much! So, if you have £1000 in bonds, you may expect to win £12.20 a year. If you invested £1000 in an ordinary savings account, paying 2.5% interest per year for example, you would get £25, which is more than twice as much – but of course you don't have the (very small) chance of winning a million pounds!

Investigate

> What is the probability of winning £1m in a year if you have £1000 in Premium Bonds?

> How does this compare with the probability of winning £1m on the lottery if you bought 500 £2 tickets?

Calculating expected returns

Probability can be used to calculate the reward you may expect to get in situations involving random events (like Premium Bonds). Here is a simple example:

WORKED EXAMPLE

In a dice game, you throw a die and receive £0 for a score of 1 or 2, £1 for a score of 3, 4 or 5 and £10 for a score of 6.

If you play the game a large number of times, what would you expect to win per game on average?

SOLUTION

The probability of getting a score of 3, 4 or 5 is one half. So, on average, you would expect to win £1 on half the games by throwing 3, 4 or 5. Therefore the expected win is $£1 \times \frac{1}{2} = 50\text{p}$.

All the expectations are shown in the table. Make sure you understand how they are calculated.

Score	Prize	Probability	Expectation
1 or 2	£0	$\frac{1}{3}$	$£0 \times \frac{1}{3} = £0$
3, 4 or 5	£1	$\frac{1}{2}$	$£1 \times \frac{1}{2} = £0.50$
6	£10	$\frac{1}{6}$	$£10 \times \frac{1}{6} = £1.6667$
		TOTAL	£2.1667

So, on average, you would expect to win £2.1667 per game.

If someone wants to make money out of getting people to play the game, they would, of course, have to charge more than the average payout to play each game.

Investigate

> Set up a spreadsheet to calculate the total Premium Bond prize payout per month, using the figures given at the start of the chapter.

WORKED EXAMPLE

Joe estimates that, in the year ahead, his business has:

- a 15% chance of making £150 000
- a 40% chance of making £80 000
- a 35% chance of covering his costs
- and a 20% chance of losing £80 000.

What is his financial expectation, or expected value for the money made, for the year ahead?

SOLUTION

$$\text{Total expectation} = 0.15 \times £150\,000 + 0.4 \times £80\,000 + 0.35 \times £0 - 0.2 \times £80\,000$$
$$= £22\,500 + £32\,000 + 0 - £16\,000$$
$$= £38\,500$$

QUESTIONS

1 The total payout per year on the £62 billion invested in Premium Bonds is £755 million. What rate of interest is this?

2 Show that the probabilities of winning prizes on Premium Bonds, plus the probability of winning nothing, add up to 1.

3 What should a fairground charge to play a game in which you roll four coins and win £5 for getting at least three heads?

4 An adviser tells a business start-up applicant that she has a 45% chance of breaking even, a 45% chance of making £100\,000 and a 10% chance of losing £50\,000. Calculate her financial expectation for the year.

5 Beth invests £750 in a crowdfunding scheme. She estimates that she has a 90% chance of her investment increasing by 20% by the end of the year, or a 10% chance of losing her money if the crowdfunding scheme fails. What is her financial expectation for the year?

E 6 A council organises a fete for each of the 6 weekends of the summer. With no rain, the expected profit each weekend is £4500, but if it rains the profit is halved. The probability of no rain on any given weekend is 0.8.

   a Work out the probability of rain on a weekend.

   b Calculate the expected profit over the 6 weekends.

CHAPTER
> ## 2.33

ARE YOU SURE ABOUT THAT?

How do I decide?

> ## Should I go to university?

I have been offered a job straight from school.

Will I earn enough to pay back the costs of university?

WHAT YOU NEED TO KNOW ALREADY

→ Understanding the difference between theoretical and experimental probability

→ Calculating simple estimates

There are many times in life when we have to make a decision without enough information. This is usually because we need to know what will happen in the future and it is not possible to know that. The best we can do is look at how likely the different scenarios are, from our current perspective, and how damaging it would be if the less likely scenarios occurred. Many aspects of this depend on our own personality and circumstances. This chapter considers some of the decisions you may have to face in life.

MATHS HELP

p322 Chapter 3.8 Representing and analysing data

p356 Chapter 3.14 Approximation

p409 Chapter 3.29 Using probability for expected amounts

PROCESS SKILLS

→ Estimating risks and costs

→ Comparing options and decisions

→ Justifying decisions

→ Criticising claims

Look at the information given below about young people who do, or do not, decide to go to university:

> **In 2013, the average salary of an employed 21-year-old (qualified with A-Levels) was £17 500; the average salary of a graduate after five years of employment was £25 563.**
>
> A typical student spends approximately £9000 per year on living expenses.
> In the 2017 to 2018 academic year, students can borrow up to a maximum of £8200 a year (£10 702 in London) to help pay for these expenses. This money is only paid back once you are earning £21 000 or more.
> Students can also spend up to £9250 per year on tuition fees, although these do not need to be paid up front.

Average salary figures from: http://www.ons.gov.uk/employmentandlabourmarket/ peopleinwork/employmentandemployeetypes/articles/ graduatesintheuklabourmarket/2013-11-19

Student loan data from:

https://www.gov.uk/student-finance/new-fulltime-students

Discuss

> Is it fair to compare those two options? Are there other things that should be considered?

Investigate

> How do the salaries of people who went to university and those who did not compare? What other information would you want to know to be able to better compare the salaries? How much uncertainty is there about these salaries?

> How much uncertainty is there in the costs of university? Which cost is the most uncertain? Why?

> What other factors might affect your decision to go to university other than the financial factors above?

You may wish to work out some of the following amounts to inform your discussion and investigation.

> Calculate an estimate of the total costs for a university student over a three-year course.

> Estimate the salary at the age of 30 of a graduate (who went to university) and of a person who did not go to university.

> Explain whether the difference between the two salaries you have estimated justifies the costs of university or not.

Things to think about

It is important to realise when a number or fact that someone is giving you is uncertain; if it is uncertain, then you also need to know what the range of possible values or outcomes is and how likely they are.

You can then consider the risks and costs of making a decision.

Sometimes people accidentally make claims without taking account of uncertainty, and othertimes they are deliberately misleading. Either way, it's important to be able to spot when a claim has been made that doesn't factor in all the uncertainties it should – you could be being ripped off!

One of the most important aspects to be critical of is 'What's in it for them?' when someone is telling you what you should do. Financial institutions no longer actually give advice; they give you information but stop short of saying what you should do. Other organisations do not have this constraint, and individuals, for the best or worst of motives, are offering advice from their own perspective.

WORKED EXAMPLE

Jessie has a lump sum of £20 000 that she wants to invest for a year.

She has a choice of two investments:

Investment A Variable growth The average growth of the investment was 6% per year over the last five years.	Investment B Fixed payment of £1000 if you invest for a year.

a   Calculate which of these options is a better investment for Jessie.

State any assumptions you have made.

b   Suggest other information that Jessie could collect to help her reduce the uncertainty in this decision.

c   A friend tells Jessie she should go for Investment A because she could get £1200 each year, better than the £1000 of Investment B. Criticise this argument.

SOLUTION

a   Let's start by looking at Investment A:

The evidence from the last five years suggest that Jessie could expect a 6% return.

6% × £20 000 = £1200, so if the investment performed like the average of the last five years then Jessie would earn £1200.

If we compare this with Investment B, where Jessie would earn £1000 regardless, then Investment A is the better option because £1200 > £1000.

We have assumed that:

● Investment A will perform in line with the average of the last five years
● Investment B is guaranteed to pay out £1000 (the company is secure).

b Jessie could collect the actual growth data for each of the last five years (or beyond) to analyse how variable the growth is.

She could also research whether the financial situation this year is likely to be similar to the last five years to decide whether 6% represents a likely outcome.

c The 6% growth is not guaranteed. The investment could give a lower return than the 5% offered in investment B. The £1000 is guaranteed, so Jessie can be certain of the extra money.

There is more information available for some decisions than others. Nobody can tell you for certain about how monetary products will perform over time, and so you have to make the best guess that you can. Nobody can tell you for certain how a particular decision will work out in the future. You should be very sceptical of someone who claims that they can!

QUESTIONS

1 Adam is involved in a contract dispute. He is claiming compensation of £125 000 in court.

Adam's lawyer believes he has a strong case and estimates that there is a 90% chance of winning.

A loss management firm offers to take Adam's case; they will give him £100 000 now but keep anything won in the case when it goes to court.

What would you advise Adam to do?

Explain your reasoning.

2 Jim is buying car insurance. He is offered legal cover that provides up to £50 000 of legal fees in the event of an accident that goes to court.

The cost of the legal cover is £24.99.

What information does Jim need to decide whether legal cover is a good investment?

3 Mary is having a baby and wants to find out what her baby will look like.

She sees this advert:

> For just £50, I can predict whether your baby will have light or dark hair.
>
> If I am wrong, I will give you back £20 and the original fee.

Comment on whether Mary should take up this offer, and whether it is a good business model.

E 4 Davina will shortly be going travelling. She needs to convert pounds into dollars to take with her.

However, Davina is uncertain whether to buy her dollars now or next week because there is a political decision happening tomorrow that may hugely affect the exchange rate.

There are two possibilities:

a The decision will cause the exchange rate to fall by about 5%.

b The decision will cause the exchange rate to rise by about 5%.

It is very unlikely that the exchange rate will be the same after the decision.

What would you advise Davina to do?

Explain your reasoning.

5 Joe has a business and has to choose between two loans:

Loan A: Borrow £100 000 and pay back £2000 per month for 6 years.

Loan B: Borrow £120 000 and pay back a variable percentage between 1% and 3% per month for 6 years.

What would you advise Joe to do? Explain your reasoning.

CHAPTER

> ## 2.34

PEACE OF MIND

> Is it more expensive abroad?

> ## Can I afford to hire a car when I go on holiday?

> Will the insurance cost a lot?

> I hope I won't have an accident.

WHAT YOU NEED TO KNOW ALREADY

→ Calculating probabilities

→ Calculating expected outcomes and values

MATHS HELP

p400 Chapter 3.27 Calculating probabilities

p409 Chapter 3.29 Using probability for expected amounts

PROCESS SKILLS

→ Evaluating options

→ Comparing decisions

→ Justifying decisions

Risk is something we all experience every day. Leaving home puts us at risk of a traffic accident; staying at home puts us at risk of the many accidents that can happen around the home. Sometimes the risk involves danger to life, sometimes damage to property. Usually it involves a bill of some kind. We can insure against having to pay out a large amount by buying insurance. Other risks cannot be offset in this way – we have to be careful and pay attention. Third party car insurance is something we have to buy in order to drive a car. This chapter looks at the costs and benefits of insurance and how other control measures can reduce risk.

Many people hire a car when travelling abroad.

Hire cars usually come with basic insurance that covers you for a major crash but leaves you with the bill for the first £500–£1000 of damage – the 'excess'.

You can buy extra insurance to prevent you having to pay this bill – it is called collision damage waiver (CDW) insurance.

Hire Car Deal!
Hire a car for only £20 per day or £120 for a full week!
Includes insurance *

* £1000 excess – this can be eliminated with our top
value Collision Damage Waiver Insurance,
just £90 for a 7-day rental.

Discuss

> What would be the total bill for hiring the car for a week including collision damage waiver insurance?

> What other additional costs would you need to factor in when deciding whether to hire a car?

> How likely do you think you are to have an accident or cause damage in a (hire) car? What factors would increase or decrease this probability?

> What are the advantages and disadvantages of purchasing the CDW insurance?

> Do you think £90 is a good price for the CDW insurance? Write a short report to justify the expense.

> Which other factors would affect your decision about whether to take out CDW insurance?

> What other information would you want to know about the insurance policy before taking it up or shopping around?

Investigate

> The probability of having an accident in the hire car over the week is 0.12. What is the expected charge?

> What is the difference between the value of this expected charge and the cost of the CDW insurance to prevent it?

Another company is offering CDW insurance for your hire car for £25, but it will only cover £800 of the potential £1000 excess charge if you have an accident.

> Would you prefer this insurance? Explain why or why not.

DID YOU KNOW?

There were 146 000 accidents on UK roads in 2012.

In the same year vehicles travelled a total of 487 billion kilometres on UK roads.

There are many other things for which you can take out insurance. Many of them are not important until you have a place of your own and other people you are responsible for. Life insurance, and buildings and contents insurance, are examples.

WORKED EXAMPLE

Jon has just bought a new 12-month mobile phone contract for £20 per month.

As part of this he has been provided with a phone with a value of £180.

He has been offered mobile phone insurance for £5 per month that will repair or replace the phone if it gets damaged or lost.

Without insurance, he will have to replace the phone himself.

a   What is the total cost of the contract:
  –   without insurance
  –   with insurance?

b   The probability of the phone being damaged or lost is 0.15.
   Calculate whether it makes financial sense for Jon to take out the insurance.

c   State any two additional factors that may affect Jon's decision whether to take out insurance.

SOLUTION

a   –   Without insurance, we work out the cost of the contract over 12 months, which is
      12 × £20 = £240.
   –   With insurance, there will be an extra £5 per month to pay, so 12 × £5 = £60. This means
      that the total cost will be £240 + £60 = £300.

b   The probability that the phone is lost or damaged is 0.15 and if this happens Jon will need to pay
   £180 for a new phone.
   So the expected cost = 0.15 × £180 = £27.

   We already know from part a that the insurance costs £60, which is a lot more than the expected
   cost of £27. Therefore, it does not make financial sense for Jon to take out the insurance.

c   Jon's cash flow may affect his decision. If he will struggle to pay £180 if the phone is damaged
   or lost then he may choose to take out the insurance, even though it is likely to cost him more
   overall.

   Jon may also be able to research other cheaper insurance providers to reduce the cost of the
   insurance to closer to the £27.

Whether to buy insurance or not is a decision you will meet at some time, whether as a householder, an employee or a volunteer organising an event.

> Write a short report explaining the advantages and disadvantages of taking out a form of insurance of your choice and justify the decision you make.

Risk assessments

Health and safety officers make checks to ensure that working environments are safe for employees and for customers.

A health and safety officer has found these concerns in a supermarket and suggested a solution for each.

Concern	Risk	Probability of this outcome if not addressed	Solution
Chiller cabinets leak water onto floor	Customer or staff member slips and falls	0.12	Allocate staff member to mop up spillages and position safety signs warning of risk
Chip and Pin machines spread bacteria	Customer falls ill with stomach bug	0.005	Give each customer a pair of latex gloves to wear before using machine
Goods on high shelves cannot be reached easily so customers and staff stretch	Customer or staff member sustains an injury	0.01	Avoid using high shelves and install an electronic platform for staff where they are used
Pedestrians in the car park have to cross traffic routes	Customer gets knocked over	0.02	Install zebra crossings in car park

> How severe is each risk above? Which would you be most and least worried about?

> What do you think the costs of each of the solutions would be?

> Which solutions would you accept and which would you dispute? Justify your decision by referring to the severity of each risk as well as the costs of the solution.

> If you had to recommend only one of the options above, which would you select? Why?

The health and safety officer has to write a report summarising the findings and recommending a course of action.

> What should they include in this report?

WORKED EXAMPLE

A TV company is producing a new series.

The series must be produced on time or the company will be fined £25000.

It is estimated by the TV company that the probability of producing the series on time is 0.84.

An insurer offers the TV company insurance costing £5000.

If the series is delayed, the insurer will pay the £25000 fine for the TV company.

a   Calculate the expected cost of delay and use this calculation to decide whether the TV company should take out the insurance.

b   The TV company could buy some improved software at a cost of £8000.

This would increase the probability of producing the series on time to 0.95.

Should the TV company invest in the improved software? State any assumptions you have made.

SOLUTION

a   The probability that the series is produced in time is 0.84.

This means that the probability of a delay will be $1 - 0.84 = 0.16$.

We can work out the expected cost of a delay by multiplying the probability by the fine.

So the expected cost of delay $= 0.16 \times £25000 = £4000$.

The insurance costs £5000, which is £1000 more than the expected cost of delay.

So, theoretically, the TV company should NOT take out the insurance because it costs more than the expected cost of delay.

However, this decision will depend on whether they can afford to pay out £25000 if the project is delayed. If not, then they may choose to take out the insurance because it is more affordable in terms of cash flow and the difference between the costs is only £1000.

b   If the TV company buy the software then the probability of delay will reduce to

$1 - 0.95 = 0.05$.

As before, we can work out the expected cost of a delay by multiplying the probability by the fine.

So now the expected cost of delay $= 0.05 \times £25000 = £1250$.

The investment of £8000 has reduced the expected cost of delay from £4000 to £1250 (a total reduction of £2750).

Therefore this investment does not make financial sense and the TV company should not invest.

We have assumed here that the company cannot use the software on future projects and so the full cost is borne by this project.

We have also assumed that the TV company can afford to pay the £25000 fine if required. If they cannot, then spending £8000 on software may be helpful if they can afford it.

QUESTIONS

1 Lisa has booked a holiday costing £750. She is now looking for cancellation insurance.

She estimates that the probability that she will need to cancel the holiday is 0.1.

a What is the expected cost of cancellation?

b Lisa is offered a cancellation insurance policy for £100. Should she buy this policy? Explain your answer.

2 A company sells washing machines for £400 with a 2-year warranty.

The warranty entitles the customer to a repair or a replacement washing machine free of charge.

For an additional £59, customers can extend their warranty to 5 years.

The probability of a washing machine breaking down in each year is shown in the table below:

	Probability of breakdown
In year 1	0.04
In year 2	0.08
In year 3	0.11
In year 4	0.12
In year 5	0.12
After 5 years	0.53

a Calculate the probability that the washing machine breaks down in years 3–5.

b Hence, calculate whether the additional warranty is good value.

3 Ali is booking a trip on a ferry. The cost of the specific ferry that Ali plans to use is £85.

If Ali needs to change the time of his ferry, he will be charged a £25 fee.

The probability that Ali will need to change the time of his ferry is approximately 0.2.

The ferry company have offered a flexible ticket instead for £92.50. With this ticket, any changes to the times of Ali's ferry will not be charged.

Should Ali buy the flexible ticket? Justify your answer with suitable calculations and explanations.

4 Mel is moving into a student house. Altogether, Mel's possessions are worth approximately £3000.

Mel has been offered contents insurance that will replace any items damaged in a fire or stolen during a burglary. The probability of a fire in a student property is 0.006, while the probability of a burglary is 0.03.

The cost of the insurance is £60 for the year.

Should Mel take out the contents insurance? Explain your answer with suitable calculations and explanations.

E 5 Jim has an ice cream van. He has a choice of two venues for the August bank holiday– he can sell ice cream at a country show or at a food exhibition.

For the last 20 years, it has rained on 7 of the August bank holidays.

His expected takings are as follows:

County show	£550 if it rains £1200 if it does not rain
Food exhibition	£850

a Jim says 'I reckon that my takings will be higher at the County show'.

Is Jim right? Justify your answer with appropriate calculations.

Jim is offered insurance for the county show. The insurance costs £75.

If it rains on the bank holiday, then Jim will receive a payment of £400.

b Calculate Jim's expected profit if he chooses the county show and takes out the insurance.

STARTING A BUSINESS

> Over half a million new businesses are started every year in the UK

Would you want to do this?

What if it fails?

It would be nice to be my own boss.

WHAT YOU NEED TO KNOW ALREADY

→ Calculating percentages of amounts
→ Using a spreadsheet

MATHS HELP

p304 Chapter 3.2 Spreadsheet formulae
p359 Chapter 3.15 Percentages
p409 Chapter 3.29 Using probability for expected amounts

PROCESS SKILLS

→ Modelling with graphs
→ Estimating costs and income
→ Representing using spreadsheets and graphs
→ Evaluating annual finances
→ Criticising estimates of costs and income

There are many reasons why people choose to start their own business. They may live in a place where there are no jobs available. They may have constraints around their availability or other things that make them unattractive to potential employers. Some may feel that, because they spend so much of their time working, they wish to do something they enjoy. Perhaps they dream of starting a very successful business that allows them to retire early! Many simply want to be their own boss. However, it is not an easy option and it is important to get advice and plan thoroughly.

Discuss

What are the advantages and disadvantages of running your own business?

There is plenty information and support available (for example, on gov.uk) for those wanting to start a new business. In this chapter, we consider some of the many aspects.

- For example, it is likely that someone starting up their own business would need to borrow money for premises, equipment or training costs. In Chapter 2.13 and Chapter 3.16 there is information about how much it costs to borrow money.

- It is also necessary to have insurance to cover accidents to you or to others and their property, and for loss or damage to business tools and equipment.

- Businesses that have a turnover of over about £80 000 per year have to register to collect VAT and pass it on to the government, but this is not required for smaller businesses.

- The value of business assets such as computers is usually written off over a number of years. So, a computer may be useful for five years before needing replacement. If it cost £2000, its value at the end of the first year is reduced by one-fifth, then it decreases by the same amount each year so that after five years it is considered to be of no value.

Discuss

> Suppose you want to set up your own business in garden maintenance. What personal qualities would you need to help you to be successful?

> Now discuss the costs of the business. Which of these are one-off set-up costs and which are recurring, every year, month or week?

> How will you generate income?

> Did you think about storage, tools, training and transport?

Now is your chance to begin to write a business plan!

> Start by choosing your business name and writing a short paragraph to say what makes your business idea worth supporting.

Now to the finances! The spreadsheet below is the start of an income and expenditure plan for your business.

> Set up the spreadsheet (use a different format if that suits you better) and fill in your estimated costs and your anticipated income. Make sensible estimates of costs and income, using your common sense and internet research. Compare your ideas with others to refine your estimates.

Ryan's Garden Services

	Expenditure			Income		
	Item	Monthly cost	Annual cost	Item	Monthly income	Annual income
Loan repayment						
Insurance						
Purchase of tools						
Website set-up						
Annual totals						
Expected annual profit						

> ❯ Do you show a profit for your first year of trading?

> ❯ If not, what can you change so that you do make a profit?

Expected returns

Those lending you money may expect you to show in your business plan what your expected profit for the year may be. You could break even, lose all or some of your money or make a profit. As you saw in Chapter 2.32, you can use probabilities to estimate your likely return. Remember that the total of the probabilities of all the outcomes must be 1.

WORKED EXAMPLE

Your possible business outcomes have the following probabilities:

Event	Profit of £10000	Profit of £5000	Break even	Lose £2000
Probability	0.2	0.6	0.15	0.05

What is your expected return?

SOLUTION

Your expected return is $= £(10\,000 \times 0.2 + 5000 \times 0.6 + 0 \times 0.15 - 2000 \times 0.05)$

$$= £(2000 + 3000 + 0 - 100)$$

$$= £4900$$

Investigate

Choose a business start-up that you are interested in and estimate possible outcomes at the end of the year and their probabilities. Calculate your expected profit.

QUESTIONS

1 Mark, Pete and Amber each set up a new business. Is it true to say that, by the end of one year, one of them will have gone out of business? Explain your answer.

2 Joe plans to set up a garden maintenance business. He estimates it will cost him £5000 to get the necessary equipment and an old van. Insurance is £1000 per year. He estimates that he can work 30 hours per week at £12 an hour. His business plan states estimated profit at $5000 + 1000 + 12 \times 30 \times 52 = £24720$.

   Criticise his plan.

3 Design a spreadsheet to draw graphs to show how a business computer (cost £2000, useful life five years), a lawnmower (cost £300, useful life seven years) and a mobile phone (cost £150, useful life two years) reduce in asset value over their useful lifetime. Write down the equation of each of your graphs.

4 Adapt your gardening maintenance spreadsheet for a different business – for example, app development, hairdressing or tech support. Can you show a profit? Write a paragraph to persuade investors to support your business.

5 You want to borrow £10000 to buy a van for your gardening business, to be paid back over three years. The total interest paid is 12%. How much is repaid each year?

E 6 Chris sets up a mobile hairdressing business. He puts together a business plan and estimates that, in the first year, he has a 10% chance of making £12000, 20% chance of making £5000, 40% chance of breaking even and 30% chance of a £2000 loss. What is his expected gain?

7 A building contractor has taken on a contract worth £30000.

   The probability of there being a delay is 0.15. If there is a delay the building contractor will be fined £5000.

   a Calculate the expected value of the contract.

   The building contractor has been offered insurance against delay.

   The cost of the insurance is £800.

   If the contract is delayed, the building contractor will receive a payment of £5000 to cover the fine.

   b Should the building contractor take out this insurance? Justify your answer with suitable calculations.

E 8 A transport company has won a contract worth £14 million.

   The contract has a 0.35 probability of delay.

   The penalty for delay is 15% of the contract value.

   Taking either of the following actions now would reduce the risk of delay:

   › employing additional engineers at a cost of £150000 would reduce the probability of delay to 0.25

   › increasing the payouts to members of the public affected by the project at a cost of £750000 would reduce the probability of delay to 0.08.

   The transport company wants to reduce the risk of delay.

   State, with detailed justification, which one of the actions you would recommend to the transport company. Include the option of choosing both.

PRACTICE QUESTIONS: PAPER 2B

1   A charity is making an international aid payment of £5 million to a country that has challenging circumstances.

The charity is concerned that there is a risk that the money will not reach the targeted people due to corruption issues with the officials in the organisations dealing with the money.

The charity estimates that the proportion of the money at risk of not reaching the targeted people is 20%.

The charity can reduce this proportion by taking one of these actions:

● employing a team of aid workers to oversee the process at a cost of £300 000 would reduce the proportion at risk to 3%

● completing quarterly inspections at a cost of £60 000 would reduce the proportion at risk to 10%.

The charity is keen to reduce the proportion of the money at risk.

State which control measure you would recommend to the charity.

Justify your recommendation.   **[5 marks]**

2   A farmer estimates that she has a yearly turnover of £600 000 from crops grown on the farm.

This turnover will be reduced by 40% in a year with bad growing conditions.

The probability of bad growing conditions in a given year is 0.35.

Taking either of the following actions now would reduce the impact of the bad weather conditions:

● buying an irrigation system at a cost of £12 000 would lessen the reduction in turnover to 30%

● using a polytunnel at a cost of £35 000 would lessen the reduction in turnover to 15%.

a   Calculate the expected turnover if no action is taken to reduce the impact of bad weather conditions.   **[3 marks]**

b   The farmer wants to lessen the risk of reduction in turnover if there are bad weather conditions.

    i   Calculate the expected turnover for each of the actions shown above.   **[3 marks]**

    ii State which one of the actions you would recommend to the farmer.
        Fully justify your recommendation.   **[4 marks]**

3   During the Rugby World Cup, at the semi-final stage there are four teams remaining. England plays Australia in one match, with probability 0.6 of winning. South Africa plays Wales in the other match with probability of 0.75 of winning.

Calculate the probability of a final that does not involve a British team.   **[4 marks]**

4   Charlie is planning to redecorate a bedroom.

The work involved has been divided into a number of activities.

The table shows:

● The activities

● The immediate predecessors of each activity

● The number of days needed to complete each activity.

Activity	Immediate predecessor	Duration (days)
A Choose colour scheme	–	1
B Buy paint	A	1
C Order curtains	A	6
D Order carpet	A	10
E Order furniture and fittings	A	15
F Strip room of old décor	–	3
G Prepare surfaces	F	2
H Paint ceiling and woodwork	B, G	3
I Paint walls	H	2
J Put up curtains	C, I	0.5
K Lay carpet	D, I	0.5
L Install furniture	E, J, K	1

    a Construct an activity network for this project.

       Show the earliest start time and latest finish time for each activity.   **[8 marks]**

    b List the critical path.   **[1 mark]**

    c Which of the tasks have 4.5 days of float time?   **[1 mark]**

    d Draw a Gantt chart for the process.   **[4 marks]**

5 A group of 40 students plan a meal out. They have various dietary requirements.

   The Venn diagram shows how many students have dietary requirements.

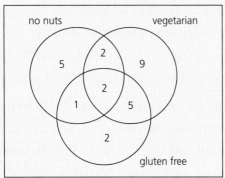

    a What is the probability that a randomly chosen student has no dietary requirements? **[1 mark]**

    b What is the probability that a vegetarian will not eat nuts?   **[2 marks]**

    c What is the probability that a randomly chosen student requires gluten free food and does not eat nuts?   **[2 marks]**

    d The results for these students are used to plan the menu for a special Christmas meal for 150 students. Work out how many gluten free meals should be provided.   **[2 marks]**

CHAPTER

❯ 2.36

THROWING A JAVELIN

❯ What has maths got to do with throwing a javelin?

She's not doing maths.

Where are the numbers?

How can you make a model with maths?

WHAT YOU NEED TO KNOW ALREADY

→ Plotting and sketching straight line graphs
→ Using the formula $y = mx + c$
→ Substituting into formulae

MATHS HELP

p301 Chapter 3.1 Straight line graphs
p352 Chapter 3.13 Formulae and calculation
p368 Chapter 3.17 Graphs with a financial context
p412 Chapter 3.30 Graph sketching and plotting

PROCESS SKILLS

→ Representing mathematical functions
→ Comparing families of functions
→ Deducing the shape of a graph

There are a number of ways to make a model using mathematics. A graph is one way and using an equation or several equations is another way. Once you have the graph or equations that describe something, you can 'play' with them to help make decisions or predict what will happen in the future. In this case, you can consider changing the angle and speed at which the javelin is thrown to get a greater distance. This chapter looks at different types of graph, the equations that describe them and the situations they can be used to model.

Kelly throws a javelin.

Her throw can be modelled using the equation $y = -0.02x^2 + x + 2$ where x is the horizontal distance travelled in metres and y is the height of the javelin in metres.

The graph shows the path of the tip of Kelly's javelin during the throw.

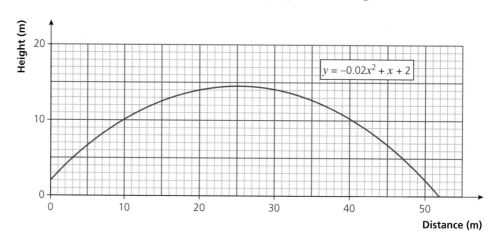

Discuss

> Describe the journey of Kelly's javelin. What happens after the javelin has travelled 25 m? Why is that?

> What type of curve is the graph? Can you tell this from the equation? How?

> Roughly how tall is Kelly? Use the graph and the equation to justify your estimate.

> Use the equation $y = -0.02x^2 + x + 2$ to find the value of y when $x = 10$. How can you check your result using the graph?

> How far does Kelly throw the javelin? How can you use the equation $y = -0.02x^2 + x + 2$ to check the accuracy of your answer?

> What would the graph look like if the equation was $y = 0.02x^2 + x + 2$? Could this new equation represent a javelin throw?

The graph is the picture of the function, and its equation tells you how the graph looks if you know what clues to look for. The next task helps you explore the clues.

HINT

Some assumptions have been made to make the model a perfect quadratic. What do you think they might be?

Investigate

> Use the equation $y = -0.02x^2 + x + 2$ to copy and complete this table of values for the javelin.

x	20	23	45
y			

> Which of the following equations will produce graphs that are linear, quadratic or neither?

$$y = x^2 - x - 4 \qquad y = 3x + 1 \qquad y = 12 - 2x$$

$$y = x^3 + 2x \qquad y = \frac{1}{x} \qquad y = -2x^2 + 1$$

> Identify where each of these curves crosses the y-axis:

$$y = x^2 + 3x + 1 \qquad y = 2x^3 + 4 \qquad y = 7 - 2x$$

Summarise what you have found so far; the ideas are developed further on the next few pages.

WORKED EXAMPLE

Plot the graph of $y = x^2 - 6x + 8$ for $-1 \leqslant x \leqslant 5$ on axes as shown below.

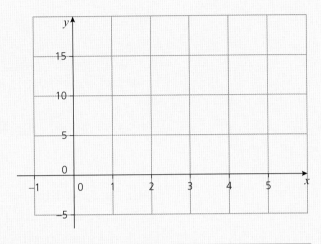

SOLUTION

We need to construct a table of values for x and y for values x of from -1 to 5.

For each value of x, we can calculate the corresponding value of y using $y = x^2 - 6x + 8$.

x	-1	0	1	2	3	4	5
y	15	8	3	0	-1	0	3

When $x = 0$, $y = 0^2 - 6 \times 0 + 8 = 0 - 0 + 8 = 8$

When $x = -1$, $y = (-1)^2 - (6 \times -1) + 8 = 1 + 6 + 8 = 15$.

Plot each pair of x- and y-values as coordinates and join the points with a smooth curve to produce the graph.

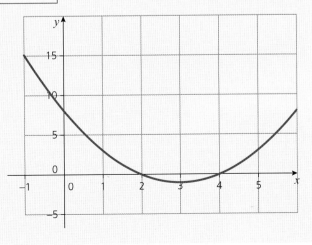

When you plot a graph it helps to fill in a table of values first so that you know what values you need on the y-axis. It also reveals patterns in the numbers that help to check accuracy.

> Plot a graph of each of these equations for $-3 \leqslant x \leqslant 3$.

1 $y = x^2 + 3x - 4$ **2** $y = \dfrac{3x + 1}{2}$ **3** $y = x^3 - 5x$ **4** $y = 2x^2 - 10$

5 $y = x^3 - x^2$ **6** $y = \dfrac{3}{x}$ **7** $y = 2^x$ **8** $y = 0.25^x$

It is helpful when plotting a graph to have some idea of what shape it will be. You can identify any errors in calculating the coordinates if you get unexpected 'bumps' in the curve.

The shape of a graph depends on its equation.

We can identify a **quadratic** graph from an equation that contains an x^2 term as its highest power of x. These graphs will follow the same shape of the graph of $y = x^2$.

We call this shape a **parabola.**

When the x^2 term is negative, the parabola will be inverted.

An equation that contains an x^3 term as its highest power of x will have a **cubic** graph shaped like the graph shown. Just like quadratic graphs, when the x^3 term is negative, the curve will be inverted.

In equations such as $y = 4^x$, the variable x is used as the power.

These equations and graphs are called **exponential** and are shaped like this:

WORKED EXAMPLE

Sketch the graph of $y = x^3 - 2$.

> **HINT**
>
> The dominant term in an expression is the one with the highest power as it contributes most to the value of the expression as x increases.

SOLUTION

This graph is a cubic because the dominant term is x^3.

Therefore, the basic shape will be the same as $y = x^3$.

When $x = 0$, $y = 0^3 - 2 = -2$ so the y-intercept is at $(0, -2)$.

Therefore the graph is:

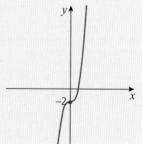

DID YOU KNOW?

A cubic graph generally has two turning points. In this case they are both at the same point producing a **point of inflection.**

Investigate

Start by sketching these 'by hand' and use graphing software or a graphical calculator to check your responses.

❯ Sketch the following graphs:

(Your sketch must show any key points clearly marked.)

$y = x^2 - 1$ $y = -x^3$

$y = 2^x$ $y = 10 - x^2$

$y = 4x - 3$ $y = 10^x + 50$

$y = 7 - 2x$ $y = (x + 3)(x - 4)$

❯ The graphing software packages include a constant controller facility that allows you to input the equation of a function with a constant replaced by a letter, for example $y = kx$. You then vary the constant to see what effect it has. Try it out with some of the functions above.

Discuss

❯ Look at these graphs that have been sketched from their equations and match up each graph with the correct equation.

$y = -x^2 + x + 1$ $y = 4^x$ $y = x^2 - x^3 + 1$ $y = 3x + 1$ $y = x^2 - 1$

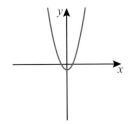 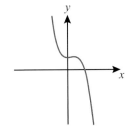 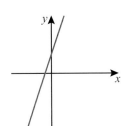 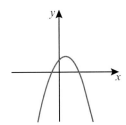 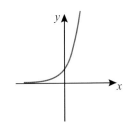

Investigate

So far we have seen four main shapes of graph:

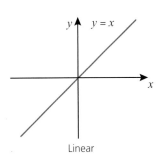 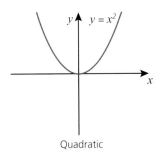 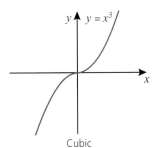 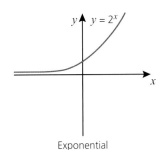

Linear   Quadratic   Cubic   Exponential

DID YOU KNOW?

Coefficient means 'the number multiplying'.

> What happens to these graphs when the **coefficient** of x, or the term with the highest power of x, is negative?

For example, how would you sketch the graphs of the functions $y = -x$, $y = -x^2$ and $y = -x^3$?

> What happens to the exponential graph when the coefficient of x is negative?

For example, what is the shape of the graph for $y = 2^{-x}$?

> What happens to the exponential graph when the base number is less than 1?

For example, what is the shape of the graph for $y = 0.5^x$?

> What happens to the standard graphs when there is a multiplier before the main term?

For example, how would you sketch the graph of $y = 3x^2$ or $y = 0.5x^3$?

> What about $y = 3 \times 2^x$?

It is useful to think about how the negative sign affects the value of the y-coordinate, and to describe the effect on the graph in terms of transformations in an informal way. The negative in front of the whole expression 'flips it to the other side of the x-axis'. The number multiplying the expression for the basic graph 'stretches it up (and down)' so it grows more quickly or more slowly.

> What effect does it have when you add a number to the basic expression for the graph?

> What effect does it have when you add other terms to the basic graphs? eg $x^2 + 5x + 6$ or $y = x^3 - 3x^2 + 6x - 5$

WORKED EXAMPLE

An economist is modelling the profit from an investment using the equation $P = 2t^2 + 1000$

P represents the profit in pounds at t weeks after the initial investment.

Sketch a graph to show how the profit would vary over time.

Show the coordinates of any points where the curve crosses an axis.

SOLUTION

The equation $P = 2t^2 + 1000$ is quadratic with a positive coefficient of t^2.

When $t = 0$, $P = 1000$. Therefore, the graph will cross the P-axis at $(0, 1000)$.

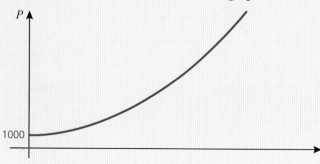

In the example above there is only one point marked on the scale so the effect of the multiplier (2) is not apparent.

Notice the effect of adding 1000 is to 'move the graph up' by 1000.

QUESTIONS

1 A biologist is modelling the number of cells in an experiment.

The number of cells, y, after t days is modelled using the equation $y = 50 \times 1.5^t$.

Sketch a graph to show how the number of cells varies over time.

Show the coordinates of any points where the curve crosses an axis.

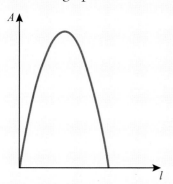

2 The relationship between the area (A) of a rectangle with a fixed perimeter, and the length (l) of the rectangle can be represented with the graph shown.

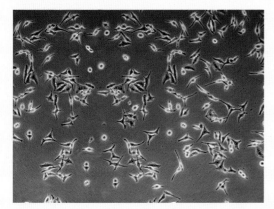

Suggest a possible equation that could model the relationship between A and l.

3 The equation $h = -4t^2 + 36$ models the height (h) of an object after t seconds.

Plot a graph to show how the height will vary with time for $0 \leqslant t \leqslant 3$.

4 The relationship between the area of a shape (A) and its height (h) can be modelled by an equation of the type $A = \alpha h^2 + \beta$.

α and β are both positive numbers.

Sketch a graph to show how A varies with h.

Show the coordinates of any points where the curve crosses an axis.

5 The population of a city during a time of expansion can be modelled by the equation $P = 5000 \times 1.1^t$.

P represents the population after t years.

Plot a graph to show how the population will vary with time for $0 \leqslant t \leqslant 3$.

6 The relationship between the number (N) of radioactive particles in an object and the time (t) in days can be modelled using the equation $N = 1000 \times 0.5^t$.

Sketch a graph to show how N varies with t.

Show the coordinates of any points where the curve crosses an axis.

7 The volume (V) of a cuboid can be modelled using the equation $V = x^3 - x^2$, where x represents the length of one of the sides of the cuboid.

a Show that the graph of $V = x^3 - x^2$ will go through the origin.

b Copy and complete the table of values for x and V.

x	0.5	1	1.5	2
V				

c Use the table of values and the equation to sketch a graph to show how V varies with x for $0 \leqslant x \leqslant 2$.

8 Esme is comparing the cost (C) of two mobile phone contracts.

She considers the cost (C) with respect to the amount of data included (d).

Contract A can be modelled as $C = 0.5d + 12$.

Contract B can be modelled as $C = -d^2 + 16$.

On the same set of axes, sketch how C varies with d for contract A and contract B.

9 Zack is modelling the velocity (v) of an object over time (t) under unusual conditions.

He plots his initial data and obtains the following graph:

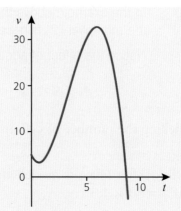

a Which of the following models would best represent Zack's data?

$v = A - Bt^3$	$v = A + Bt^2$	$v = A \times 2^t$	$v = A + Bt$

Explain your choice.

b For the model you have selected, estimate the value of A.

CHAPTER

> 2.37

THE ADVANCE OF TECHNOLOGY

> ## Can the past predict the future?

How much faster is the latest processor?

Are there more smartphones than laptops?

How come devices keep getting smaller?

WHAT YOU NEED TO KNOW ALREADY

→ Plotting graphs
→ Reading and interpreting charts and diagrams
→ Using vertical line charts

MATHS HELP

p412 Chapter 3.30 Graph sketching and plotting

PROCESS SKILLS

→ Justifying decisions
→ Interpreting graphs
→ Inferring behaviour of models
→ Estimating future values
→ Representing using graphs
→ Comparing data

Thirty years ago, a computer took up a whole room and the word internet was barely known. Thanks to continual technological advancements, communication is very different now compared with how it was then. Most business communication took three days using the postal system and if your train was cancelled you would have to find a telephone box, and, provided you had the correct change, you could call home. Email and mobile phones have changed our lives. Changes continue to happen as processing power increases and that power becomes achievable with smaller and smaller devices. Graphs are very useful for showing how things have developed. This chapter will look at trends in technological development and considers what this might mean for the future.

Number of global internet users

Year	No of users (millions)
1994	25.4
1995	44.9
1996	77.6
1997	121.0
1998	188.5
1999	281.5
2000	414.8
2001	502.3
2002	665.1
2003	781.4
2004	913.3
2005	1030.1
2006	1162.9

The figures in the table show the number of people using the internet from 1994 to 2006.

A **time series graph** allows you to look at trends over time. Always plot time on the *x*-axis and the quantity you are investigating on the *y*-axis.

Discuss

> Which of the three graphs below would you use to represent this data?

A

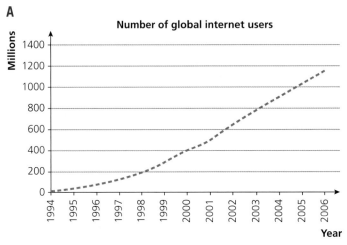

B

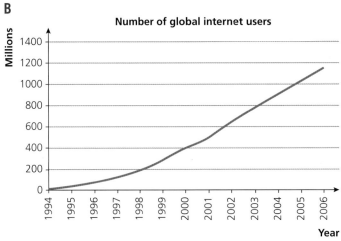

C

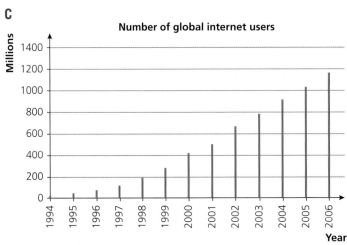

HINT

Although time is a **continuous variable**, a point between two years makes little sense (for example, 2005.5), although this date could perhaps represent 1 July each year, or the halfway point between data collections. It is therefore incorrect to draw a smooth line joining two consecutive years. Instead, a dotted line (A) or alternatively draw a vertical line chart (C). This lets the reader know that they should not use **interpolation** (the process of reading off values between the points that have been plotted) to find values in between those given.

DID YOU KNOW?

One of the best examples of **exponential growth** is observed in bacteria. It takes bacteria roughly an hour to reproduce through prokaryotic fission. If you placed 100 bacteria in an environment and recorded the population size each hour, you would observe exponential growth. There would be 200 at the start of the second hour, 400 at the start of the third, 800 at the start of the fourth, and so on.

Technological trends, like population sizes, often show **exponential growth**. This is when a measure increases at a constantly growing rate. On a graph, this is represented by a curve with a constantly increasing gradient.

When a measure decreases exponentially, it is known as **exponential decay**.

Discuss

› Describe the trend the graphs above show.

Extrapolation is the process of extending a graph beyond the range of the points that have been plotted in order to make predictions. It should be used with caution. There is no certainty that the relationship demonstrated by the data will continue outside the data range.

Discuss

The graph below shows two possible ways to extrapolate the graphs above to predict the number of internet users in later years.

› Which do you think would produce results closest to the actual data, A or B?

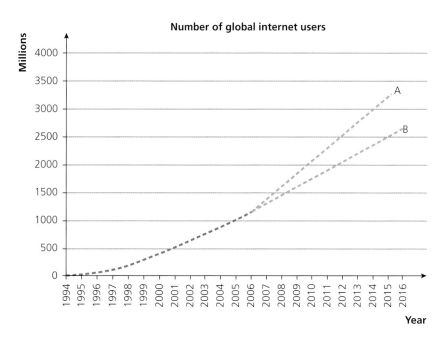

WORKED EXAMPLE

Smartphone sales have risen exponentially since 2005. Company A sold 40 560 smartphones in 2005, 51 250 in 2006, 78 250 in 2007 and 102 150 in 2008.

a   Draw a graph to represent this information.

b   Use extrapolation to estimate:
   –   the number of sales in 2009
   –   the number of sales in 2012.
   In both cases, explain how reliable you think your estimate is.

SOLUTION

a   The orange line shows the data that has been plotted.

b   The green line is the extrapolation of the orange line. The graph has been extrapolated to show exponential growth. From the graph:

   –   Estimated sales for 2009 are 150 000. Although extrapolation is a very risky thing to do, smartphone sales have continued to grow, so this estimate, being only one year further on, is probably quite close.

   –   Estimated sales for 2012 are 580 000. Even though we know smartphone sales have risen exponentially, it is not possible to know the exact steepness of the curve.
   Also, there will come a time when almost everyone has a smartphone and then sales will begin to plateau. This could be before 2012. Therefore, this estimate is very unreliable.

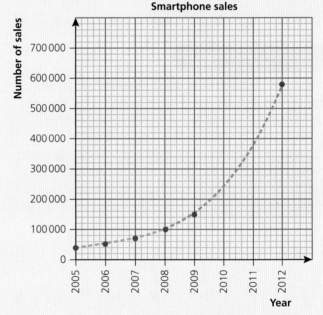

DID YOU KNOW?

No population can grow exponentially forever. A population will grow when resources are in surplus, decline when resources are scarce, and stabilise when the population is at the maximum level that can be sustained. This is known as carrying capacity.

DID YOU KNOW?

The number of transistors that can be placed on an integrated circuit doubles approximately every two years. This trend has continued in a smooth and predictable curve for over half a century and is expected to continue beyond 2020. It is known as **Moore's Law**. Processing speed, memory capacity and even the pixels in digital cameras are all linked to Moore's Law, and are increasing at exponential rates too.

Date of microprocessor	Number of transistors on the integrated circuit (000s)
1980	29
1985	275
1990	1 180
1995	5 500
2000	21 000

Investigate

> The data in the table shows the typical number of transistors on a circuit board in the years given. Plot a graph to represent this information.

> Explain how your graph supports Moore's Law.

> The Intel 18-core Xeon Haswell-E5 processor produced in 2014 had 5 560 000 000 transistors on its integrated circuit. Does this support Moore's Law? Explain your answer.

WORKED EXAMPLE

Microfast and Superquick produce processors for laptops. In January 2015, Microfast produced an integrated circuit board that contained 6 billion transistors. They expect the number of transistors to double every two years. In January 2014, Superquick produced an integrated circuit board with 8 billion transistors, but they only expect this number to double every three years.

When would you expect both companies to have the same number of transistors on their circuit boards?

SOLUTION

Putting the information into a table:

Drawing a graph:

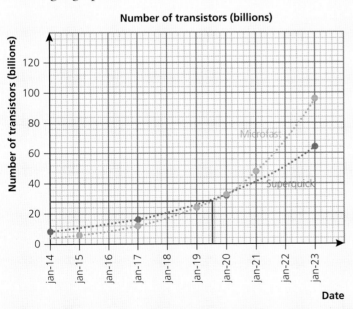

| | Number of transistors (billions) | |
Date	Microfast	Superquick
Jan-14		8
Jan-15	6	
Jan-16		
Jan-17	12	16
Jan-18		
Jan-19	24	
Jan-20		32
Jan-21	48	
Jan-22		
Jan-23	96	64

The two companies have the same number of transistors (28 billion) on their circuit boards where the two graphs cross in July 2019.

WORKED EXAMPLE

Sam is comparing mobile phone contracts. He wants the most up-to-date technology and is likely to upgrade relatively quickly.

Contract A costs £30.99 per month

Contract B costs £23.99 per month with an initial payment of £95

a   Express the cost, £C, of each contract as a function of t, where t is the number of months since it started.

b   Plot the two functions on a graph.

c   Use your graph to estimate when the two contracts cost the same

d   Solve an equation to determine this time exactly.

e   Advise Sam which contract is likely to be better for him.

SOLUTION

a   A: $C = 30.99t$ B: $C = 95 + 23.99t$

b
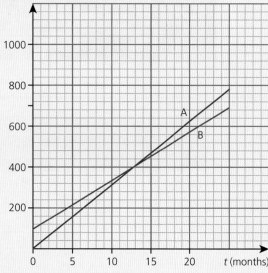

c   When $t \approx 13.5$ months the two amounts are the same.

d $30.99t = 95 + 23.99t$

$7t = 95$

$t = 95 \div 7 = 13\frac{4}{7}$ so after 13 months and 17 days

e   Although Sam is likely to upgrade fairly quickly 13 months is not likely to be long enough for a substantial jump in the technological capability to have happened, so contract B is better for him.

QUESTIONS

1 Typical microchip transistor sizes between 2000 and 2013 are given in the table below.

Year	Transistor size (nm)
2000	130
2002	90
2006	65
2008	45
2010	32
2013	14

DID YOU KNOW?

nm stands for nanometre;
1 nm = 1×10^{-9} metres

a Show this data on a graph. What does your graph tell you about the size of transistors?

b Extrapolate your graph to estimate the size transistors will be in 2020. Explain how reliable you think your estimate is.

2 a The world population between 2000 and 2014 is given in the table below.

Year	World population (billions)
2000	6.13
2002	6.28
2004	6.44
2006	6.60
2008	6.76
2010	6.93
2012	7.10
2014	7.27

i Represent the data on a graph. What does your graph tell you about the world population?

ii Extrapolate your graph to estimate the world population in 2020. How reliable do you think your estimate is?

b The world population between 1900 and 2000 is given in the table below.

Year	World population (billions)
1900	1.65
1920	1.86
1940	2.30
1960	3.02
1980	4.44
2000	6.13

i Represent the data on a graph. What does this graph tell you about the world population?

ii Extrapolate your graph to estimate the world population in 2020.

c Which graph do you think gives a better estimate for the world population in 2020? Explain why.

E 3 The table below shows the death rates per 1000 people in the UK and Japan between 1960 and 2010.

	1960	1965	1970	1975	1980	1985
UK	11.5	11.6	11.8	11.8	11.7	11.9
Japan	7.6	7.1	6.9	6.2	6.1	6.2

	1990	1995	2000	2005	2010
UK	11.2	11.1	10.3	9.6	8.9
Japan	6.7	7.4	7.7	8.5	9.5

a Represent both countries on the same graph. Comment on the trends in death rates in the two countries.

b Use your graph to estimate the year in which the UK death rate became lower than the death rate in Japan. How reliable is your estimate? Explain your reasoning.

c Use your graph to explain what you think the relationship between the death rate in these two countries will be between 2020 and 2030. How reliable is your prediction? Explain your reasoning.

CHAPTER

> **2.38**

HOW GRAPHS AND EQUATIONS MAKE MODELS

> How can an equation crash a bank?

> ## The mathematical equation that caused the banks to crash

Who makes up these equations?

Does nobody check them?

WHAT YOU NEED TO KNOW ALREADY

→ Using spreadsheets
→ Interpreting equations and graphs

MATHS HELP

p301 Chapter 3.1 Straight line graphs
p352 Chapter 3.13 Formulae and calculation
p416 Chapter 3.31 Rates of change

Mathematical models are used everywhere, providing essential information for policymakers on population sizes, weather, life expectancy and so on. But, as the headline above shows, if mathematical models are inappropriately applied the results can be disastrous. (In the story above the equation was applied to market conditions for which the model was not appropriate.)

PROCESS SKILLS

→ Modelling using linear models
→ Comparing models with data
→ Interpreting linear graphs
→ Estimating from graphs
→ Representing data using graphs
→ Criticising the accuracy of the model

We'll start with something simpler! A journey by car from Great Harwood in Lancashire to South Petherton in Somerset is 255 miles.

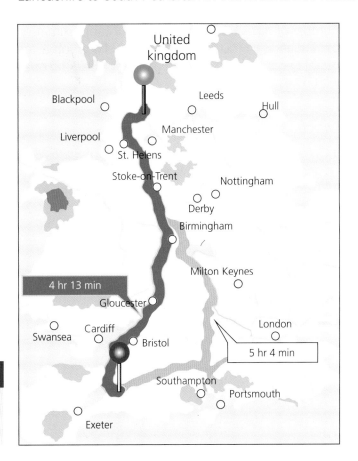

HINT

Google Maps says the journey will take 4 hours 13 minutes.

A simple model for the journey would be to assume that you are travelling at the average speed throughout the journey. The result is shown in this distance–time graph.

You could use the graph to see how far you will travel in the first hour, or when you will reach the halfway point.

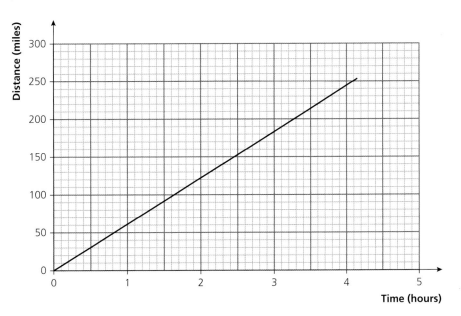

The graph is a straight line graph through the origin, indicating a relationship of direct proportion. The equation of the graph is of the form $y = mx$, where the constant m is the gradient of the graph, which in this case is the average number of miles travelled in one hour (approximately 60.5). So the graph has the equation $y = 60.5x$. The gradient of a graph is the rate of change of one quantity with respect to another, in this case the rate at which distance changes with time.

Discuss

> What are the limitations of this linear model? What is the underlying assumption? Does it give you an accurate measure of where you are at a particular time on the journey? Could the model be improved?

Many straight line graphs do model information exactly.

WORKED EXAMPLE

This graph shows the cost of petrol (at £1.04 a litre), which is directly proportional to the amount you buy.

a   Why is the graph a continuous straight line?

b   What is the gradient of the graph?

c   What does the gradient represent?

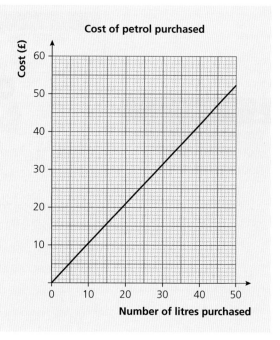

SOLUTION

a   You can buy any amount of petrol, not just whole litres, so a continuous straight line is appropriate.

b   The graph has a gradient of 1.04.

c   It represents the rate of change of the price with respect to the number of litres.

Discuss

> What examples can you find where the price you pay for something is *not* directly proportional to the amount you buy?

> What do the graphs look like?

In Chapter 2.18 you considered some **conversion graphs**. These model the situation exactly. Most conversion graphs are straight line graphs through the origin; there is a directly proportional relationship between, say, a number of miles and the equivalent number of kilometres. An exception is the graph to convert degrees Fahrenheit to degrees Celsius, which is a straight line but it does not go through the origin.

> Why is this?

At GCSE you considered correlation. Quantities such as people's height and weight are correlated – in general, the taller a person is, the more they weigh – but, as the graph below shows, this is not a close correlation.

However, you can draw by hand (or get the computer to draw for you) **a line of best fit**, which shows an approximate linear relationship between height and weight.

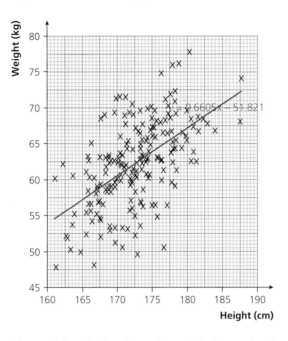

The red line is the line of best fit drawn by the computer. It models the approximate connection between heights and weights.

The equation of the line is $y = 0.6605x - 51.821$

So, if a person's height is 175 cm, the model says that his or her weight is

$(0.6605 \times 175 - 51.821)$ kg

≈ 63.8 kg

Discuss

> What are the limitations of this linear model? Could it be improved?

Population

The prediction of population is important for many reasons, and the mathematical model chosen for the prediction very much affects the results.

Mathematicians use information about current and past populations and other relevant factors to devise a model to predict future populations. It is a challenging task!

The diagram below shows world population up to the present and then how the number of people will change over the next century according to three different models, shown in red, orange and green.

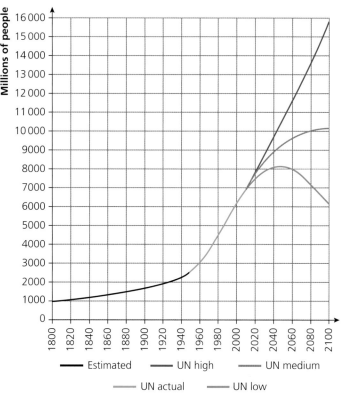

The red line (which continues from the blue line showing actual figures) shows linear growth in population, with an annual rate of increase of approximately 100 million.

Note the enormous difference between this prediction and the orange line showing a much slower (and decreasing rate of growth) and the green line, which shows a decrease in population from the middle of the century. But they are all respectable UN mathematical models!

We do not know which model will best predict future populations, but we can see that the linear model is not a good match for the population before 1950, as shown by the black line; a different linear graph, with a smaller gradient, would be a better fit.

〉 What is the rate of change for the black line?

〉 Why does it change after 1950?

It is typical of mathematical models that they apply only to certain conditions or for a particular time period. If this is not taken into account unexpected results will occur, as the story at the beginning of the chapter shows.

Using **interpolation** to estimate values within the time period or circumstances of the model usually gives reliable results. Using **extrapolation** to predict values outside the time period or circumstances gives very dubious outcomes. It is appropriate to be very critical of values obtained through extrapolation.

Discuss

〉 Why is the black line an estimate?

WORKED EXAMPLE

a How is the estimate of an annual rate of increase for the blue line on the previous graph obtained?

b Is 'approximately 100 million' supported by your calculation?

SOLUTION

a The time increases by 60 years from 1950 to 2010.

The population increases from 2500 million to 7000 million, which is 4500 million.

The rate of increase, or gradient = 4500 million ÷ 60

= 75 million a year.

b 100 million is the nearest power of ten to 75 million and so is the same order of magnitude. Yes, it is supported by the calculation.

QUESTIONS

1 The equation $y = 60.5x$ describes the graph of distance against time for the journey from Great Harwood to South Petherton. Kyle says that means they must have been travelling on dual carriageways all the way as they were travelling at more than 60 mph. Explain why Kyle is wrong.

2 a Use the facts that 0 degrees Celsius = 32 degrees Fahrenheit and 100 degrees Celsius = 212 degrees Fahrenheit to draw a graph to convert temperatures in degrees Celsius to temperatures in degrees Fahrenheit.

b What is the equation of your graph?

c A simple 'rule of thumb' for converting Fahrenheit temperatures to Celsius is to halve the Fahrenheit temperature. Draw a graph of this model on your conversion graph and comment on its accuracy.

3 a Use the linear model $y = 0.6605x - 51.821$, where y is weight in kg and x is height in cm, to estimate the height of a person who weighs 70 kg.

b Nicola says that this means her scales must be wrong as she weighs 70 kg and her height is 165 cm. Explain why Nicola's scales may be correct.

4 Find the equation of the straight line graph that fits the black line in the world population graph on page 266. What would be the population now if this trend had continued?

E 5 The path of a golf ball hit at an angle of $\theta°$ can be modelled by the linear function $y = x \tan \theta$, where x is the horizontal distance in metres of the golf ball from the tee and y is its height in metres above the ground.

a Calculate the height of the ball hit at an angle of 40° when it is 30 m horizontally from the tee.

b Explain why the linear model is not appropriate for the whole of the path of the ball.

CHAPTER

> ## 2.39

SUNRISE AND SUNSET

> ## The evenings are drawing in so fast at the moment

It's so nice to get up in the light.

The dark winter evenings seem to last forever.

I can't wait for the long summer evenings.

WHAT YOU NEED TO KNOW ALREADY

→ Calculating the gradient of a line segment

MATHS HELP

p301 Chapter 3.1 Straight line graphs
p416 Chapter 3.31 Rates of change
p420 Chapter 3.32 Distance, speed and time

PROCESS SKILLS

→ Estimating gradients
→ Interpreting gradients
→ Deducing maximum and minimum points

If you can measure where things are and how they are changing, you can predict to some extent where they will be in the future. This is clearly a very useful thing to be able to do. You do, however, need to be very careful when extrapolating in this way as there is no guarantee that the rate of change will be the same beyond the moment for which it is calculated. This chapter looks at the idea of instantaneous rate of change (the rate of change at a moment) for something that is modelled by a curve.

Straight lines have a constant rate of change that is easy to determine as it is equal to the gradient of the line. However, for a curve, the gradient is constantly changing so we estimate the gradient by drawing a **tangent** to the curve at the point where we wish to know the rate of change of the function. A tangent is a straight line with the same gradient as the curve at the point of contact.

Investigate

The graphs below show how sunrise and sunset times in London change through the year. Look carefully at the graphs and see what you notice.

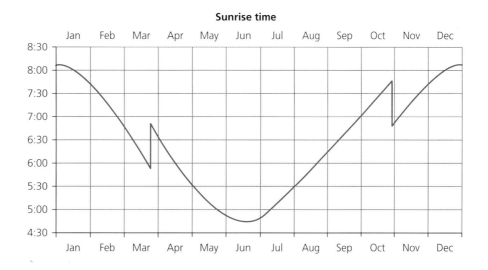

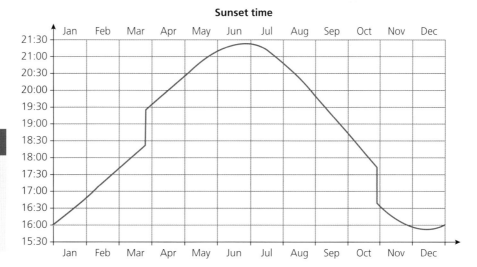

DID YOU KNOW?

Solstice means 'sun stands still'. Identify the winter and summer solstices on each graph.

DID YOU KNOW?

Equinox means 'equal night (and day)'. Identify the vernal (spring) and autumnal equinoxes on each graph.

Discuss

> Why isn't the graph a smooth curve?

> What do the highest and lowest points on the graphs represent?

> What does the gradient represent?

> At what times are the sunrise and sunset times changing fastest?

The clocks go back and forward each year making the sunrise and sunset times suddenly jump one hour. This happens close to the equinoxes where the times are changing fastest. You can see the gradient of the graph is greatest at these times. You may also notice that the earliest sunrise time doesn't quite match the latest sunset time. The maximum and minimum times for each are when the rate of change is instantaneously zero.

All of these things can be identified 'by eye' and no measurements are required to answer the questions above.

However, it is sometimes helpful or even necessary to quantify these rates of change by calculating an estimate of the gradient.

Investigate

› Tangents are drawn at the points A, B and C of this curve. Identify which tangent has a positive gradient, which tangent has a negative gradient and which tangent has a zero gradient.

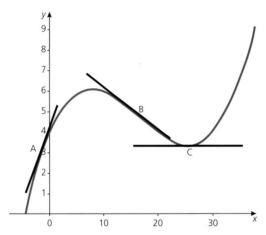

› Can you identify a pair of points where the gradients are equal in size but opposite in sign?

WORKED EXAMPLE

This curve shows the concentration in ml/kg of a drug in someone's body over a period of 24 hours.

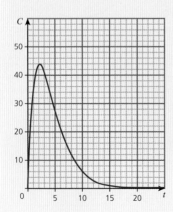

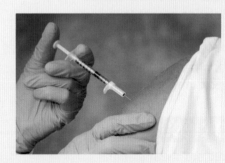

a   Calculate the instantaneous rate of elimination at 7 hours.

b   Explain what is happening at 2 hours.

c   Describe what is happening to the concentration of the drug over 24 hours.

SOLUTION

a The instantaneous rate of elimination is obtained by estimating the gradient of the curve at 7 hours. Draw a tangent to the curve where the time is 7 hours. The gradient of the tangent at any point is the measure of the instantaneous gradient of the curve at that point.

You can see the triangle drawn against the tangent to work out its gradient.

Gradient is $\dfrac{-40}{(10-2)} = \dfrac{-40}{8}$

$= -5$

The instantaneous rate of elimination at 7 hours is 5 ml/kg per hour.

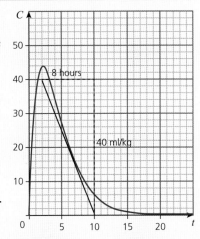

b At 2 hours the instantaneous gradient of the curve is zero. This means the rate of change is zero. The drug has been absorbed completely and will now start to be eliminated.

c The concentration increases from zero until the drug has been absorbed completely. It is then gradually eliminated until, after about 18 hours, there is none left.

Discuss

› James and Jemina are discussing gradients. James says that the gradient is always constant for his graph and Jemina says the gradient is always changing for her graph. Explain, using examples, how both James and Jemina can be correct.

WORKED EXAMPLE

The following graph indicates the number of 16- and 17-year-olds in the UK between 1992 and 2015 (original data from www.ons.gov). Compare the rate of change of the population between the periods 1994–1996 and 2008–2015.

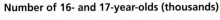

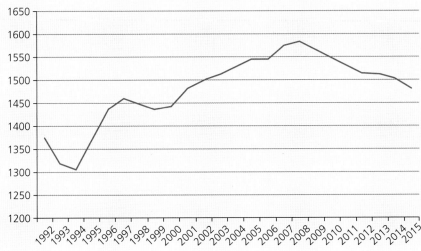

SOLUTION

Between 1994 and 1997 the number of 16- and 17-year-olds is increasing, whereas between 2008 and 2015 the number is decreasing. This is indicated by a positive gradient on the curve in the first period and a negative gradient in the second period. It is better to use an average measure of rate of change rather than an instantaneous rate of change because the population changes unevenly with time. Between 1994 and 1996 the number increased from 1 300 000 to 1 440 000, an average growth of 70 000 per year. Between 2008 and 2015 the population dropped from 1 580 000 to 1 480 000. This is an average reduction of approximately 15 000 per year.

QUESTIONS

1 This graph shows the heights of water at Felixstowe during a day.

   a Use the graph to estimate the times when the water is at low tide.

   b When is the tide rising most quickly?

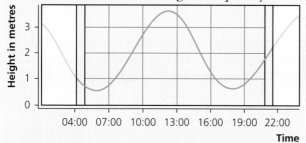

2 Use a tangent to estimate the instantaneous gradient of this curve when the value on the horizontal axis is 1.

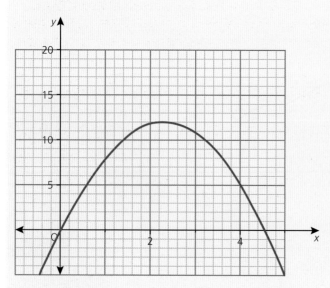

3 The kinetic energy (KE) of a particle of mass m kg and velocity v m s^{-1} is given by the equation KE $= 0.5 \times m \times v^2$.

   a Plot a graph of KE against velocity for a mass of 5 kg and velocity from 0 m s^{-1} to 10 m s^{-1}.

   b Use the graph to estimate the instantaneous rate of change of kinetic energy when the velocity is: (i) 4 m s^{-1} and (ii) 8 m s^{-1}.

E 4 A population of bacteria is growing. The times (in minutes) that the population reaches 100, 200, 300 and 400 are plotted on the figure below.

   a When is the rate of change of the population greatest?

   b Estimate, with reasons, the rate of growth at 30 minutes.

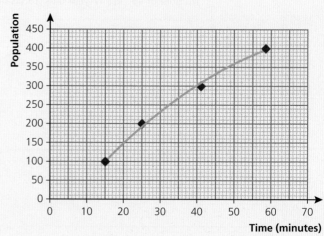

HOLIDAY FLIGHTS

> How long does it take to get there?

> What is the maximum speed of the plane?

> At what speed does an aeroplane travel?

I hate take-offs – I'm afraid we won't be going fast enough to leave the ground.

Does it go at the same speed for the whole flight?

WHAT YOU NEED TO KNOW ALREADY

→ Plotting and interpreting graphs
→ Drawing tangents
→ Finding the gradient of a straight line
→ Rearranging and substituting into formulae
→ Converting between units of measurement

Cruising speed is the average speed of a plane once it has reached its cruising altitude. (On short flights, this may never happen!) It doesn't mean that the plane will always be travelling at that speed! The cruising speed is key to the length of your holiday flight. It will depend on many factors such as the type of plane, the weight of the plane and the weather conditions. This chapter looks at different aspects of speed.

MATHS HELP

p301 Chapter 3.1 Straight line graphs
p412 Chapter 3.30 Graph sketching and plotting
p416 Chapter 3.31 Rates of change
p420 Chapter 3.32 Distance, speed and time

PROCESS SKILLS

→ Modelling motion
→ Checking results
→ Interpreting graphs
→ Estimating gradients
→ Representing

The table below shows outbound and return flights with EasyGo Airline from Manchester to Barcelona.

EasyGo Airline	Outbound: 14:45–18:00	Return: 07:15–08:30

Discuss

> Why do the outbound and return flight times appear to be different?

> In both cases, passengers will be on the plane for the same number of hours. How long is this?

> The outbound flight says it starts its journey at 14:45. How long do you think it is before it takes off?

The graph below shows the **ground speed** of an EasyGo plane during its journey from Manchester to Barcelona.

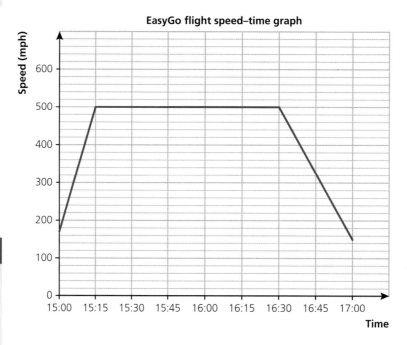

EasyGo flight speed–time graph

This is called **a speed–time graph.** It is sometimes also referred to as a **velocity–time graph.** Velocity is just speed in a given direction.

> **DID YOU KNOW?**
>
> Outbound and return flights are not always scheduled to take the same amount of time. The routes taken are sometimes different so that one can take longer than the other.

> **DID YOU KNOW?**
>
> **Ground speed** is an aviation term. It is equivalent to the speed of any vehicle travelling along the ground.

WORKED EXAMPLE

Look at the graph above.

a   What information does the graph give you?

This flight left the departure gate on time at 14:45 UK time. It taxied to the runway and took off at 15:00. It landed 2 hours later.

b   What was the plane's speed at take-off?

c   What was the plane's speed when it landed?

SOLUTION

a At 15:00, the plane was travelling at a speed of 170 mph. The speed increased at a constant rate until 15:15 when its speed was 500 mph. Its speed then remained constant until 16:30 when it began to reduce at a constant rate until 17:00 when its speed was 150 mph.

b The plane took off at 15:00. Its speed was 170 mph.

c The plane landed at 17:00. Its speed was 150 mph.

When an object's speed is increasing, the object is said to be **accelerating.** When an object's speed is decreasing, the object is said to be **decelerating**. The **gradient** of a speed–time graph tells you the **amount of acceleration**.

WORKED EXAMPLE

This graph shows the speed of an EasyGo plane during its journey from Manchester to Barcelona.

a What is its acceleration between 15:00 and 15:15?

b What is its acceleration between 15:15 and 16:30?

c What is its acceleration between 16:30 and 17:00?

SOLUTION

a Find the **gradient** of the graph between 15:00 and 15:15.

Acceleration = $\frac{\text{change in speed}}{\text{time taken}}$

= $\frac{330 \text{ mph}}{0.25 \text{ h}}$

= 1320 miles/h²

b Between 15.15 and 16.30 the speed is constant, so there is **no acceleration.**

c Between 16.30 and 17.00, the plane is **decelerating**.

Deceleration = $\frac{\text{change in speed}}{\text{time taken}}$

= $\frac{-350 \text{ miles / h}}{0.5 \text{ h}}$

= −700 miles/h²

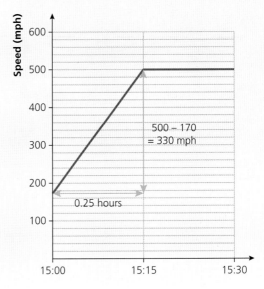

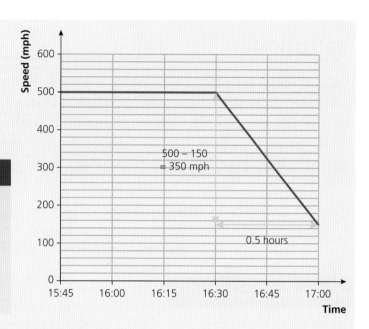

> **HINT**
>
> There are many ways to write units of speed, acceleration, distance and time. The main thing is to check that they all match. If time is in seconds, then the speed and acceleration must be relative to seconds. If the distance is in metres, then the speed and acceleration must be in metres too.

The table below shows the common units and how they are written.

Distance	Time	Speed	Acceleration
Metres (m)	Seconds (s)	m/s or $m\,s^{-1}$	m/s^2 or $m\,s^{-2}$
Kilometres (km)	Hours (h)	km/h or $km\,h^{-1}$	km/h^2 or $km\,h^{-2}$
Miles	Hours (h)	mph or miles/h	$miles/h^2$

The **average speed** of any journey can be found using the formula:

$$\text{Average speed} = \frac{\text{total distance travelled}}{\text{total time taken}}$$

The formula can be rearranged to find distance or time:

$$\text{Total distance travelled} = \text{average speed} \times \text{time taken}$$

$$\text{Time taken} = \frac{\text{total distance travelled}}{\text{average speed}}$$

WORKED EXAMPLE

During the first 15 minutes of its journey from Manchester to Barcelona, the EasyGo plane travelled 84 miles. What was its average speed?

SOLUTION

$$\text{Average speed} = \frac{\text{total distance travelled}}{\text{total time taken}}$$

$$= \frac{84 \text{ miles}}{15 \text{ minutes}}$$

$$= \frac{84 \text{ miles}}{0.25 \text{ hours}}$$

$$= 336 \text{ miles per hour (mph)}.$$

Check:

We know that during the first 15 minutes, the speed increased from 170 mph to 500 mph.

$$\frac{170 + 500}{2} = \frac{670}{2} = 335 \text{ mph } ✓$$

(The difference is due to rounding errors; all numbers are given to the nearest whole number.)

This **distance–time graph** shows the distance of the EasyGo flight from Manchester throughout its journey.

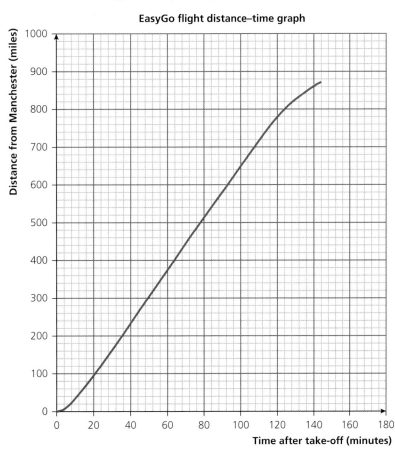

EasyGo flight distance–time graph

Speed can also be found by calculating the **gradient on a distance–time graph**.

WORKED EXAMPLE

Use the distance–time graph to estimate the average speed of the plane **during the first 15 minutes** of its flight.

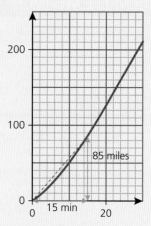

SOLUTION

Find the gradient of the **chord** to the curve between 0 minutes and 15 minutes.

$$\text{Average speed} = \frac{\text{change in distance}}{\text{time taken}}$$

$$= \frac{85 \text{ miles}}{15 \text{ minutes}}$$

$$= \frac{85 \text{ miles}}{0.25 \text{ hours}}$$

$$= 340 \text{ miles per hour (mph)}.$$

Note: It is not possible to read values accurately from a graph, so the average speed found from the gradient is an estimate.

DID YOU KNOW?

A straight line joining two points on a curve is called a **chord**.

WORKED EXAMPLE

Use the distance–time graph on page 277 to estimate the speed of the plane 10 minutes after take-off.

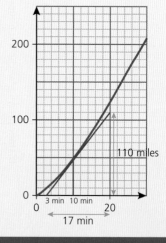

SOLUTION

Find the gradient of the tangent to the curve at 10 minutes.

$$\text{Speed} = \frac{\text{change in distance}}{\text{time taken}}$$

$$= \frac{110 \text{ miles}}{17 \text{ minutes}}$$

$$= \frac{110 \text{ miles}}{0.28 \text{ hours}}$$

$$= 393 \text{ miles per hour (mph)}$$

DID YOU KNOW?

A straight line that touches a curve at one point only is called a **tangent**.

Investigate

> Use the graph to estimate the average speed of the plane between 100 and 120 minutes into the flight.

> Use the graph to estimate the speed of the plane 105 minutes into the flight.

Discuss

On the return flight from Barcelona to Manchester, the plane took off at 7:45 and landed one hour 45 minutes later, taking the same path of 874 miles.

> Which of the following calculations is correct for calculating the average speed of this holiday flight?

a Average speed $= \frac{\text{total distance}}{\text{total time}} = \frac{874}{1.45} = 602.8$ mph

b Average speed $= \frac{\text{total distance}}{\text{total time}} = \frac{874}{1.75} = 499.4$ mph

> What would you estimate the cruising speed to be for this flight?

Discuss

The graph below shows the distance–time graph for a flight from London Heathrow to Sydney, Australia.

> Explain what is happening to the flight between 12.5 and 14 hours into the flight.

> This is common on long-haul flights. Why?

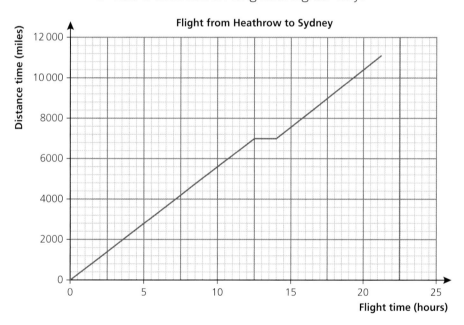

Flight from Heathrow to Sydney

Investigate

> Calculate the speed of the plane 5 hours into the flight.

> What is the total distance travelled by the plane?

> What is the plane's average speed for the entire journey?

QUESTIONS

1 The speed–time graph shows a train journey.

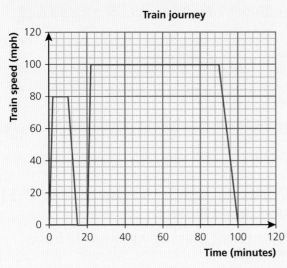

a What does the graph tell you about the journey?

b Use the graph to estimate the final deceleration of the train.

c A second train travelled 60 miles at an average speed of 45 mph. How long did the journey take?

2 The distance–time graph shows a commuter train journey.

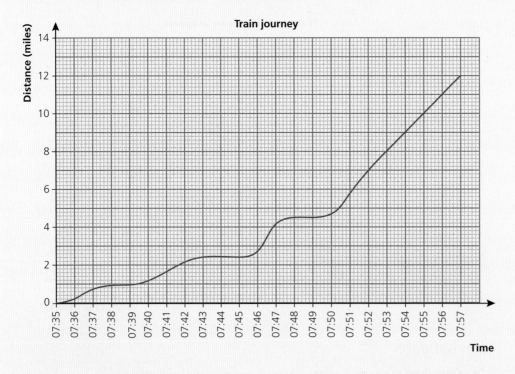

a How long does the journey last?

b How many times does the train stop before it reaches its destination?

c Estimate the maximum speed the train reaches in miles per hour.

3 **a** Amerdeep cycles for 20 minutes at an average speed of 5 metres per second. How far does he cycle?

b One Tuesday, Amerdeep records his ride as shown in the table.

Total distance cycled	Time from start
1 km	5 minutes
4 km	10 minutes
8 km	15 minutes
12 km	25 minutes

i Draw a distance–time graph to represent his journey.

ii On which part of the journey was Amerdeep fastest?

iii On which part of the journey was he slowest?

iv Calculate his average speed for the entire journey in metres per second.

E **4** A car's satellite navigation system gives three route options:

route A   152 km   2 hours 10 minutes

route B   160 km   1 hour 55 minutes

route C   148 km   2 hours 5 minutes

a Calculate the average speed on each route.

b On another journey, the satellite navigation system calculated a journey using the following data.

Distance travelled	Speed over that distance
5 km	40 km h^{-1}
15 km	55 km h^{-1}
85 km	100 km h^{-1}
10 km	40 km h^{-1}

c Write a short report on how the distance and speed vary on the journey. You should include graphs in your report.

★★★

CHAPTER

> 2.41

THE NUMBER 'e'

> ## What's so special about this number?

Why is it on my calculator?

It seems to go on forever.

You're about to find out!

WHAT YOU NEED TO KNOW ALREADY

→ Using indices
→ Understanding gradients
→ The diagram below shows part of the graphs of $y = 2^x$ and $y = 3^x$ (the solid lines).

The irrational number e appears to be part of abstract advanced mathematics (and indeed it is), but it has some surprising and important connections with, among other things, interest rates and the growth (or decline) in populations. Its graph also has some interesting properties as you will discover in this chapter.

MATHS HELP

p412 Chapter 3.30 Graph sketching and plotting
p416 Chapter 3.31 Rates of change
p422 Chapter 3.33 Exponentials

PROCESS SKILLS

→ Modelling growth
→ Deducing the equation of one graph from another
→ Representing exponential functions
→ Evaluating values of functions
→ Comparing graphs

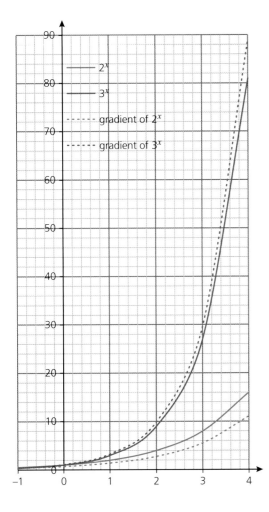

Discuss

> What is the same and what is different about the two solid graph lines?

> How would the graphs continue for higher values of x? Or for lower values of x?

The dotted graphs are the gradients of the two original graphs.

Discuss

> What is the same and what is different about the two dotted graphs?

Note that all four graphs are increasing functions with increasingly large gradients.

These are all graphs of **exponential functions**.

The dotted graphs are both of a similar shape to $y = e^x$.

The gradient function for $y = 2^x$ has lower values than $y = 2^x$; it lies below $y = 2^x$.

The gradient function for $y = 3^x$ has higher values than $y = 3^x$; it lies above $y = 3^x$.

This indicates that there is a function of the form $y = P^x$, where P is between 2 and 3, for which the gradient function exactly matches the function itself. This is shown in the diagram below; the graph shown by the grey line is matched exactly by the black dotted line, its gradient.

The value of P for which this is true is 2.71828182845904523536…
– normally known as **e** or **Euler's number** after the mathematician Leonhard Euler.

The function $y = e^x$ is the only exponential function to have the property that its gradient function is the same as the function itself.

So the gradient at any point on $y = e^x$ is the same as the y value at that point.

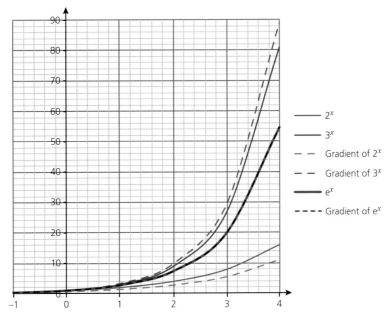

Investigate

> Find out more about Euler and his work. Write a short report on his contributions to mathematics.

The number e also occurs in another, apparently entirely different, mathematical situation – payment of compound interest.

If you invest £1 at 100% interest, at the end of the year you will have £2 – the original £1 plus £1 interest.

If the interest is paid twice a year instead of once, you will get 50% of the interest after six months and another 50% (of the original investment plus interest) at the end of the year. You will then have £1 × $(1 + \frac{1}{2})^2$ = £2.25.

If the interest is paid every month, at the end of the year you will have

$$£1\left(1+\frac{1}{12}\right)^{12} \approx £2.6130$$

If the interest is paid every week, at the end of the year you will have

$$£1\left(1+\frac{1}{52}\right)^{52} \approx £2.6926$$

If the interest is paid every day, at the end of the year you will have

$$£1\left(1+\frac{1}{365}\right)^{365} \approx £2.7146$$

If the interest is paid every hour, at the end of the year you will have

$$£1\left(1+\frac{1}{8760}\right)^{8760} \approx £2.7181$$

which is beginning to look at bit like e.

Investigate

Continue this to find the result when the interest is paid every minute, then every second. How close to e do you get?

In some situations interest is compounded continually. The interest on £1 at the end of the year is £e.

So e is an intriguing number that is fundamental to some important parts of mathematics. As the calculations above indicate, it is the limit of the function $f(n)=\left(1+\frac{1}{n}\right)^n$ as n tends to infinity.

It is an **irrational number** (meaning that it cannot be expressed as a fraction) and a **transcendental number**.

Investigate

> What is a transcendental number?

Because of its particular properties, e is used as the standard base for exponential functions, which can all be expressed in the form $y = A e^{bx}$, where A and b are constants. Different values of the constants A and b produce functions that model continuous growth (if b is positive) and decay (if b is negative). The rate of change of the function is directly proportional to the value of the function itself.

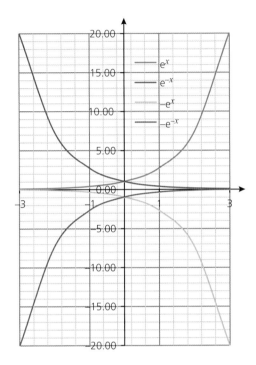

This diagram shows the graph of $y = e^x$ and three transformations of it that make the graphs:

$$y = e^{-x}$$

$$y = -e^x$$

$$y = -e^{-x}$$

Make sure you understand how the graph of $y = e^x$ has been transformed to make each of the other three graphs above.

DID YOU KNOW?

The function can be used to draw a **logarithmic** spiral like the one shown. It uses polar coordinates which are a different coordinate system from the Cartesian coordinates you are used to.

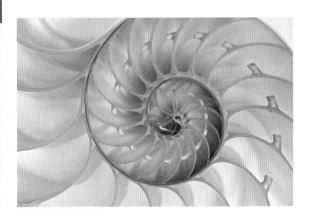

QUESTIONS

1 What irrational numbers do you know? What transcendental numbers do you know?

2 The diagram below shows the graph of $y = e^x$ (in black) and the tangent to the graph at the point where $x = 2$ (in red). The gradient of the tangent is the same as the gradient of the curve at the point where $x = 2$.

  a Calculate the gradient of the tangent and confirm that is equal to e^2.

  b Find the equation of the tangent.

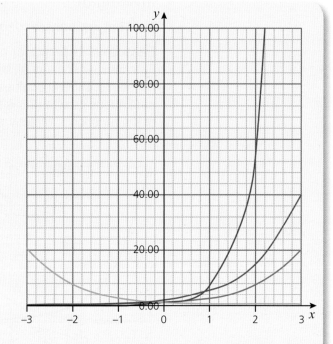

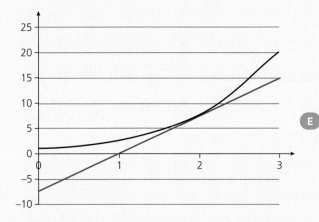

3 The following diagram shows the graphs of:

$y = e^x$

$y = e^{2x}$

$y = e^{-x}$

$y = 2e^x$

Identify which graph is which and explain how you know.

E 4 Under certain conditions, the number of blue-green algae in a population grows according to the model $y = 1000e^{0.087x}$, where y is the number of blue-green algae after x hours.

  a State the number of algae at the start.

  b Work out how many algae there are after 4 hours.

  c Work out how long it takes for the number of algae in the population to reach 10 000.

CHAPTER

> 2.42

GERMS GALORE

> ## Wedding party hit by food poisoning outbreak

How many were infected?

Which food caused it?

Could it have been prevented?

WHAT YOU NEED TO KNOW ALREADY

→ Substituting into formulae, including exponential examples

→ Solving simple linear equations

→ Sketching the graphs of exponential functions

The growth of any population is modelled by multiplying the initial population by a scale factor. This is because each individual (or pair) in the population produces, on average, several offspring. This multiplicative effect is referred to as exponential growth. Mathematical modelling often involves the growth of populations, whether of humans, animals, bacteria or insects. This chapter looks at exponential growth and the equations that describe it.

MATHS HELP

p352 Chapter 3.13 Formulae and calculation
p412 Chapter 3.30 Graph sketching and plotting
p422 Chapter 3.33 Exponentials
p425 Chapter 3.34 Exponential growth and decay

PROCESS SKILLS

→ Deducing values and equations

→ Modelling real life scenarios

→ Interpreting equations

The graph shows how bacteria increase over a period of 45 hours. You can see how quickly the population is growing.

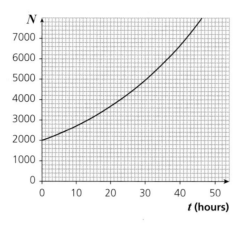

Bacteria in food multiply rapidly over time.

The growth of a particular bacteria population at room temperature can be modelled by the function:

$$N = 2000e^{0.03t}$$

where N represents the number of bacteria after t minutes.

Investigate

› How can you use the graph to find the initial population of the bacteria?

› How could you have found this directly from the equation?

› How can you find the population of the bacteria after 100 hours? Do you use the equation or the graph for this?

› How many hours does it take for the bacteria population to reach 5000?

› Is there a way of finding out how many hours it will take for the bacteria population to reach 20 000?

› Do you think the population of bacteria will continue to grow like this forever?

It is useful to be able to answer these questions to help prevent disease and to evaluate measures taken to inhibit the growth of the bacteria.

Investigate

› Find the number of bacteria after:

  – 13 hours   – 65 hours   – 200 hours

› Use trial and improvement to find out roughly how many hours will have passed when the bacteria population reaches 10 000.

Discuss

› What equation do we need to solve to find the number of hours that have passed when the bacteria population reaches 50 000?

HINT

The laws of indices state that when one power is raised to another you multiply the powers

It is time-consuming to solve equations where the unknown that we are seeking is part of a power.

We can, however, use **logarithms** to solve the equation.

Logarithms are the inverse of exponentials. This means that applying both in succession results in being back where you started.

We use **log** as a shorthand notation for logarithm and **ln** as a shorthand notation for 'natural logarithm', so the natural logarithm of 3^x would be written as $\ln(3^x)$.

In particular $\ln(e^x) = x$ and $e^{\ln x} = x$ as the ln and e functions combine to take you back to x.

And so $e^{\ln a} = a$ and $(e^{\ln a})^x = a^x$

$(e^{\ln a})^x = e^{x \ln a} = a^x$

So, $\ln(3^x) = \ln(e^{x \ln 3}) = x \ln 3$ as the e and ln combine to leave $x \ln 3$.

Your calculator can work out the logarithm of any number, so you can find that the value of $\ln 3 = 1.0986$.

As with all equations, remember that if you apply the ln function to one side, you must do the same to the other.

WORKED EXAMPLE

Solve the equation $20 = 2^x$.

HINT

You can use the result $\ln(3^x) = x \ln 3$, or the version with any number in place of 3, in your calculations.

SOLUTION

The equation has the unknown x as a power.

Therefore, we will take the logarithm of each side of the equation:

$\ln 20 = \ln\left(2^x\right)$

We know that $\ln\left(2^x\right) = x \ln 2$ so we can write our equation as

$\ln 20 = x \ln 2$

Dividing both sides by $\ln 2$ gives us that

$x = \frac{\ln 20}{\ln 2} = 4.32$ to 3 significant figures.

WORKED EXAMPLE

An economist is modelling the value of an investment using the equation $y = 10e^{2x}$. Find the value of x when $y = 10\,000$.

SOLUTION

When $y = 10\,000$, the equation is $10\,000 = 10e^{2x}$

First, divide both sides by 10 to leave just the exponential term on its own: $1000 = e^{2x}$

Take the logarithm of both sides to get

$$\ln 1000 = \ln\left(e^{2x}\right)$$

$$\ln 1000 = 2x$$

$$x = \frac{\ln 1000}{2} = 3.454$$

Methods for solving equations like these follow the principle of keeping the equation balanced by applying the same operation to both sides. The only difference is when you are simplifying the part with the logarithm.

Solve the following equations to find the value of x:

> $100 = 3^x$ > $1000 = 2^x$ > $5000 = 1000 \times 5^x$

> $48 = e^x$ > $200 = e^{3x}$ > $96 = 8e^{2x}$

> $10 = 2 + e^{5x}$ > $1250 = 375 + 5e^x$

Sometimes we know that a function is exponential but we do not have the exact equation for its model.

In this situation we can use some known values to find the missing parts of the equation.

WORKED EXAMPLE

The value of an online business (y) after x years can be modelled using an equation of the form $y = Ce^{kx}$.

When $x = 0$, $y = 200$ and when $x = 8$, $y = 1000$.

Formulate an equation connecting x and y.

SOLUTION

When $x = 0$, $y = 200$ so $200 = C \times e^{k \times 0} = C \times e^0 = C$.

So the equation will be of the form $y = 200e^{kx}$.

When $x = 8$, $y = 1000$ so $1000 = 200 \times e^{8k}$

Dividing both sides by 200 gives us

$$5 = e^{8k}$$

Take the logarithm of each side to give

$\ln 5 = 8k$

So $k = \frac{\ln 5}{8} = 0.201$

Therefore the equation is

$y = 200\mathrm{e}^{0.201x}$

Investigate

Explore how to use your calculator to answer these questions.

> An exponential function is of the form $y = C\mathrm{e}^x$.

Find the value of C if

a $y = 50$ when $x = 0$ **b** $y = 50$ when $x = 8$ **c** $y = 28$ when $x = 2.5$

> An exponential function is of the form $y = C\mathrm{e}^{kx}$.

Find the value of C and k if

a $y = 100$ when $x = 0$ and $y = 150$ when $x = 2$

b $y = 16$ when $x = 0$ and $y = 128$ when $x = 5$

c $y = 12\,000$ when $x = 0$ and $y = 12\,500$ when $x = 2$

Discuss

Since the Second World War, the Gross Domestic Product (GDP) of the US economy has been growing by approximately 2% per year.

In 1945 the GDP was approximately $250 billion.

How can we write an equation to find out the year in which the GDP of the USA was 10 times that of 1945?

20 times? 100 times?!

WORKED EXAMPLE

A nuclear reactor produces thermal energy (E) exponentially at the very start of its reaction process.

This energy emission (E) after t milliseconds can be modelled using an equation of the form $E = C\mathrm{e}^{kt}$

At time $t = 0$, $E = 5000$. After 2 milliseconds, $E = 7459$.

a Formulate an equation connecting E and t.

b How long will it take for the reactor to release a million times more heat than its initial output?

SOLUTION

a   When $t = 0$, $E = 5000$ so $5000 = Ce^{k \times 0} = Ce^0 = C$.

Therefore, the equation is of the form $E = 5000e^{kt}$.

When $t = 2$, $E = 7459$ so substitute to give $7459 = 5000e^{2k}$

Dividing both sides by 5000 $\frac{7459}{5000} = e^{2k}$

Taking a logarithm of each side $\ln\left(\frac{7459}{5000}\right) = 2k$

So $k = \ln\left(\frac{7459}{5000}\right) \div 2 = 0.2$

Therefore, the equation is $E = 5000e^{0.2t}$

b   The initial output at time $t = 0$ was 5000. A million times this figure is 5 000 000 000. So we need to solve $5\,000\,000\,000 = 5000e^{0.2t}$

Divide through by 5000 to get $1\,000\,000 = e^{0.2t}$

Now take a logarithm of each side $\ln(1\,000\,000) = 0.2t$

So $t = \dfrac{\ln(1\,000\,000)}{0.2} = 69$ milliseconds.

QUESTIONS

1 An economist is modelling the value (N) of an investment after t years using the function

$N = 2000 \times 1.025^t$.

  a Find the value of the investment after 10 years.

  b Calculate the number of years for which the investment will need to operate before the initial value of the investment has been doubled.

2 A physicist is modelling the breakdown of an avalanche using an equation of the form $y = Ae^{kx}$.

When $x = 0$, $y = 1000$ and when $x = 10$, $y = 3000$.

Formulate an equation connecting x and y.

3 Josie invests £20 000 at a fixed compound interest rate of 1.5% per annum.

  a Write an equation to model the value (V) of Josie's investment after n years.

  b Use your equation to find the amount of time it will take for Josie to double her investment.

4 A mathematician is modelling a geometric sequence using the formula $T = 2 \times 3^n$ where T represents the value of the nth term of the sequence.

  a Find the first five numbers in the sequence.

  b How many terms of the sequence have a value less than 10 000?

E 5 The relationship between the carbon dioxide in the atmosphere and the global temperature has been modelled by an environmental scientist using a function of the form $C = Ae^{bt}$ where C is the number of tonnes of carbon and t is the global temperature in Celsius.

When $t = 0$, $C = 10\,000$.

When $t = 10$, $C = 18\,221$.

  a Find the values of A and b and hence formulate an equation for C in terms of t.

  b Work out the temperature when the amount of carbon exceeds a billion tonnes.

CHAPTER

> ## 2.43

CARBON DATING

> ## How old is it?

How can you find out?

It looks really ancient!

I'd love to know their history.

WHAT YOU NEED TO KNOW ALREADY

→ Using a calculator
→ Exponential growth
→ Interpreting graphs
→ Manipulating formulae

MATHS HELP

p352 Chapter 3.13 Formulae and calculation
p412 Chapter 3.30 Graph sketching and plotting
p422 Chapter 3.33 Exponentials
p425 Chapter 3.34 Exponential growth and decay

PROCESS SKILLS

→ Modelling with exponential functions
→ Estimating from graphs
→ Representing using equations and graphs
→ Evaluating the accuracy of models
→ Comparing different models

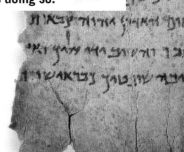

The Dead Sea Scrolls were discovered in the middle of the twentieth century. They are of great historical significance, as they contain material later included in the Bible. The age of the manuscripts was determined by a combination of factors, including carbon dating – an important archaeological technique developed by Willard Libby in the 1940s. Libby was awarded the Nobel Prize in Chemistry for his work.

As the previous chapters show, exponential functions provide useful models for various aspects of life. This chapter looks at radioactive decay and its application to finding the age of ancient objects, as well as other situations where numbers are decreasing in constant proportion or can be modelled as doing so.

The basis of carbon dating is **exponential decay**. In the last two chapters you learnt about functions of the form $y = a^x$ and, in particular, the function $y = e^x$, with the special property that its gradient function is also $y = e^x$. These exponential functions can model growth, as you have seen, and they can also model decay, which we consider in this chapter.

Carbon-14 is a radioactive isotope of carbon that is continuously being produced in the atmosphere, at a rate that matches its natural decay rate, to maintain an approximately constant amount. Carbon-14 enters all living things via the food chain. While living, a plant or animal contains the same proportion of carbon-14 as the atmosphere. When the plant or animal dies, it no longer takes in carbon-14. The carbon-14 in the plant or animal experiences radioactive decay, so the older the plant or animal the less carbon-14 it contains. Determining the percentage that remains of the original amount of carbon-14 allows calculation of the date that the plant or animal died.

Investigate

> Find out more about carbon dating and how it is used.

The graph shows how the percentage of carbon-14 decreases over time.

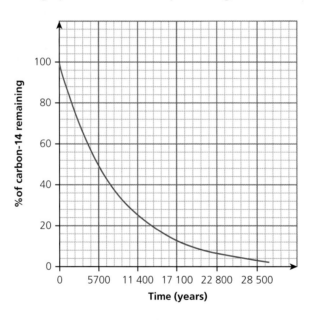

The unusual horizontal scale shows an important property of exponential decay. The time taken to decay by a half (from 100% to 50%, then from 50% to 25%, then from 25% to 12.5% and so on) is always the same. In the case of carbon-14, this time is 5700 years.

If a test of a sample of ancient material (for example, a bone) shows that 30% of the original amount of carbon-14 remains, we can see from the graph that the sample is approximately 10000 years old.

The age can also be found by solving an equation, as you learnt in the last chapter.

WORKED EXAMPLE

Work out the percentage of carbon-14 remaining in the sample of ancient material after 10 000 years using an appropriate equation.

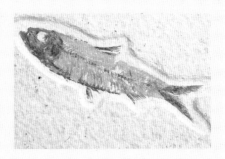

SOLUTION

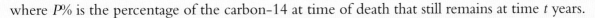

The formula representing radioactive decay can be expressed in the form

$$P = 100\mathrm{e}^{-1.216 \times t \times 10^{-4}}$$

where $P\%$ is the percentage of the carbon-14 at time of death that still remains at time t years.

So, after 10 000 years, the percentage remaining is

$$100\mathrm{e}^{-1.216 \times 10\,000 \times 10^{-4}}\%$$

$$= 100\mathrm{e}^{-1.216}\%$$

$$= 100 \times 0.296\%$$

$$\approx 30\%$$

as we found from the graph.

To find the year if we know the percentage, we can use the logarithm function that you met in the last chapter.

Find the age of the fossil using the equation.

If we know that 70% of the carbon-14 remains in a bone sample and we want to know how old the bone from the fossil is, we need to solve the equation

$$70 = 100\mathrm{e}^{-1.216 \times t \times 10^{-4}}$$

First, we divide both sides of the equation by 100

$$0.7 = \mathrm{e}^{-1.216 \times t \times 10^{-4}}$$

then take the logarithm of each side

$$\ln 0.7 = \ln\left(\mathrm{e}^{-1.216 \times t \times 10^{-4}}\right)$$

so

$$\ln 0.7 = -1.216 \times t \times 10^{-4}$$

so

$$-0.3567 = -1.216 \times t \times 10^{-4}$$

therefore

$$0.3567 = 1.216 \times t \times 10^{-4}$$

$$t = \frac{0.3567}{1.216 \times 10^{-4}}$$

$$\approx 2900$$

So the bone is about 2900 years old; this can be confirmed by looking at the graph.

Discuss

> Make sure you can understand all the steps of this calculation.

Other examples of exponential decay

Where else could exponential models of decay apply? You may have heard of Newton's law of cooling. This law states that the rate of cooling of a hot object is directly proportional to the difference in temperature between the object and its surroundings, so the hotter the object in relation to its surroundings, the faster it cools.

The proportional relationship indicates an exponential function. The formula for the temperature T_t in degrees Celsius after time t minutes is

$$T_t = T_s + (T_0 - T_s)e^{-\lambda t}$$

where T_s is the temperature of the surroundings, T_0 is the initial temperature of the object and λ is the cooling constant.

Here is a specific example of cooling. A cup of coffee, initially at a temperature of 95°C, is left to cool in a room whose temperature is 20°C. The cooling constant in this case is 0.056.

DID YOU KNOW?

You can use this technique to work out the time of death for a body that is found some hours later.

Discuss

> What is the temperature of the coffee after half an hour?
> How long would it take the coffee to cool down to 70°C?

WORKED EXAMPLE

A cup of coffee cools according to the model

$$T = A + Ce^{-0.04t},$$

where T is the temperature of the coffee, in degrees Celsius, t minutes after it is made, and A and C are constants.

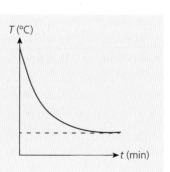

Room temperature is 18°C and the initial temperature of the coffee is 88°C.

a State the values of A and C.

b Give a reason why the initial temperature is not 100°C.

c Assuming the coffee is drinkable when it reaches 60°C, how long do you need to wait to drink your coffee?

SOLUTION

a $A = 18$ (when t becomes very large the second term becomes negligible so $T \approx A$).

   $C = 70$ (when $t = 0$, $T = 88 = 70 + 18$)

b There may be milk in the coffee; the cup will have cooled the water immediately it makes contact.

c Solve: $60 = 18 + 70e^{-0.04t}$

$$42 = 70e^{-0.04t}$$
$$e^{-0.04t} = \frac{42}{70}$$
$$-0.04t = \ln\left(\frac{42}{70}\right)$$
$$t = \ln\left(\frac{42}{70}\right) \div (-0.04) = 12.77 \text{ minutes}$$

QUESTIONS

1 Use the graph of the decay of carbon-14 to estimate the age of an artefact in which 10% of the original amount of carbon-14 remains. Substitute your answer into the formula $P = 100e^{-1.216 \times t \times 10^{-4}}$ to check its accuracy.

2 Calculate the age of a garment in which 40% of the original carbon-14 remains. Check on the graph that your answer is reasonable.

3 The crosses on this diagram show the numbers of local authority properties rented out in the years 2002–2010. The line graph is an exponential line of best fit plotted by the computer. Its equation is $y = 3.9748e^{-0.061x}$

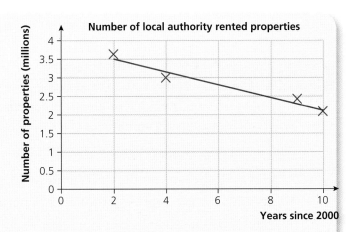

a Use the equation to estimate the number of local authority rented properties in 2017.

b Solve the equation to estimate when the number of local authority rented properties will fall below one million.

c Comment on the likely accuracy of the results obtained by using this exponential function to model the number of properties.

d Find a straight line to model the data and use that to answer the same questions. What are the advantages and disadvantages of each model?

4 A clay pot is fired in a kiln and reaches a temperature of 300°C. It is brought out of the kiln and left to cool in a workshop whose temperature is 25°C. Use Newton's law of cooling, $T_t = T_s + (T_0 - T_s)e^{-\lambda t}$, where T_t is the temperature after t hours, T_s is the temperature of the surroundings, T_0 is the initial temperature of the object and λ is the cooling constant (1.299 in this case).

a What is the temperature of the clay pot after half an hour?

b The pot can be comfortably handled when it reaches a temperature of 35°C. How long will it take to reach this temperature?

E 5 The depreciation in value of a car can be modelled by the formula $P = P_0 \times 2^{-kt}$, where P is the value of the car in pounds after t years, P_0 is its value when new and k is a constant.

A car that cost £17 000 when new is worth £12 000 after three years. Find k and use the formula to estimate the value of the car after five years.

6 The population of tigers in the wild declined from approximately 100 000 in 1900 to approximately 3500 in 2000.

Assuming that the population can be modelled by the formula $N = N_0 e^{-kt}$, where N is the population after t years, N_0 is the original population and k is a constant, show that k is approximately 3.4×10^{-2} and use the formula to estimate when the population of tigers will fall below 1000.

PRACTICE QUESTIONS: PAPER 2C

1 The population of a colony of insects is modelled using an equation of the form $N = Ce^{kt}$ where N represents the number of insects after t hours.

The initial population of insects is 3000 and after 10 hours it has grown to 3154.

a Work out the number of insects in the colony after 2 hours. [4 marks]

b After how many hours will the insect population reach 8000? [3 marks]

2 A popular website experienced rapid growth in its first phase of release.

The number of visitors (V) after t days since launch was modelled by the equation

$V = 500 + 200e^{0.5t}$

a Sketch a graph showing how the number of visitors varies over time. [2 marks]

b Calculate the number of visitors after 14 days. [2 marks]

c Work out how long it takes for the number of visitors to exceed a million. [3 marks]

3   The Forestry Commission monitors the woodland bird population. They can never know the exact population, but they can monitor increases and decreases in a variety of locations.

Assuming a population index of 100 in the year 2000, they calculate indices for other years based on the increases and decreases they have recorded.

Some of these indices are shown below.

Year	Woodland bird population index (year 2000 = 100)
1970	122.7
1975	134.6
1980	125.4
1985	121.0
1990	113.4
1995	95.3
2000	100.0
2005	99.4
2010	99.3

a   Draw a graph to show these indices.

Write a short report about what the graph shows.   **[4 marks]**

b   Estimate the woodland bird population index in 1987 and in 2020.

Use your graph to help you.

Suja says, 'The 1987 estimate is more reliable.' Give detailed reasons why Suja is right.   **[3 marks]**

4   Jane is planning to take a taxi. She considers two firms, A and B.

The charges for each firm are:

A: £4.50 as a fixed charge and then £1.30 per mile after the first mile

B: Cost = £2.5 + 1.4m, where m is the number of miles

a   Write down the rate per mile for firm B.   **[1 mark]**

b   Draw a graph to show the rates for each firm for $0 \leqslant m \leqslant 10$ **[2 marks]**

c   Jane says, 'I shall use firm B for journeys up to 8 miles and firm A for longer journeys as that is the cheapest way.' Is Jane right?

You **must** show your working.   **[3 marks]**

→

5 A car accelerates from rest at some traffic lights.

The graph shows its speed for the first ten seconds.

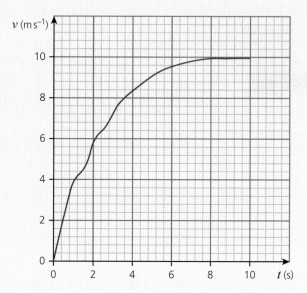

a Write a short report describing the car's motion during that ten seconds.
You should give details of what is happening and at what time. **[4 marks]**

b Find the car's acceleration 6 seconds after it starts. **[3 marks]**

c The car travels 77 m during this time. Calculate its average speed. **[2 marks]**

6 The number of plants of the hairy bittercress weed is modelled by $N = e^t$, where t is the
number of months the plants are left undisturbed and N is the number of plants on 1st March.

a Sketch a graph to show the number of plants over a period of 6 months. **[2 marks]**

b Write down the rate at which the number of plants is increasing on 1st May. **[2 marks]**

CHAPTER
> 3.1

STRAIGHT LINE GRAPHS

Finding the gradient of a straight line

To find the gradient of any line, first choose two points on the line.

HINT

Remember / has a positive gradient and \ has a negative gradient.

To find the gradient between the two points

Think of the line between the two points as the hypotenuse of a right-angled triangle.

Find the lengths of the vertical and horizontal lines of the triangle.

The gradient of the line between the two points $= \dfrac{\text{vertical distance}}{\text{horizontal distance}}$

WORKED EXAMPLE

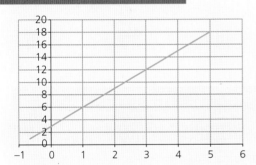

Find the gradient of the line above.

SOLUTION

Possible points on the line to choose are $(1, 6)$ and $(5, 18)$.

$$\text{Gradient of the line} = \frac{\text{Vertical distance}}{\text{Horizontal distance}}$$

$$= \frac{12}{4}$$

$$= 3$$

HINT

It is easier to choose two points that are as far away from each other as possible and whose coordinates are easy to determine.

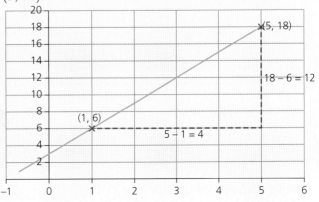

Equation of a straight line

The equation of a straight line is of the form $y = mx + c$

where m is the gradient of the line and c is the y-intercept.

Note: Vertical lines are an exception to this rule. A vertical line going through $(a, 0)$ has an equation $x = a$.

WORKED EXAMPLE

Plot the graph of $y = 3x + 2$.

SOLUTION

This is a straight line graph with gradient 3 and y-intercept at $(0, 2)$.

Plot the graph by finding three points on the line (you need two to draw the line plus a check point).

x	0	2	4
$y = 3x + 2$	2	8	14

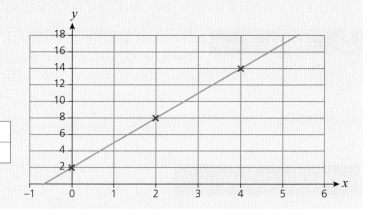

WORKED EXAMPLE

Find the equation of this line.

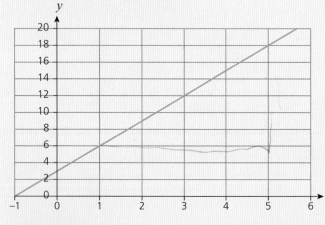

> ### HINT
>
> If the y-intercept is not visible on the graph, take any point on the line and substitute the coordinates into $y = 3x + c$
>
> e.g. The point $(1, 6)$ lies on the line, so
>
> $6 = 3 \times 1 + c$
>
> $6 = 3 + c$
>
> $c = 3$

SOLUTION

To find the equation of the line we need to know its gradient and y-intercept.

As in the example on page 301, the gradient of the line is 3.

By inspection, looking at the graph, the line crosses the y-axis when $y = 3$. This is the y-intercept.

So the equation of the line is $y = 3x + 3$.

QUESTIONS

1 a Using x- and y-axes from −2 to 6, draw the line $y = x - 1$.

b State the gradient of this line.

2 a Find the gradient of this line.

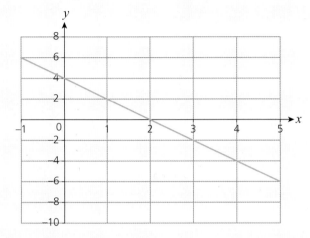

b State the equation of the line.

3 a Plot the points (−4, 1) and (6, 11) on a graph.

b Find the gradient of the line joining the two points.

c On the same graph, draw the line $y = 1 - x$.

d What do you notice about the two lines?

4 Here are the equations of five lines.

A $y = 4x + 4$

B $y = 4 - 4x$

C $y = 4 + 5x$

D $y = 5x - 4$

E $y = 4$

a Which lines have the same y-intercept?

b i Which lines have the same gradient?

ii Plot the lines with the same gradient on the same graph.

iii What do you notice about these lines?

c What is the equation of the line at right angles to $y = 4$ passing through the point (3, 2)?

5 ABCD is a parallelogram. Find the equations of lines AB, BC, CD and DA.

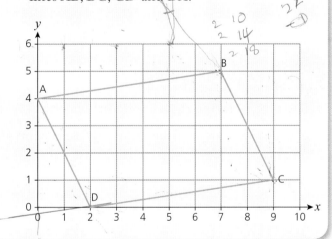

SPREADSHEET FORMULAE

You can type formulae into cells in a spreadsheet one symbol at a time. There are some shortcuts you can take.

Click on the cell rather than typing in its reference (A2 for example).

Copy and paste a formula into a new position and the cell references automatically update to the new position.

A formula will not work unless there is an = sign at the start.

It is best to edit a formula in the status bar, rather than in the cell.

* is the symbol for multiplication and / is the symbol for division.

WORKED EXAMPLE

The picture on the right shows part of a spreadsheet.

The average (mean) number of hours worked in weeks 1, 2 and 3 is 39.67.

What formula should go into cell B5 to work out the mean?

	A	B
1	Week	Hours worked
2	1	45
3	2	35
4	3	39
5		39.67

SOLUTION

The formula in cell B5 could be any of these:

= (B2+B3+B4)/3

= AVERAGE(B2:B4)

= SUM(B2:B4)/3

= B2/3+B3/3+B4/3

= 1/3*(B2+B3+B4)

WORKED EXAMPLE

This table shows how much four people earn and how much they spend on housing.

	A	B	C	D
1	Name	Monthly income	Monthly housing cost	Percentage
2	Alan	£1,325	£435	33
3	Bev	£1,993	£605	30
4	Charlie	£934	£345	37
5	Dot	£2,012	£820	41

The percentage of his income that Alan spends on housing is $\frac{435}{1325} \times 100\% = 33\%$

What is the spreadsheet formula needed to do this calculation?

SOLUTION

The spreadsheet formula needed to do this calculation is C2/B2*100.

The formula goes in cell D2 and is copied into cells D3 to D5.

QUESTIONS

	A	B	C
1	Week	Hours worked	Earnings
2	1	45	£342.00
3	2	35	£266.00
4	3	39	£296.40
5		39.67	£301.47

1 Rewrite each of the five formulae on page 304 so that it will calculate the mean of the earnings for weeks 1, 2 and 3, as shown in this table.

2 Work out what formula was used in cell C2 of the table above (and replicated in C3 and C4) to calculate the earnings each week.

3

	A	B	C
1	Hours paid at £7.60	Hours paid at £8.10	Total earnings
2	37	4	

What formula should go in cell C2 of this table to calculate total earnings?

4 What formula would go in D6 in the table in the worked example above to calculate the mean percentage of income spent on housing?

5 This table shows the numbers of males and females in some college classes.

	A	B	C	D
1	Class	Number of males	Number of females	Percentage of class that are females
2	C1	12	8	
3	C2	5	5	
4	C3	8	20	
5	C4	15	13	

a What formula goes in cell B6 to calculate the total number of males in these classes?

b What formula goes in cell D2 to calculate the percentage of class C1 that are female?

c Set up the spreadsheet yourself to test out the formula in cell D2, replicate it, and find the average percentage of females in these four classes.

CHAPTER

> 3.3

PERIMETER AND AREA

Circles

Circumference of a circle = $2\pi r$

Area of a circle = πr^2

WORKED EXAMPLE

A biologist needs to estimate the area and circumference of this bacteria sample.

The biologist takes several measurements of the distance across the sample in different directions.

By modelling the sample as a circle:

a   Calculate an estimate for the area of the bacteria sample to the nearest square micrometre.

b   Calculate an estimate for the circumference of the bacteria sample to the nearest micrometre.

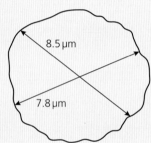

8.5 μm

7.8 μm

DID YOU KNOW?

μm means micrometre and is 10^{-6} metres

SOLUTION

Since we have two different measurements for the distance across the sample we begin by finding the average distance.

$$\frac{8.5 + 7.9}{2} = 8.2\,\mu m$$

For the estimates of the area and the circumference we will use this value as an estimate for the diameter of the bacteria sample.

a   Diameter of the sample   = 8.2 μm

Radius of the sample   = 4.1 μm

Area of a circle   = πr^2

= $\pi \times 4.1^2 = 52.8\,\mu m^2$

To the nearest square micrometre the estimate for the area would be 53 μm².

b   Diameter of the sample   = 8.2 μm

Circumference of a circle = $\pi d = \pi \times 8.2 = 25.8\,\mu m$

To the nearest micrometre the estimate for the circumference would be 26 μm

Composite shapes

A composite shape can usually be split up in several different ways.

Some ways may involve adding areas.

Other ways may involve subtracting two areas.

The simplest is usually the best!

WORKED EXAMPLE

Mark has recently grown a new lawn in his back garden.
The diagram shows some of the dimensions of Mark's garden.

A 750 ml bottle of lawn feed will cover 75 m² of lawn. Mark will use
the lawn feed once a week for 6 weeks. How many bottles should Mark buy?

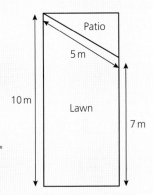

SOLUTION

Mark's lawn can be considered to be a composite shape; it can be split into
a triangle and a rectangle as follows:

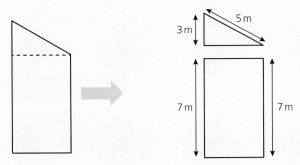

> **HINT**
>
> It can also be considered as a trapezium.

Using the triangle and Pythagoras' theorem we can work out the width of the lawn.

$$x^2 = 5^2 - 3^2 = 16$$

$$x^2 = \sqrt{16} = 4 \, \text{m}$$

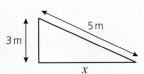

The width of the lawn is 4 m.

Now we can calculate the area of the individual components that make up Mark's lawn:

Area of triangular section $= \frac{3 \times 4}{2} = 6 \, \text{m}^2$

Area of rectangular section $= 7 \times 4 = 28 \, \text{m}^2$

Total area of lawn $= 34 \, \text{m}^2$

Mark needs to use lawn feed once a week for 6 weeks therefore he will need to cover a total area of

$34 \times 6 = 204\,m^2$

Each bottle will cover $75\,m^2$

$204 \div 75 = 2.72$

Therefore Mark will need to buy 3 bottles of lawn feed.

QUESTIONS

1 Alice is making jam tarts. Her pre-rolled pastry is a rectangle measuring 0.5 m by 2 m. Alice needs to cut circles of diameter 6 cm from the pastry for her jam tarts. Any remaining pastry will be thrown away.

Calculate the area of pastry that will be thrown away. Give your answer as a percentage of the original amount.

2 A church is replacing one of their stained glass windows. The window is made from a rectangle and a semi-circle.

a Work out the area of glass needed.

The church will also add a border to their stained glass window once it has been fitted.

b Calculate the perimeter of the glass window.

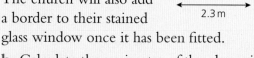

3 Emily is decorating a cake for an upcoming wedding. She wants to put a ribbon around the cake.

The cake was baked in a round 6-inch tin. Emily wants to wrap the ribbon around the cake twice.

What is the smallest length of ribbon she can use? Give your answer in centimetres (1 inch = 2.5 cm).

4 Katie is painting the back wall in her garage. The diagram shows the dimensions of the wall.

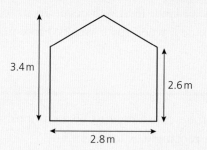

A litre of paint covers 9 square metres. Will it be enough to cover the wall?

5 Oscar is tiling a swimming pool. He still has this wall to tile and 90 m² of tiles left to complete it with. Does Oscar have enough tiles? Justify your answer.

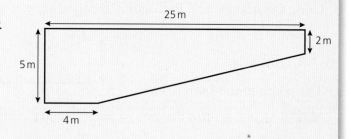

SURFACE AREAS

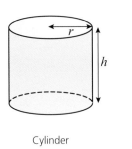

Cylinder

Surface area = $2\pi r^2 + 2\pi rh$

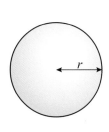

Sphere

Surface area = $4\pi r^2$

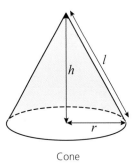

Cone

Surface area = $\pi r^2 + \pi rl$

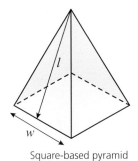

Square-based pyramid

Surface area = $w^2 + 2wl$

> **HINT**
>
> In some contexts, one face of a three-dimensional shape might be missing. For example, when calculating the area of a bucket, there is no lid on the cylinder. Be careful not to include it in the total surface area.

A **hemisphere** is half a sphere.

A **frustum** of a cone is a cone with its top cut off.

A regular tetrahedron has four identical faces. Each is an equilateral triangle.

> **HINT**
>
> Sometimes you can leave your answer in terms of π. This can give you an exact answer rather than a rounded one.

WORKED EXAMPLE

A trinket box is made from a cylindrical container with a lid in the shape of a hemisphere. The outside is coated in silver.

Find the area of the silver parts.

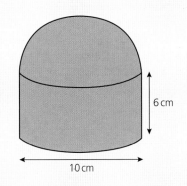

6 cm

10 cm

SOLUTION

Diameter of the cylinder = 10 cm

Radius of the cylinder = 10 ÷ 2 = 5 cm

Surface area of the cylinder = area of base + area of curved surface

$$= \pi r^2 + 2\pi rh$$

$$= \pi \times 5^2 + 2 \times \pi \times 5 \times 6$$

$$= 25\pi + 60\pi$$

$$= 85\pi$$

$$= 267.04 \text{ cm}^2$$

Radius of the sphere = 5 cm

Surface area of sphere = $4\pi r^2$

$$= 4 \times \pi \times 5^2$$

$$= 100\pi$$

$$= 314.16 \text{ cm}^2$$

Surface area of hemisphere = $\frac{1}{2} \times$ surface area of a sphere

$$= \frac{1}{2} \times 314.16$$

$$= 157.08 \text{ cm}^2$$

Area of silver parts = surface area of cylinder + surface area of hemisphere

$$= 267.04 + 157.08$$

$$= 424 \text{ cm}^2$$

QUESTIONS

1 Find the surface area of this regular tetrahedron.

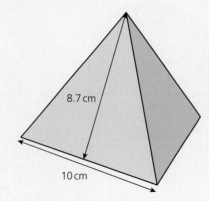

8.7 cm

10 cm

2 Find the area of sheet metal needed to make this pig trough.

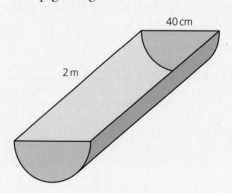

40 cm

2 m

3 Find the external surface area of this thermos cup and lid.

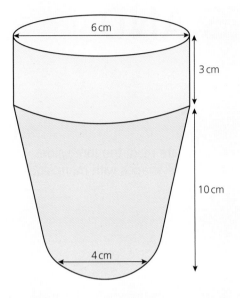

4 Two buttons are in the shape of a hemisphere and a pyramid with a square base. Which of these buttons has the larger surface area? Show all your working.

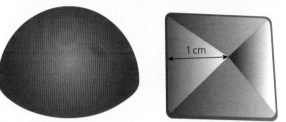

Diameter = 1.5 cm   Base is 1.5 cm × 1.5 cm square

TRIANGLES AND SIMILARITY

Triangles are important shapes in structures as they are rigid; the three sides define the shape so that it cannot be deformed, unlike shapes with more sides.

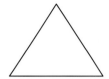

Triangles can be of many shapes and sizes – but all equilateral triangles (like all circles and other regular polygons) are the same shape as they are all enlargements of each other.

Triangles that are the same shape but not necessarily the same size are called **similar**.

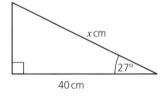 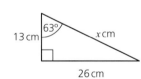

Corresponding angles in similar shapes are **equal**.

Corresponding lengths are **in the same ratio**.

These two triangles are similar as their corresponding angles are equal.

The sides of the triangles are in the same ratio, so the height of the bigger triangle is 20 cm.

Pythagoras' theorem states that, in a right-angled triangle, the square on the hypotenuse equals the sum of the squares on the other two sides.

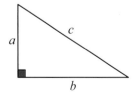

or $a^2 + b^2 = c^2$

Pythagoras' theorem also works in three dimensions to work out the length of the diagonal of a cuboid.

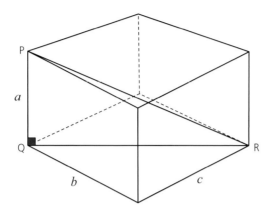

PR is the diagonal of the cuboid. $PR^2 = a^2 + QR^2 = a^2 + b^2 + c^2$

WORKED EXAMPLE

Work out the third side, marked x, in the smaller triangle on page 312.

Then use similarity to work out the third side of the larger triangle.

SOLUTION

The third side (the hypotenuse) can be found using Pythagoras' theorem

$x^2 = 13^2 + 26^2$

$= 169 + 676 = 845$

$x = \sqrt{845} \approx 29.1$

So the third side of the small triangle is 29.1 cm.

The third side of the large triangle is therefore $29.1 \, \text{cm} \times \frac{40}{26} \approx 44.7 \, \text{cm}$.

WORKED EXAMPLE

The diagram shows a tree and a person and their distances from a point O.

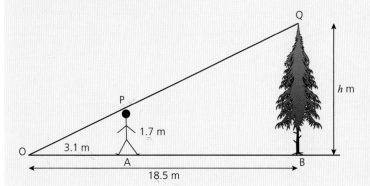

Find the height (h) of the tree in metres.

SOLUTION

The triangles OPA and OQB are similar (one is an enlargement of the other)

$$\therefore \frac{h}{1.7} = \frac{18.5}{3.1}$$

$$\therefore \quad h = \frac{18.5}{3.1} \times 1.7 \approx 10.1$$

So the height of the tree is 10.1 metres.

(Scale drawing can also be used to solve this problem and others like it — you may like to try it.)

When a shape is enlarged to a similar one with double the dimensions, what happens to the area?

The obvious (but wrong!) answer is that the area is doubled, but these diagrams show that this is not the case.

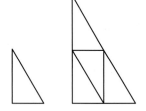

In each case, doubling the measurements results in quadrupling the area.

Suppose the measurements are tripled?

What does this diagram show?

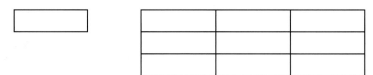

The area of the enlarged rectangle is nine times the area of the original rectangle.

In general, we can say that an enlargement with scale factor f results in the area being multiplied by f^2.

This can be extended to volumes too – if a solid is enlarged with scale factor f, the volume is multiplied by f^3.

This mathematical fact also affects the relationship of linear, square and cubic units.

A square centimetre is the same as 100 square millimetres, as the diagram shows.

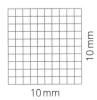

10 mm

10 mm

WORKED EXAMPLE

A map has a scale of 1 cm to 100 m.

What is the area of a rectangular park that has an area of 0.7 cm by 1.2 cm on the map?

SOLUTION

The area of the park on the map is 0.7 cm × 1.2 cm = 8.4 cm²

Every square centimetre on the map represents 100 m by 100 m in real life.

So the area of the park in real life is 8.4 × 10 000 m² = 84 000 m² or 8.4 hectares.

> **DID YOU KNOW?**
>
> The area of 1 hectare is 10 000 m²

QUESTIONS

1 Explain why all right-angled isosceles triangles are similar.

A mobile phone mast has a shadow 22.5 m long while at the same time of day the shadow of a 1.4 m child is 2.1 m.

Find the height of the mobile phone mast.

2 Find the length of
   a the diagonal of a 10 cm square
   b the diagonal of a 10 cm cube.

3 The Louvre Pyramid in Paris is a glass and steel pyramid on a 35 m by 35 m square base. Its height is 22 m.

What is the length of each of its sloping edges? (Remember that the height is measured vertically from the centre of the base – don't confuse it with the height of the sloping triangular faces.)

4 On a map drawn to a scale of 1 cm to 1 km, the area of a lake is 0.75 cm².

What is the area of the lake in real life?

5 A model car, made on a scale of 1 : 10, has a fuel tank with capacity 0.05 litres.

How many litres does the tank of the full-size car hold?

The mass of the full-size car is 1.3 tonnes.

What is the mass of the model?

What assumptions do you have to make to answer these questions?

> **DID YOU KNOW?**
>
> A tonne is 1000 kg and is approximately equal to 1 ton, the imperial version.

PERSONAL FINANCE

Earning money

When you get a job you can be paid weekly or monthly. If you are paid an hourly rate, you are often paid weekly and this is called your **wages**. Many people are paid an annual rate, called a **salary,** and this is often paid monthly.

You may get asked to do **overtime**, which may be paid at time-and-a-half or even double-time. Time-and-a-half means you are paid 50% more per hour for those hours of work.

Most people have to pay **income tax** on their earnings. How much you have to pay depends on how much you earn. The money you pay in income tax goes to the Government to provide revenue for public spending.

As an employee, as well as paying tax, you pay **National Insurance** contributions. National Insurance contributions are specifically for the Government to use to fund state pensions and other social benefits.

Employers deduct tax and National Insurance from wages and salaries before paying their employees.

Bank accounts

Most adults have a bank account and their salary is usually paid into their account by their employer. This is a **credit** to the account. It is money going in to the account. It may well be the only credit to an account, but there will be many **debits**. People use their salaries to pay their bills, buy their food and enjoy themselves! Money they spend is a debit on their account.

Most people keep track of their accounts online, but paper statements are also available. As well as the debits and credits, statements show the **balance** on your account. That is the amount in the account after any transactions.

WORKED EXAMPLE

Date	Description	Credit	Debit	Balance
25 Jan 2016	Salary	£1,560.00		£2033.00
28 Jan 2016	Gas bill		£30.00	
	Cash machine		£50.00	£1953.00
29 Jan 2016	Electricity bill		£30.00	£1923.00
31 Jan 2016	Rent		£450.00	£1473.00
1 Feb 2016	Pizza takeaway		£8.45	£1464.55

Here are the entries from Jack's bank statement between 25 Jan and 1 Feb. Add the following transactions to his account:

- A cheque from his mum, Amelia Taylor, on 3 Feb for £35.00 towards a new pair of jeans
- Purchase of a pair of jeans on 7 Feb for £44.99
- Withdrawal of £50 cash on 6 Feb

SOLUTION

Date	Description	Credit	Debit	Balance
25 Jan 2016	Salary	£1,560.00		£2033.00
28 Jan 2016	Gas bill		£30.00	
	Cash machine		£50.00	£1953.00
29 Jan 2016	Electricity bill		£30.00	£1923.00
31 Jan 2016	Rent		£450.00	£1473.00
1 Feb 2016	Pizza takeaway		£8.45	£1464.55
3 Feb 2016	Cheque from A. Taylor	£35.00		£1499.55
6 Feb 2016	Cash machine		£50.00	£1449.55
7 Feb 2016	Purchase in fashion store		£44.99	£1404.56

Interest rates

The bank account described above is called a current account. It is designed for as many withdrawals and credits as you wish. If you want to save some money for a particular project or for a rainy day, you can open a **savings account**. There is usually a limit on the number of withdrawals you can make but the bank will pay you **interest** for leaving your money with them. A typical interest rate might be 0.5% per annum. Sometimes the interest is paid into the savings account and sometimes it is paid via a cheque to the account holder.

The cost of goods and services

The cost of goods in the shops is made up of several elements:

- the cost of the item to the shopkeeper (the **cost price)**
- a contribution towards the shop expenses and the staff salaries
- **profit** (the amount the shopkeeper 'makes' after all expenses are paid)
- **VAT** (Value Added Tax).

Adding together all the elements above gives you the **selling price** of the item. That is the price on the label in the shop.

Shopkeepers do not always make a profit. If goods do not sell, they may need to set the selling price below the cost price. In this case they will make a **loss**.

WORKED EXAMPLE

John makes up goodie bags to sell at the sports club summer fayre.

He buys

- An assortment of sweets sold in bulk £84.00
- 100 cellophane bags £10.00
- 2 reels of ribbon £4.80

a He uses all the sweets and makes up 100 bags. He plans to sell each bag for £1.50. What profit will he make if he sells all the bags?

b He only sells 57 bags. What loss does he make?

SOLUTION

a Total cost price = £84.00 + £10.00 + £4.80 = £98.80

Total selling price = 100 × £1.50 = £150

Profit = £150 − £98.80 = £51.20

b Total selling price = 57 × £1.50 = £85.50

Loss = £98.80 − £85.50 = £13.30

VAT

VAT is a sales tax that provides the Government with revenue for public spending. It has already been included in the price of goods and services you buy. The seller has to add it to the price they wish to charge. Sometimes it is shown as a separate line on an invoice or receipt.

QUESTIONS

1 Part of Arif's bank statement is shown below.

Date	Description	Credit	Debit	Balance
24 June				£73.50
26 June	Gas bill		£30.00	
	Cash machine		£50.00	-£6.50
27 June	Salary	£1010.00		£1003.50
1 July	Rent		£350.00	£653.50
4 July	Online payment		£28.35	£625.15
7 July	Mobile phone direct debit		£35.00	
	Retail refund	£20.00		£610.15
8 July	Cash machine		£30.00	£580.15

a Explain what happened on 26 June.

b On 9 July Arif had a dentist appointment which cost £35 and a grocery shop of £45.13. What was her final balance on 9 July?

2 Jess renovates computers. He sells a laptop for £135 + VAT. The laptop cost him £30.

His costs including wages for staff were £73.

What profit did he make?

3 Erin makes 6 cakes for a charity day. She cuts each cake into 12 slices and sells each slice for 75p. She makes a profit of £36. What was the cost price of each cake?

4 Camille organised a local charity fun day. She had to hire some premises and equipment as well as buy prizes and goods to sell. She spent £120 on the premises, £42 on equipment and £94 on prizes and goods to sell.

On the day, she took the following money:

–   £115 on entrance tickets
–   £165 on activities
–   £53.30 on refreshments
–   £62.40 on sale of other goods

What profit did she make?

5 Jim works 40 hours at £8.50 an hour. Last week he did 5 hours of overtime at time-and-a-half and 6 hours at double-time. What was his total gross pay for the week?

MATHS HELP

p 359 Chapter 3.15 Percentages

CHAPTER > 3.7

STATISTICAL TERMS

Types of data

In this chapter, we consider what types of data there are – qualitative or quantitative; continuous or discrete?

Qualitative data is something that describes a quality such as someone's gender. It is a category, not a number or measurement.

Quantitative data is information such as height or shoe size – it is a number or measurement on a scale. Quantitative data can be **discrete** – such as shoe size, for which only certain values are possible – or **continuous**, such as weight, which can be anywhere on a continuous scale.

WORKED EXAMPLE

Here are examples of the personal data that may be kept on someone (you, perhaps?).

Name, address, age, date of birth, place of birth, passport number, National Insurance number, gender, medical history, disability/no disability, blood group, height, weight, waist measurement, educational qualifications, ethnic group, marital status, trade union membership, work history, religious affiliation. Such data of course has to be treated with care and according to the law.

Sort the data above into groups, as in the diagram.

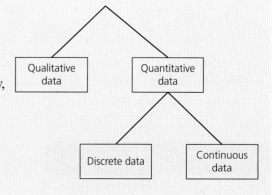

SOLUTION

Qualitative:

Name, address, place of birth, National Insurance number (it contains letters too), gender, medical history, disability/no disability, blood group, educational qualifications, ethnic group, marital status, trade union membership, work history, religious affiliation.

Quantitative discrete:

Passport number, date of birth.

Quantitative continuous:

Age, height, waist measurement.

Primary and secondary data

Data that you collect yourself, or that someone collects on your behalf, for a particular investigation is **primary data**. Data that you get from elsewhere, that was collected for a different purpose, is **secondary data**.

Populations and samples

Most statistics are generated by collecting information from a sample of people or things. For example, when pollsters are trying to estimate people's voting intentions before an election, they ask a sample of perhaps 2000 people and use their answers to make deductions about the population as a whole. The **population** is the whole group (in this case, all the people of voting age in the UK) about which information is needed; the **sample** is a subset of this population. Unless the sample is **representative** of the population as a whole, the results may be **biased**. For example, if the pollsters chose the sample from only a particular part of the country, or from a particular age group, the sample would be biased.

The population does not have to be people – it could be, for example, all the cars in the country, or all the bees in a hive.

QUESTIONS

1 Which of these are examples of continuous quantitative data?

 Age, number of siblings, distance from home to college, bank account number, GCSE Mathematics grade.

2 Write down three examples of qualitative data.

3 Explain the difference between primary and secondary data using examples.

4 Explain why a survey about people's shopping habits would be biased if the researcher stood at the entrance to a supermarket and asked customers to complete a survey.

5 Suppose you want to find the average height of boys in a school. How could you select a 10% representative sample from the whole population of boys in the school?

CHAPTER 3.8

REPRESENTING AND ANALYSING DATA

Drawing and interpreting diagrams

Categorical data is best displayed using bar charts, pie charts or pictograms.

Bar charts may be horizontal or vertical, with the height or length of the bar being equal to the frequency of its category.

Two bar charts may be combined to create a dual bar chart where the bars for each category are placed side by side or otherwise combined.

Pictograms use an icon to represent several items and they are arranged in a similar way to a bar chart.

Pie charts represent the proportion in a category, using a sector of a circle with an angle that is the same proportion of 360°.

Ungrouped discrete data is best displayed using a **vertical line chart**, which is as the name suggests!

HINT

Bars in bar charts should all be the same width with equal gaps between them.

HINT

Pictograms should be drawn so that the icons line up properly and are able to be cut down clearly to show fractions of the amount each icon represents.

WORKED EXAMPLE

Alan conducts a survey on where a group of 18-year-olds intend to be living over the next year.

Here are the results:

With a parent or carer	With a parent or carer	In rented accommodation	In student housing	In student housing
With a parent or carer	In student housing	With a parent or carer	In student housing	In student housing
In rented accommodation	With a parent or carer	In rented accommodation	In own property	With a parent or carer

Construct a frequency table and a pie chart to show the results of the survey.

→

SOLUTION

Begin by listing the categories in the first column of the table.

We then tally or count the number of responses to find the frequency.

So we get:

Type of accommodation	Frequency
With a parent or carer	6
In student housing	5
In rented accommodation	3
In own property	1

To produce a pie chart, find the total frequency = 6 + 5 + 3 + 1 = 15.

A pie chart contains 360° and we have a total frequency of 15.

We need to split the circle into 15 parts which makes each part 24° (because 360 ÷ 15 = 24).

Each frequency is multiplied by 24 to get the angle for the pie chart for that category.

We include these angles as an extra column in the table:

Type of accommodation	Frequency	Angle (°)
With a parent or carer	6	144
In student housing	5	120
In rented accommodation	3	72
In own property	1	24

Draw the pie chart with these four angles to get:

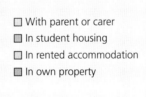

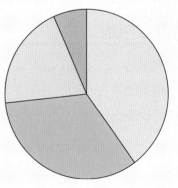

Analysing data by calculating averages and range

There are three **averages** in common use: mode, median and mean.

The **mode** is the most frequent item for categorical or ungrouped data.

The **modal class** is the one with the highest **frequency density** for grouped quantitative data. This is the same as the one with the highest frequency if the class widths are equal.

The **median** is the middle item of discrete data when the data is written in order.

The **mean** is calculated by adding all the data and dividing by the number of items of data.

The **spread** of the data, or how **variable** it is, can be measured using the range.

The **range** is the difference between the greatest and the smallest items of data.

An **outlier** is an item of data which is much larger or smaller than the rest of the data. It may be an error or a genuine item of data.

WORKED EXAMPLE

Ali is investigating the cost of holidays during term time.

He surveys people taking one-week UK holidays and collects the cost per person.

Here are the results of his survey:

£380	£310	£250	£400	£220	£240	£235	£295
£300	£275	£375	£360	£280	£240	£250	£240

a   Calculate the mean cost of a holiday in term time.

b   Find the range of the data.

A newspaper states that the cost of a week's UK holiday taken in the school holidays varies from £350 to £750 per person, with an average of £460.

c   Compare the costs of holidays in term time and those in the school holiday period.

SOLUTION

a   The mean is the sum of all the data divided by the number of items of data, which is 16 for this set.

$$\text{So mean} = \frac{380 + 310 + 250 + 400 + 220 + 240 + 235 + 295 + 300 + 275 + 375 + 360 + 280 + 240 + 250 + 240}{16}$$

$$= £291 \text{ (to the nearest pound)}.$$

b   The lowest value in the list is £220 and the highest value in the list is £400.

So the range = £400 − £220 = £180.

c   The difference between the two averages (the one from Ali's survey and the one from the newspaper) is £460 − £291 = £169.

The range of costs from the newspaper is £750 − £350 = £400, compared with £180 in Ali's data above.

So we can say:

The average cost of a holiday in the UK is £169 more expensive in the school holidays than in term time.

There is much more variation in costs for holidays in the school holidays (up to £400) than for those in term time, which vary by up to only £180.

WORKED EXAMPLE

The table shows the results of an experiment which measures the time taken to stop a car driving at 30 mph in normal conditions.

a State the modal class.

b State the class interval in which the median stopping distance lies.

c Estimate the range of the stopping distances.

The frequency polygon shows the time taken to stop a car driving at 30 mph in rainy conditions.

d Compare the stopping distances in normal conditions and in rainy conditions.

Stopping distance (m)	Frequency
$15 \leqslant x < 18$	2
$18 \leqslant x < 20$	8
$20 \leqslant x < 22$	24
$22 \leqslant x < 25$	27
$25 \leqslant x < 30$	9

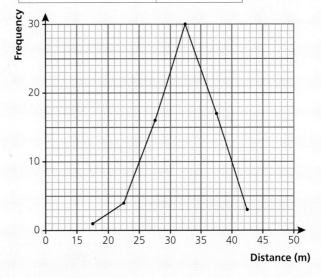

SOLUTION

a The modal class is the category with the highest frequency density.

Stopping distance (m)	Frequency	Frequency density
$15 \leqslant x < 18$	2	0.7
$18 \leqslant x < 20$	8	4
$20 \leqslant x < 22$	24	12
$22 \leqslant x < 25$	27	9
$25 \leqslant x < 30$	9	1.8

In this case, that is $20 \leqslant x < 22$.

b The total frequency $= 2 + 8 + 24 + 27 + 9 = 70$, so the median is in the category where the 35th item of data lies.

There are two items of data in the first interval, then 8 more in the second, which takes us to 10 altogether. There are 24 items of data in the next category, which takes us to a total so far of 34. Therefore, the 35th item of data must be in the next interval so the median lies in $22 \leqslant x < 25$.

c The lowest possible data value is 15 and the highest possible data value is 30.

Therefore, an estimate of the range is $30 - 15 = 15$ m.

d We can see from the frequency polygon that the modal class lies around 32.5 (the highest point of the graph). Although we cannot work out the range fully, we know it is at least $42.5 - 17.5 = 25$ m.

If we total up all the frequencies we get 1 + 4 + 16 + 30 + 17 + 3 = 71 so the median will be at the 36th data value.

This happens in the 4th class interval because 1 + 4 + 16 = 21 is less than 36 and 1 + 4 + 16 + 30 = 51 is more than 36. So the median lies in the class interval with midpoint 32.5.

So we can say:

On average it takes longer to stop in rainy conditions than in normal conditions (because the median and the mode are higher).

There is more variation in the stopping distance in rainy conditions than in normal conditions (because the range is higher).

Frequency polygons are not explicitly mentioned in the specification. They are included here as an example of a statistical diagram that could appear and require interpretation.

Scatter graphs

WORKED EXAMPLE

The scatter graph shows some information about the height and arm span of ten basketball players.

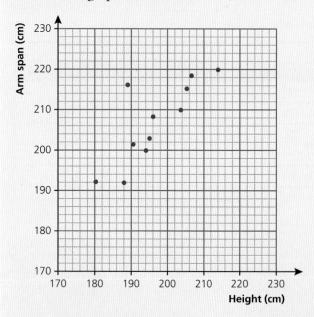

a   Additional data for two further basketball players below could be included on the scatter graph. Do they follow the trend for the other players?

	Player A	Player B
Height (cm)	198 cm	210 cm
Arm span (cm)	208 cm	218 cm

b   Identify any outlier data points on the graph.

c   Describe the correlation of the graph.

d   Explain the relationship between height and arm span that the graph shows.

SOLUTION

a The additional players do follow the trend of the graph.

b There are no significant outliers; the most outlying point is circled.

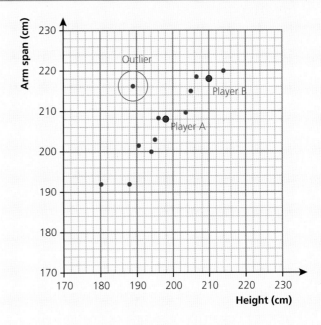

c The scatter graph shows positive correlation.

d The greater the height of a basketball player, the larger his or her arm span.

QUESTIONS

1 The table shows the number of medals won by the European competitors in the 2010 Gymnastics World Championships:

Medal	Gold	Silver	Bronze
Frequency	7	8	5

a Construct a pie chart to represent the table.

The pie chart below shows the medals won by European competitors in the 2014 World Championships.

2014 Championships
- Gold
- Silver
- Bronze

b There were 16 medals awarded to European competitors overall in 2014.

Estimate the number of gold medals won by Europeans at the 2014 World Championships.

c Compare the performance of European gymnasts in 2010 and 2014.

2 The table below shows the number of visitors to two websites over a week (to the nearest hundred).

	Monday	Tuesday	Wednesday	Thursday	Friday	Saturday	Sunday
Website A	3000	3200	3600	3500	3100	2900	2700
Website B	2800	2900	3100	3400	3900	4200	4000

a Construct a dual bar chart to represent the data in the table.

b Compare the number of visitors to website A and website B over the week.

3 Eva is analysing the performance of her team.

She has graded each team member's performance from 1 to 5, where 1 is excellent performance and 5 is inadequate performance.

Here are the performance grades for Eva's team:

1	1	4	3	2	2	2	2	5	3	3	2

a Construct a vertical line chart to represent Eva's performance data.

b Describe the performance of Eva's team.

4 Jay is a supermarket buyer for the vegetable section.

He is comparing two suppliers of cauliflowers to decide which one to use.

Jay needs to pick a supplier with a good average cauliflower size and good consistency of size.

Supplier A provides a sample of cauliflowers with the following diameters (to the nearest cm):

17 cm	15 cm	16 cm	14 cm	13 cm
20 cm	18 cm	17 cm	15 cm	16 cm
14 cm	14 cm	16 cm	18 cm	19 cm
16 cm	17 cm	13 cm	13 cm	18 cm

a Construct a frequency table for the cauliflowers from Supplier A.

b For supplier A's cauliflowers, calculate:

i the mean

ii the median

iii the mode

iv the range

Supplier B's sample of cauliflowers had diameters with the following averages and range:

Mean	14.5 cm
Median	15 cm
Mode	15 cm
Range	3 cm

c Which supplier should Jay select based on this data? Why? What else may affect his decision?

5 The sales of vinyl records in a specialist music shop over six years are shown in the table below:

Year	2010	2011	2012	2013	2014	2015
Sales	3500	4100	3900	5000	5200	5500

a Construct a line graph to represent the data in the table.

b Describe what has happened to sales of vinyl records over the time period of the graph.

6 Jamie is a farmer who sells eggs. He wants to investigate the relationship between a chicken's age and the number of eggs it produces. He samples ten of his chickens, recording their age in years and the number of eggs they produce in a fortnight.

Here are the results:

Age of chicken	2	1.5	1	1.25	2	3	2.75	5	0.75	3.5
Number of eggs	12	13	14	12	11	10	11	6	14	9

a Construct a scatter graph to represent the data.

b Describe the correlation of the graph.

c Suggest an explanation for the relationship that the graph shows between the ages of chickens and their egg production.

7 A scientist is conducting an experiment using springs to see how the mass a spring is supporting affects its extension.

She collects the following data values:

Mass (g)	35	50	60	75	100	120	125
Extension (cm)	2.2	2.9	4	4.5	6.2	7.5	7.8

a Construct a scatter graph to represent this data.

b Describe the relationship between the mass and the extension of the spring.

→

Height (m)	172	184	145	164	153	161	148	157	162
Average GCSE points score	49.2	51.2	38.8	57.5	48.6	29.9	32.2	45.1	38.0

8 A sociologist believes that there is a relationship between people's heights and their academic achievement.

To test the hypothesis, the sociologist collects heights and GCSE points scores from a sample of people.

The results are shown in the table above:

a Construct a scatter graph to represent this data.

b Describe the correlation between height and academic achievement that the data shows.

MATHS HELP

p 320 Chapter 3.7 Statistical terms

SAMPLING METHODS

It is not always possible, because of time, finance or geographic location, to survey a whole **population,** so a subset of the population is used. The subset is called a **sample**. Samples can be chosen in different ways depending on the original population and the restrictions in place. The four types of sampling method are **random, cluster, quota** and **stratified** in Paper 1, and **systematic** sampling is included in Paper 2A. In all cases, the sample needs to be as representative as possible and the method needs to be efficient.

Random sampling: each member of the population has the same chance of being selected, perhaps selected by putting all the names on a piece of paper and drawing them from a hat.

Cluster sampling: a typical subgroup is chosen as the sample, perhaps one street in a town, or one town in a county.

Quota sampling: the number of people with particular characteristics is decided and then the data collected on an opportunistic basis, perhaps a certain number in each of several age groups.

Stratified sampling: the **population** is divided into layers, or strata, and the number in each layer in the sample is proportional to the number in that layer in the whole population.

Systematic sampling: the **population** is ordered into a list and every twentieth (for example) item is selected for the sample.

DID YOU KNOW?

A **population**, in statistics, is all the items you could collect data about for your investigation.

DID YOU KNOW?

In a **census** you collect data about each member of the **population**.

WORKED EXAMPLE

A fast food restaurant chain wants to interview customers in its restaurants in the UK and has recruited one interviewer to interview 100 customers chosen at random from ten restaurants in a week. What type of sampling would you recommend that they use to select the restaurants?

SOLUTION

The restaurants are spread all over the UK, so if the restaurants were chosen at random the interviewer might have to travel hundreds of miles, which would be a waste of time and money. Therefore, the preferred method for them to use is a cluster sample. This will enable a representative sample to be made but with an efficient use of time. The method divides the UK into small geographical regions that are roughly equivalent in terms of the profile of the people living there, and one region is randomly chosen. The random sample can be achieved by numbering each region then choosing the region using a random number generator. For the chosen region they would choose a random sample of 10 restaurants.

WORKED EXAMPLE

A sports club owner wants to survey the opinions of specific types of member. The survey needs twice as many men as women and two age ranges are required – from 35 to 45 and from 65 to 75. What type of sampling would you recommend the sports club owner use?

SOLUTION

A quota sample would be appropriate because a specific number of each type of club member is needed. We do not know the actual mix of the population of the sports club. If we did have this, then a stratified sample could be appropriate, especially if the sample needed to represent the membership in the right proportions. Whoever is doing the survey should be instructed to interview a set number of men and women (for example, 40 and 20) and a set number of each age (20 of the men to be aged 35–45 and 20 to be aged 65–75; 10 of the women to be aged 35–45 and 10 to be aged 65–75).

WORKED EXAMPLE

A school has 120 pupils in each of years 7 and 8, 150 in year 9 and 180 in each of years 10 and 11. The head teacher wants to sample the opinions of pupils on a new school uniform. What type of sample would the head teacher be advised to use?

SOLUTION

Since there are unequal numbers in each year, and because opinions may differ in each year, a stratified, quota or cluster sample could be used because the same mix of pupils is needed in the sample. If a stratified sample was used then this could be 20 pupils from each of years 7 and 8, 25 pupils from year 9 and 30 pupils from each of years 10 and 11. This would represent one-sixth of each year group. This could also be seen as a quota sample since the numbers from each group are given. A cluster sample is possible by randomly choosing one of the tutor groups in each year.

A systematic sample, chosen by selecting every sixth pupil on the school roll, is also an option.

Although a random sample is possible it would not guarantee the correct proportions of pupils in each year group.

WORKED EXAMPLE

A football club wants to sample the opinions of those supporters who buy a programme. What type of sample could the football club use?

SOLUTION

All programmes have a number and so a random sample is a possibility. A list of random numbers could be generated before the game and when that programme number is sold, that person could be asked to answer the survey. If there is not time before a game then the chosen supporters could be asked to send the survey back. To encourage them to do it, all returns could be entered in a draw. A systematic sample could also be used – the programme numbers would allow you to sample every 10th person.

There are usually several reasonable choices for a sampling method. Always explain why you have chosen a particular method unless the wording of the question suggests it is not necessary.

QUESTIONS

1   A tyre company has garages all over UK. The company has recruited a team of three interviewers for one week to sample the opinions of their customers. What type of sample would you recommend that the company use? Give the reasons for your choice.

2   A driving instructor wants to survey her customers. She knows that among her pupils she has three times as many females as males. What type of sample do you think she should use? Give the reasons for your choice.

3   A dentist is considering making changes to make the waiting room more friendly for children. He has a list of his patients and is considering using a quota sample. Is this a good choice?

4   A teacher wants to sample his class of 35 students. Describe an appropriate sampling method to use and give the reasons for your choice.

5   A sports club has the following members:

Age	Male	Female
2–12	10	8
13–15	18	24
16–21	75	95
22–32	40	30
33–52	80	60
53–	32	20

Recommend a suitable sampling technique that could be used to find the views of the members and give the reasons for your choice.

CHAPTER
3.10

CALCULATING STATISTICS

Quartiles and interquartile range

DID YOU KNOW?

The lower and upper quartiles are actually the medians of the lower and upper halves of the data. The values obtained using $\frac{1}{4}(n + 1)$ and $\frac{3}{4}(n + 1)$ are estimates of these values. When n is odd the estimate is the same as the quartiles, but that is not the case when n is even.

Not only is it useful to know the median and range of a set of data, it is also useful to know the **interquartile range**. To do this you need to find the **quartiles**.

– The **lower quartile** is the value of the item a quarter of the way through the data set.

– The **upper quartile** is the value of the item three-quarters of the way through the data set.

– The middle, or second, quartile is the value of the item halfway through the data set. It is better known as the **median**.

Interquartile range = upper quartile – lower quartile

The interquartile range is the range of the middle 50% of the data.

WORKED EXAMPLE

Fifteen students were timed eating their lunch. Here are the results (the times are in minutes).

15	14	9	12	13	4	18	7
13	12	7	15	16	12	5	

a   Work out the median time.

b   Work out the interquartile range of their times.

SOLUTION

a   The first step is to arrange the times in order.

4   5   7   7   9   12   12   12   13   13   14   15   15   16   18

The middle time is the 8th one.

4   5   7   7   9   12   12   (12)   13   13   14   15   15   16   18

So the median time is 12 minutes.

b The interquartile range is the difference between the lower and upper quartiles.

These are the medians of the lower half and the upper half of the data respectively.

4   5   7   (7)   9   12   12   1|2   13   13   14   (15)   15   16   18

The interquartile range = 15 − 7 = 8 minutes.

Percentiles

- The 50th percentile is the value of the item $\frac{50}{100}$ of the way through the data set.

  $\frac{50}{100} = \frac{1}{2}$. So this is the median.

- The 25th percentile is the value of the item $\frac{25}{100}$ of the way through the data set.

  $\frac{25}{100} = \frac{1}{4}$. So this is the lower quartile.

- The 75th percentile is the value of the item $\frac{75}{100}$ of the way through the data set.

  $\frac{75}{100} = \frac{3}{4}$. So this is the upper quartile.

- The 5th percentile is the value of the item $\frac{5}{100}$ of the way through the data set.

- The 90th percentile is the value of the item $\frac{90}{100}$ of the way through the data set.

The 5th and 95th percentiles are often used to eliminate the extremes as the middle 90% of the data falls between them. The 10th and 90th percentiles are also often used for the same reason – the middle 80% of the data falls between them.

WORKED EXAMPLE

40 plots of 1 square metre are chosen randomly from a cricket pitch. The number of daisies in each is counted. Here are the results.

7	5	7	8	6	4	4	2	8	5	5	8	3
3	2	4	9	8	6	1	4	8	5	4	6	5
7	3	3	9	1	3	6	14	1	4	6	2	1
8												

What is the range of the middle 80% of the data?

➡

SOLUTION

The first step is to organise the data; a frequency table is easier to compile and use than a list.

The middle 80% of the data is between the 10th and the 90th percentiles.

The 10th percentile is the 4th item of data = 1

The 90th percentile is the 36th item of data = 8

So the range of the middle 80% = 8 − 1 = 7.

Number of daisies	Tally	Frequency
1	IIII	4
2	III	3
3	IIII	5
4	IIII I	6
5	IIII	5
6	IIII	5
7	III	3
8	IIII I	6
9	II	2
14	I	1

Standard deviation

The standard deviation is the average (standard) distance (deviation) that the items of data in a data set are from the mean of the data.

– A small standard deviation means that most of the data lies close to the mean, that is, the data is fairly consistent.

– A larger standard deviation means that the data is more spread out, or less consistent.

To calculate the standard deviation of a set of values:

1 work out the mean
2 for each item of data, calculate the difference between its value and the mean
3 square all of these differences
4 find the mean of these squared differences
5 take the square root of the mean.

Scientific and graphical calculators have a statistical function that will work this out automatically. You will be expected to use this in the exam. Make sure you know how it works on your calculator. All calculators are slightly different.

HINT

Most calculators provide two options for **standard deviation**. One is marked with '$n-1$' the other with 'n'. They may be denoted using s or σ. You get credit for either in the exam. (Note that the difference is beyond the requirements for this qualification.)

WORKED EXAMPLE

Here are the weights of eight boys aged 18 months:

11.6 kg   12.4 kg   11.2 kg   11.7 kg   12.9 kg   10.9 kg   11.9 kg   11.8 kg

Find the mean and standard deviation of their weights.

➜

SOLUTION

Mean $= \frac{94.4}{8} = 11.8\,\text{kg}$

Mean of squared differences $= \frac{2.8}{8} = 0.35$

Taking square root $\sqrt{0.35} = 0.59$

So the standard deviation is $0.59\,\text{kg}$.

Weights (kg)	Difference from the mean	Square
11.6	−0.2	0.04
12.4	0.6	0.36
11.2	−0.6	0.36
11.7	−0.1	0.01
12.9	1.1	1.21
10.9	−0.9	0.81
11.9	0.1	0.01
11.8	0	0
Total	**94.4**		**2.8**

QUESTIONS

1 The distances, in miles, travelled by 18 students to get to college each week are shown below.

51   21   29   38   50   38   51   27

35   16   23   54   32   40   34   32

44   48

a Work out the median distance.

b What is the interquartile range of the distances?

2 The price of a certain laptop differs between retailers. Here are the prices found in each of six shops.

£356   £382   £407   £399   £345   £415

Find the mean price and standard deviation.

3 The amount, in £s, spent on lunch by 50 apprentices on their day in college each week is shown in the table.

£s spent on lunch	Frequency
0	1
1	3
2	10
3	10
4	7

5	8
6	7
7	3
8	1

a What is the interquartile range of their lunch spend?

b What is the range of the middle 80% of the lunch spend?

4 Two machines produce piston rings. A good fit is very important.

One ring is selected at random from each machine each day for a week and its diameter is measured.

Machine A   20.4 mm   20.4 mm   20.3 mm   20.2 mm   20.4 mm   20.1 mm   20.5 mm

Machine B   20.2 mm   20.3 mm   20.3 mm   20.4 mm   20.3 mm   20.3 mm   20.0 mm

a Find the mean and standard deviation of the diameters of the piston rings produced by each machine.

b Use your data from part **a** to decide which machine is more reliable. Give a reason for your answer.

DIAGRAMS FOR GROUPED DATA

CHAPTER

3.11

Histograms

Histograms are useful diagrams for visualising grouped discrete data and continuous data. Histograms look similar to bar charts, but they have one very significant difference – the **area** of a histogram represents the frequency of the data.

WORKED EXAMPLE

The table shows the waiting times for a rollercoaster at a major theme park during one day in summer.

The last class is assumed to have the same width as the previous class.

a   Draw a histogram to represent this data.

The theme park wants to improve waiting times. The target is for 90% of customers to wait for 60 minutes or less for this rollercoaster.

b   Did the theme park achieve the target on this particular summer day?

Length of wait (minutes)	Number of customers
$0 < x \leqslant 10$	60
$10 < x \leqslant 20$	84
$20 < x \leqslant 40$	56
$40 < x \leqslant 80$	30
$80 < x \leqslant 140$	24
$x > 140$	6

SOLUTION

a

Length of wait (minutes)	Number of customers	Lower class boundary	Upper class boundary	Class width	Frequency density
$0 < x \leqslant 10$	60	0	10	10	$60 \div 10 = 6$
$10 < x \leqslant 20$	84	10	20	10	$84 \div 10 = 8.4$
$20 < x \leqslant 40$	56	20	40	20	$56 \div 20 = 2.8$
$40 < x \leqslant 80$	30	40	80	40	$30 \div 40 = 0.75$
$80 < x \leqslant 140$	24	80	140	60	$24 \div 60 = 0.4$
$x > 140$	6	160	200	60	$6 \div 60 = 0.1$

➡

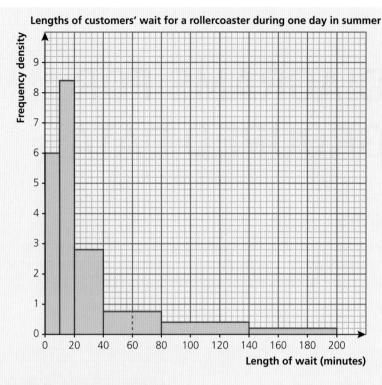

Lengths of customers' wait for a rollercoaster during one day in summer

b On this summer day, waiting times were recorded for 260 customers. The target is for 90% of customers to wait for 60 minutes or less.

90% of 260 = 234 customers.

To meet the target, 234 customers need to have waited for 60 minutes or less.

In a histogram, the area of the bars represents frequency.

The area of the bars up to and including 60 minutes is

$10 \times 6 = 60$

$10 \times 8.4 = 84$

$20 \times 2.8 = 56$

$20 \times 0.75 = 15$

In total $60 + 84 + 56 + 15 = 215$ customers waited for 60 minutes or less for this rollercoaster.

Therefore the theme park did not meet its target on this particular day.

Cumulative frequency graphs

Cumulative frequency graphs can be used to find the median, quartiles and percentiles of a data set.

Cumulative frequency gives a running total of the values in a data set.

WORKED EXAMPLE

A teacher carried out a survey on the number of hours worked each week by a group of year 13 students with part-time jobs.

a   Draw a cumulative frequency graph to present this data.

b   The teacher is concerned about how many hours the year 13 students are working at their part-time jobs. He feels that working for 12 hours or more during the week will have a negative impact on his students. Use the cumulative frequency graph to estimate how many students this teacher should be concerned about.

Time h (hours)	Number of students
$0 < h \leqslant 5$	5
$5 < h \leqslant 10$	32
$10 < h \leqslant 15$	45
$15 < h \leqslant 20$	10
$20 < h \leqslant 25$	3

SOLUTION

a

Time h (hours)	Number of students	Upper class boundary	Cumulative frequency
$0 < h \leqslant 5$	5	5	5
$5 < h \leqslant 10$	32	10	37
$10 < h \leqslant 15$	45	15	82
$15 < h \leqslant 20$	10	20	92
$20 < h \leqslant 25$	3	25	95

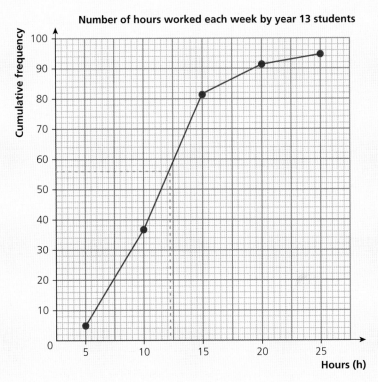

b   From the cumulative frequency graph, 12 hours gives a value of approximately 56 students.

$95 - 56 = 39$

This means that 39 students are working for 12 hours or more.

The teacher should be concerned about 39 of the students out of the 95 surveyed.

HINT

You can see how to use a cumulative frequency graph to draw a box plot in Chapter 3.12.

Box and whisker plots

Box and whisker plots (box plots) are diagrams that are plotted using the median, quartiles and highest and lowest values of a data set. These values can be found from cumulative frequency graphs or from raw data.

WORKED EXAMPLE

A survey asked a group of students how many times they checked Facebook during their lunch break.

The frequency table shows the results.

a   Find the median and quartiles.

b   Draw a box plot to illustrate this data.

Number of times	Frequency
0	4
1	6
2	8
3	5
4	1
5	1

SOLUTION

a   Total number of students = 25

For 25 values the median will be the 13th value. Therefore the median is 2.

For 25 values the lower quartile (LQ) will be the 6.5th value.

Therefore the LQ will be 1.

For 25 values the upper quartile (UQ) will be the 19.5th value.

Therefore the UQ will be 3.

b   Summary of information

Minimum value	0
LQ	1
Median	2
UQ	3
Maximum value	5

Number of times a group of students checked Facebook during their lunch break

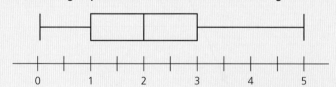

Stem-and-leaf diagrams

HINT

Numbers with more than two significant figures can also be represented in a stem-and-leaf diagram. The spread of the numbers will inform your decision about the stems and the leaves.

Stem-and-leaf diagrams are used to organise small sets of raw data. In some situations, when two sets of data are being compared, it is preferable to use a back-to-back stem-and-leaf diagram to display the data.

Steps to draw a stem-and-leaf diagram:

– Choose the stem – this is the first digit for a two-digit number.

– List the stems in ascending order vertically.

– List the leaves – this is the second digit of a two-digit number – against the appropriate stem.

– Redraw the diagram with the leaves in ascending order.

– Add a key and title.

If there are only a few stems, they can be split so that leaves 0 to 4 are on one line and then 5 to 9 are on the next line.

If there are more than about 15 stems, they can be compressed so that the leaves go from 0 to 19 on one line.

This displays the shape of the distribution more clearly.

See Chapter 2.5 page 58 for an example.

WORKED EXAMPLE

For a group of 21 people the list gives the ages at which they passed their driving test.

| 49 | 31 | 20 | 17 | 18 | 17 | 18 | 18 | 18 | 25 | 26 | 29 | 19 | 19 | 18 |
| 17 | 18 | 19 | 21 | 17 | 20 | | | | | | | | | |

a Draw a suitable stem–and–leaf diagram.

b Comment on the ages of the people when they passed their driving test.

SOLUTION

a Ages of 21 people when they passed their driving test

```
1 | 7 7 7 7 8 8 8 8 8 8 9 9 9
2 | 0 0 1 5 6 9
3 | 1
4 | 9
```

This has only a few stems so we can expand it to give:

```
1 | 7 7 7 7 8 8 8 8 8 8 9 9 9
2 | 0 0 1
2 | 5 6 9
3 | 1
3 |
4 |
4 | 9
```

Key: 1 | 7 means 17 years old

b The expanded stem-and-leaf diagram shows that most of the people pass their driving test soon after reaching the age of 17. A few wait until their twenties. The reason for so few older people could be that most people of that age have already passed or the reasons that stopped them when they were younger still apply.

QUESTIONS

1 The table shows the times taken by 45 sixth-formers to complete a charity 10 km run.

Time taken (minutes)	Number of sixth-formers
$40 < x \leqslant 45$	4
$45 < x \leqslant 50$	12
$50 < x \leqslant 60$	19
$60 < x \leqslant 80$	10

Draw a histogram to represent this data.

2 The table gives estimates for the age distribution of the UK population in June 2015.

Age (years)	Number of people (thousands)	Lower class boundary	Upper class boundary	Class width	Frequency density (thousands per year)
0–5	4835	0	6	6	4835 ÷ 6 = 806
6–20	11327	6	21	15	11327 ÷ 15 = 755
21–35	13191	21	36	15	
36–56	18341				
57–72	11193				
73–88	5521				
89+	386				

a Copy and complete the table.

b Draw a histogram to illustrate the age distribution.

HINT

Frequency density = frequency ÷ class width

3 The table shows the distances that a group of football fans travel to a match.

Distance travelled (km)	Number of football fans
$0 < x \leqslant 5$	40
$5 < x \leqslant 10$	70
$10 < x \leqslant 20$	240
$20 < x \leqslant 30$	160
$30 < x \leqslant 50$	120

Here is a histogram to illustrate this data.

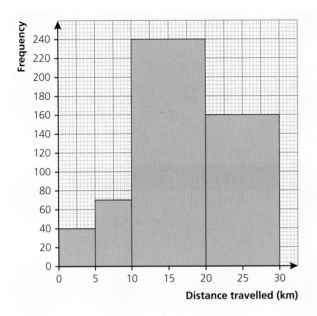

a Give reasons why this histogram is incorrect.

b Use the table to draw a correct histogram.

HINT

The area of the bar represents frequency.

4 David drives to work. The table gives the times (*t*) in minutes that it takes David to drive to work over a 100-day period.

Time taken, *t* (minutes)	Number of days
$0 < t \leqslant 10$	17
$10 < t \leqslant 20$	32
$20 < t \leqslant 30$	34
$30 < t \leqslant 40$	13
$40 < t \leqslant 50$	4

Draw a cumulative frequency graph to represent this information.

5 The table below shows the results of a survey about the ages of Instagram users in the UK.

Ages of Instagram users in the UK (years)	Percentage of the people surveyed
$0 < x \leqslant 10$	5
$10 < x \leqslant 15$	17
$15 < x \leqslant 20$	39
$20 < x \leqslant 35$	23
$35 < x \leqslant 50$	15
$50 < x \leqslant 90$	1

Draw a cumulative percentage graph to represent the results of this survey.

HINT

Cumulative frequency is always plotted against the *upper* class boundary.

6 The data below shows the time in minutes that it took a group of ten Geography students to complete an essay:

17 17 18 19 25 27 28 30 32 33

Draw a box plot to represent this information.

7 Alice organised a raffle to raise money for charity. She sold 30 tickets in total.

Number of tickets purchased	Frequency
1	5
2	6
3	8
4	0
5	10
6	1

Draw a box plot to represent this information.

8 The data below gives the reaction times in milliseconds of ten distance runners and ten hurdlers.

Distance runners	219	220	224	234	242	245	247	248	256	261
Hurdlers	215	218	222	227	230	232	239	240	241	241

a Draw a back-to-back stem-and-leaf diagram to represent this information.

b Compare the distance runners and the hurdlers.

HINT

Cross off the values as you add them to the stem-and-leaf diagram.

9 Oscar carried out a survey of the heights of trees around his college. The data below shows the heights, in metres, of 15 trees that he recorded.

4.3 5.4 11 4.8 10.6 7.3 8 9.5 4.9
7.8 3 9.5 4.6 5.4 7.2

a Draw a stem-and-leaf diagram to represent Oscar's findings.

HINT

For larger sets of data, re-order the stem-and-leaf diagram after you have listed every value.

b Calculate the median height of the trees. Comment on how representative it is of the data.

HINT

When finding the median of a back-to-back stem-and-leaf diagram, take care to count to your median in the correct order.

ESTIMATING STATISTICS FROM GRAPHS

Box plots

Box and whisker diagrams (box plots) make it easy to compare data and to identify the range of the data in relation to the range of the middle half of the data. It is a useful way to compare data sets.

Because box plots are constructed using the least and greatest values, the lower and upper quartiles and the median, these values can easily be read from the graph and associated statistics such as range and interquartile range can be calculated.

WORKED EXAMPLE

Here is a diagram showing box plots of samples of male and female heights.

The statistics are shown by the boxes and their whiskers.

Compare the male and female heights.

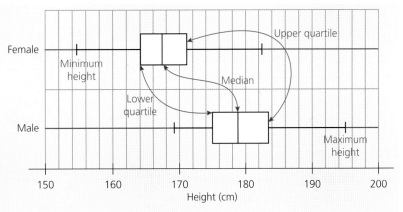

SOLUTION

We can see that the male heights are in general greater than the female heights as the median for the males is greater than that for the females. This is the same for each of the quartiles and extreme values. There is quite a bit of overlap, however, as the position of the male upper quartile shows that about three-quarters of the males are shorter than the tallest female.

The median of the female heights shows that about half the females are shorter than the shortest male.

We can also see that the range of the male heights is 195 cm − 169 cm = 26 cm, and that the range of the female heights is 181 cm − 155 cm = 26 cm, so the ranges are the same, which tells us that the variation in height is the same for males and females.

Cumulative frequency graphs

It is also straightforward to read range, median, upper and lower quartiles from cumulative frequency graphs. You locate the appropriate position on the vertical (cumulative frequency) axis and find the corresponding points on the horizontal axis.

You need to estimate the least and greatest values of the data from the graph in order to work out an estimate of the range. The estimate of the range is the difference between the most extreme values on the graph.

DID YOU KNOW?

Readings from a cumulative frequency graph can only be estimates of the quartiles and so we use the points a quarter and three-quarters of the way through the data rather than the approach we use with raw data (finding the medians of the lower and upper halves of the data).

WORKED EXAMPLE

The weights of 100 18-month-old girls were recorded and the results presented in a cumulative frequency graph.

a   Find the median and the interquartile range.

b   The median and the interquartile range for 100 18-month-old boys were found to be 11.7 kg and 1.7 kg. Compare the weights of boys and girls at 18 months old.

SOLUTION

a   Reading from the cumulative frequency diagram:

Median = 11 kg

Interquartile range
= upper quartile − lower quartile
= 11.7 − 10.5
= 1.2 kg

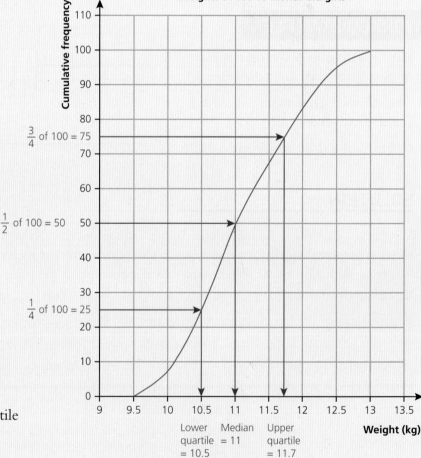

Weights of 100 18-month-old girls

$\frac{3}{4}$ of 100 = 75

$\frac{1}{2}$ of 100 = 50

$\frac{1}{4}$ of 100 = 25

Lower quartile = 10.5   Median = 11   Upper quartile = 11.7

Weight (kg)

b

	Median	Interquartile range
Boys	11.7 kg	1.7 kg
Girls	11 kg	1.2 kg

At 18 months old, boys weigh about 0.7 kg more than girls. There is also more variation in their weights. The weights of the middle 50% of boys vary by about 0.5 kg more than those for the girls.

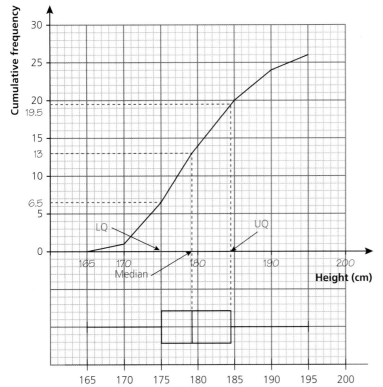

This diagram shows the cumulative frequency graph and the associated box plot for a sample of male heights.

There are 26 males in this sample, so to find the median look for 13 (half of 26) on the cumulative frequency scale, go across to meet the graph, then go down to the height scale and read off the median, 179 cm.

Similarly, the lower and upper quartiles are read off by following across from 6.5 (a quarter of 26) and 19.5 (three-quarters of 26).

Look at the start of the graph and you can see that the group with the fewest members is 165 to 170. The least data value could be anywhere in this group.

For consistency we choose the lower bound of the group, so 165 is the estimate for the least value.

Looking at the end of the graph you can see that the group with the most members is 190 to 195. The greatest value could be anywhere in this group, and again we choose the most extreme value, which is the upper bound of the group, 195.

Drawing the box plot beneath the cumulative frequency graph uses the statistical measures without writing them down explicitly.

The data for the cumulative frequency graph is grouped, so some detail is lost, which is why the upper quartile (UQ) doesn't quite equal the UQ on the box plot, which is plotted using the raw data.

If there is enough data though, the statistics read off the cumulative frequency graph will be good approximations of the exact values.

Cumulative frequency graphs can also be used to find percentiles. Just as the median and the quartiles divide the data into quarters, percentiles divide it into hundredths.

WORKED EXAMPLE

This is a cumulative frequency graph of the body mass index (BMI) of a sample of 200 males in 1993.

Estimate the 50th percentile.

Estimate the 20th percentile.

SOLUTION

To find the 50th percentile, read across from 100 (50% of 200) on the cumulative frequency axis and down to the BMI axis. The 50th percentile is 19.3.

The 20th percentile is 18.1.

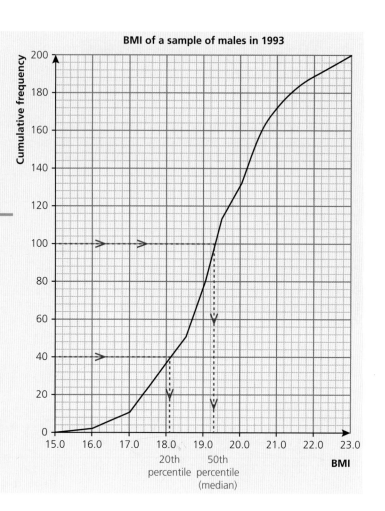

BMI of a sample of males in 1993

DID YOU KNOW?

The 50th percentile is the same as the median.

WORKED EXAMPLE

The weights of 100 18-month-old girls were recorded and the results presented in a cumulative frequency graph, as shown on page 346.

a   Find the 5th and 95th percentiles.

b   The 5th and 95th percentiles for 100 18-month-old boys were found to be 9.8 kg and 14 kg. Compare the weights of boys and girls at 18 months old.

SOLUTION

a   From the cumulative frequency graph on page 346,
    5th percentile = 9.9 kg
    95th percentile = 12.4 kg

b

	5th percentile	95th percentile
Boys	9.8 kg	14 kg
Girls	9.9 kg	12.4 kg

The 5th percentiles are very similar but the 95th percentile is about 1.6 kg heavier for 18-month-old boys. That means nearly all 18-month-olds weigh more than 9.8 kg.
The heaviest 5% of girls weigh over 12.4 kg while the heaviest 5% of boys weigh more than 14 kg.
The lightest boys and girls are similar in weight but at the top end, boys are noticeably heavier.

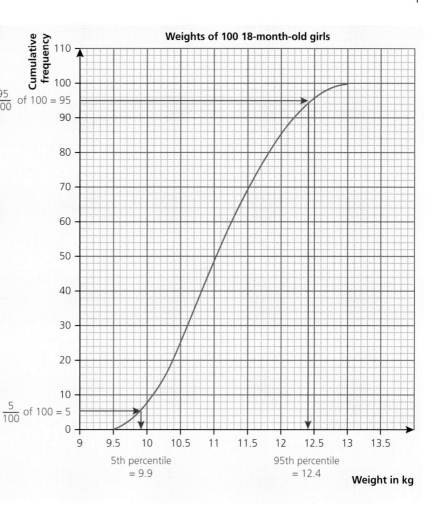

Stem-and-leaf diagrams

These are good for representing small amounts of data in an organised way without losing any of the detail.

The back-to-back format helps you to compare the distribution of male and female heights.

WORKED EXAMPLE

This diagram shows the heights, in centimetres, of a sample of 26 males and 26 females.

Key: 15 | 7 represents 157 cm

Male		Female
	15	5 7 7 8
9	16	2 3 4 4 4 5 6 6 7 7 7 8 9
9 9 9 8 8 8 7 **5** 4 3 3 3 2	17	0 0 1 2 3 4 6 6
9 8 6 5 **4** 4 2 2 1 0	18	1
5 0	19	

a   What is the modal class for the males?

b   What is the median for the males?

c   What is the interquartile range of the male height data?

SOLUTION

a   For the males, the modal class is 170–180 cm.

b   The data is arranged in order, so to find the median you find the height of the middle person – or, as 26 is an even number, the mean of the heights of the 13th and 14th people.

   For the males, the 13th and 14th numbers are both 9 (shown in green), so the median male height is 179 cm.

c   The median of the first 13 numbers (the lower half of the data) is the 7th number, so the lower quartile for the males is 175 cm (see the red number 5).

   The median for the upper half of the numbers is the 20th number, so the upper quartile for the males is 184 cm, also in red.

   So the interquartile range = 184 cm – 175 cm = 9 cm.

QUESTIONS

1   The downhill skiing times of 50 competitors in a race were recorded and the data plotted in a cumulative frequency diagram.

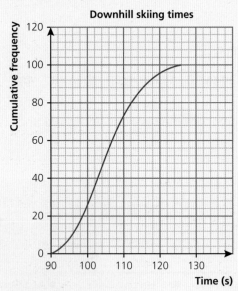

Use the graph to find:

a   the median

b   the interquartile range

c   the 10th percentile

d   the 90th percentile.

e   One competitor's result was missing.

   i   A spectator says that this result is more likely to be within the upper and lower quartiles than outside them. What reply would you give?

   ii   What is the probability that the result would be within the 10th and 90th percentiles?

2   Aisha has developed a new species of rose. She is recording the height of 200 plants every week. Here are her records for one particular week.

Height (x cm)	Frequency
$18 < x \leqslant 20$	26
$20 < x \leqslant 22$	46
$22 < x \leqslant 24$	71
$24 < x \leqslant 26$	33
$26 < x \leqslant 28$	20
$28 < x \leqslant 30$	4
	200

a Copy the table and add a cumulative frequency column to it.

b Draw a cumulative frequency diagram.

c Use your diagram to find the median and interquartile range.

d Use your diagram to find the 5th and 95th percentiles.

e Another type of rose had a median height of 24 cm and an interquartile range of 6 cm for the same age of plant. Compare the two types of plant at this age.

3 These two box plots show the results of a test taken by a group of students before and after being taught the material tested.

The test was marked out of 50.

Before

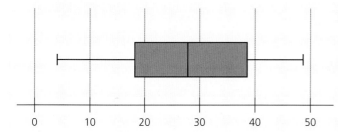

After

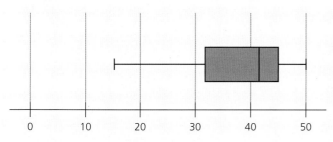

Compare the performance of the students before and after they were taught the material.

4 a What is the 75th percentile of the BMI data on page 348?

b What percentile is a BMI of 21?

c Draw a box plot of the BMI data.

5 a What is the mode for the female heights, as shown in the stem-and-leaf diagram on page 349?

b Use the stem-and-leaf diagram of male and female heights to find the interquartile ranges of the two sets of data.

What does this tell you about the distribution of the height?

c The heights of the 26 males in the stem-and-leaf diagram are also shown in the cumulative frequency diagram on page 347.

What are the advantages and disadvantages of each form of presentation?

6 The stem-and-leaf diagram shows the (systolic) blood pressure of a group of patients before and after a drug was given.

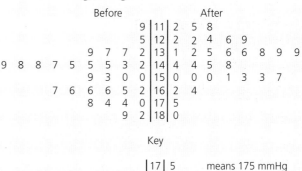

a Calculate the median blood pressure before and after the drug was given.

b Find the range of blood pressures before and after the drug was given.

c Describe the impact of the drug on the blood pressures of the patients. Does this data support the company's claim that their drug reduces blood pressure?

FORMULAE AND CALCULATIONS

Substituting into formulae is straightforward as you simply replace each letter by a numerical value. You now have a calculation to do.

Doing the calculation follows the conventions for the priority of operations.

There is a natural hierarchy that gives rise to the conventions.

Addition and subtraction are inverse operations so are equal. Multiplication/division are repeated addition/subtraction and so have priority over addition and subtraction.

To change this order, you insert brackets as the convention is that calculations inside brackets are to be done first.

Simple formulae

WORKED EXAMPLE

As you know, to find the perimeter of a two-dimensional shape you have to add up the lengths of all its sides. So, if a rectangle has length l units, width w units and the perimeter is P units, we can say that

$P = l + w + l + w$

This can also be correctly expressed as $P = 2l + 2w$ and $P = 2(l + w)$

If $l = 6.2$ and $w = 1.7$, calculate the perimeter of the rectangle in all three ways.

SOLUTION

$P = l + w + l + w$

$\quad = 6.2 + 1.7 + 6.2 + 1.7$

$\quad = 15.8$ units

$P = 2l + 2w$

$\quad = 2 \times 6.2 + 2 \times 1.7$

$\quad = 12.4 + 3.4$

$\quad = 15.8$ units

HINT

All of the operations are addition and so we calculate them from left to right.

HINT

Multiplication ... one before addition.

→

$P = 2(l + w)$

$\quad = 2 \times (6.2 + 1.7)$

$\quad = 2 \times 7.9$

$\quad = 15.8$ units

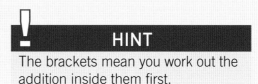

HINT

The brackets mean you work out the addition inside them first.

Notice the similarities and differences between the last three formulae. They show you some ways you can manipulate or simplify terms and expressions in algebra.

You could use a calculator to do the calculation. Scientific and graphical calculators are programmed to follow the correct order for the priority of operations in a calculation.

Division can be shown in several different ways in a calculation, as shown below.

The perimeter formula can be rearranged so that, if you know the perimeter and the width of a rectangle, you can calculate the length.

To find the length, we can use this formula, which can be written in several ways

$$l = \frac{P}{2} - w \qquad l = \frac{1}{2}P - w \qquad l = P \div 2 - w \qquad l = \frac{P - 2w}{2}$$

WORKED EXAMPLE

Use all four versions of the formula for the length of the rectangle to calculate the length when the perimeter is 7.5 units and the width is 1.25 units.

SOLUTION

$l = \frac{P}{2} - w$

$\quad = \frac{7.5}{2} - 1.25 = 3.75 - 1.25$

$\quad = 2.5$

HINT

Division is carried out before subtraction.

$l = \frac{1}{2}P - w$

$\quad = \frac{1}{2} \times 7.5 - 1.25 = 3.75 - 1.25$

$\quad = 2.5$

HINT

Multiplication by $\frac{1}{2}$ is carried out before subtraction.

$l = P \div 2 - w$

$\quad = 7.5 \div 2 - 1.25 = 3.75 - 1.25$

$\quad = 2.5$

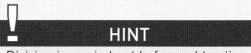

HINT

Division is carried out before subtraction.

$$l = \frac{P - 2w}{2}$$

$$= \frac{7.5 - 2 \times 1.25}{2} = \frac{7.5 - 2.5}{2}$$

$$= \frac{5}{2} = 2.5$$

HINT

The numerator is worked out first as the fraction line acts like a pair of brackets.

The multiplication is carried out before the subtraction.

Finally, the division is worked out.

Powers and roots

Powers are a way of recording repeated multiplication (or division) and so have priority over multiplication and division. Roots are also powers (fractional ones) and so have the same priority as powers. Again, brackets can be used to change the conventional priority.

WORKED EXAMPLE

A formula to find the total interest (I) on £2000 invested for n years at 1.6% compound interest is $I = 2000(1.016^n - 1)$.

Use the formula to show that the interest after three years is £97.54.

SOLUTION

$I = 2000 \times (1.016^3 - 1)$ ← Work out the brackets first.

$= 2000 \times (1.048772096\ldots - 1)$ ← The power is worked out before the subtraction.

$= 2000 \times 0.048772096\ldots$

$= £97.54$

The order that you use to work out a calculation informs the order that you deal with the terms when you solve equations. It is usually the reverse order, so the last part of the possible calculation is dealt with first.

WORKED EXAMPLE

We receive £150 interest after three years.

What is the interest rate?

SOLUTION

In this case, if the interest rate is $r\%$

$$150 = 2000\left(\left[\frac{100 + r}{100}\right]^3 - 1\right)$$ ← The 2000 is the only part outside the brackets so it would be calculated last. Here we divide both sides by 2000 as a first step.

$$\frac{150}{2000} = \left[\frac{100 + r}{100}\right]^3 - 1$$ ← The −1 is the final part as it comes after the power, so the next step is to add 1 to both sides.

$$\left[\frac{100+r}{100}\right]^3 = \frac{150}{2000}+1$$ ← The cubing is the last thing to calculate as it is outside the brackets so the next step is to cube root both sides.

Now we need to find the cube root of both sides of the equation.

$$\frac{100+r}{100} = \sqrt[3]{\frac{150}{2000}+1}$$ ← We calculate the answer to the cube root of the right-hand side.

$$\frac{100+r}{100} \approx 1.0244$$ ← The numerator is enclosed by brackets as shown by the fraction line so the division by 100 is the final step of the calculation so we now multiply both sides by 100.

$$100 + r \approx 102.44$$

$$r \approx 2.44$$

So the interest rate is 2.44%

Reciprocals

The reciprocal of a number is 1 divided by the number, so the reciprocal of 4 is $\frac{1}{4}$ or 0.25. Your calculator has a button for it, labelled $1/x$ or x^{-1}.

When dividing by something, you can multiply by its reciprocal; to divide by 2 you multiply by $\frac{1}{2}$, and to divide by $\frac{3}{4}$ you multiply by $\frac{4}{3}$.

So $\frac{5}{8} \div \frac{3}{4} = \frac{5}{8} \times \frac{4}{3} = \frac{20}{24} = \frac{5}{6}$

QUESTIONS

1 A formula to convert temperature in degrees Fahrenheit to Celsius is

$C = \frac{5}{9}(F-32)$, where C and F are temperatures in degrees Celsius and Fahrenheit respectively.

Work out the Celsius equivalent of 98 degrees Fahrenheit.

2 The equation of a straight line is given by $y = 5 - \frac{3x}{4}$.

Work out the value of y when $x = 10$.

3 Pete is paid travel expenses for travelling to meetings. The first 50 miles are paid at a rate of 40p per mile and the rest at 25p per mile. Which of these formulae will correctly calculate the expenses (E) for a journey of n miles, where n is greater than 50?

$E = (n-50) \times 0.25 + 50 \times 0.4$

$E = 0.25(n+3)$

$E = 0.25n - 50 + 20$

$E = \frac{n}{4} + 25 \times 0.3$

$E = \frac{n}{4} - \frac{50}{4} + \frac{50 \times 4}{10}$

4 The period of oscillation (T) of a simple pendulum is given by

$$T = 2\pi\sqrt{\frac{L}{g}}$$

where L is the length of the pendulum in metres and g is the acceleration due to gravity.

Work out the period of oscillation of a pendulum length 1.2 m with $g = 9.8\,\text{ms}^{-2}$.

5 The volume of a sphere is given by the formula $V = \frac{4}{3}\pi r^3$.

Work out the volume of a sphere with radius 4.3 cm.

6 Find the reciprocal of each of the following numbers

10, 1.5, $\frac{1}{5}$, 0.01.

7 Calculate five-ninths divided by two-thirds.

CHAPTER

> 3.14

APPROXIMATION

When something is measured to the nearest whole number of units, then the error can be ± half the unit of measurement.

Calculations with money are often **rounded** (or **truncated**) to the nearest penny at each stage to give an amount in pounds and pence.

Truncation means the amount is rounded down, even if you would expect it to be rounded up because it is nearer the higher amount. So £51.897 becomes £51.89.

This may often have little or no effect on the final result. Sometimes, however, when there are many stages to the process, or when the amounts are large or repeated for many accounts, the differences build up.

WORKED EXAMPLE

£50 000 is invested at 9% annual interest, compounded monthly, for one year. What is the effect of truncation on the amount at the end of the year?

SOLUTION

9% annual interest is equivalent to 9 ÷ 12 = 0.75% monthly interest.

The amount with no truncation = £50 000 × 1.0075^{12}

$\qquad\qquad\qquad\qquad\quad$ = £54 690.34

The amount with truncation is given by this table, which shows the amount at the end of each month after truncation. The truncated amount is then used to calculate the next month's amount.

This gives a final amount of £54 690.29 which is 5p less than the other method.

Months	Amount in £ when truncated
1	50 375.00
2	50 752.81
3	51 133.45
4	51 516.95
5	51 903.32
6	52 292.59
7	52 684.78
8	53 079.91
9	53 478.00
10	53 879.08
11	54 283.17
12	54 690.29

Amounts are often quoted to 1 or 2 significant figures when giving information about yields from investments.

This may mean there is a large range of possible outcomes.

WORKED EXAMPLE

£12 000 invested at 3% interest for (exactly) 5 years will be worth £15 000.

Comment on this claim.

SOLUTION

£12 000 may only be accurate to 2 significant figures so it is between £11 500 and £12 500.

3% is between 2.5% and 3.5%

The minimum yield $= 11\,500 \times 1.025^5$

$\qquad\qquad\quad = £13\,011.19$

The maximum yield $= 12\,500 \times 1.035^5$

$\qquad\qquad\quad = £14\,846.08$

This rounds to £15 000, so the claim is not entirely untrue but it is misleading, with a more likely yield of £14 000 (and it could be as low as £13 000).

WORKED EXAMPLE

An endowment policy to repay a mortgage quotes a likely return on the investment of between 2% and 5% for the remaining 10 years of the mortgage. The current value of the investment is £43 654.

Will it be enough to repay the mortgage of £65 000?

SOLUTION

Lowest amount $\quad = £43\,654 \times 1.02^{10}$

$\qquad\qquad\qquad = £53\,213.98$

Highest amount $\quad = £43\,654 \times 1.05^{10}$

$\qquad\qquad\qquad = £71\,107.77$

£65 000 is nearer the upper boundary than the lower one so it is more likely that it will not be enough to repay the mortgage.

QUESTIONS

1 £60 000 is invested at an annual interest rate of 5%.

What is the effect of truncation after 6 years if the interest is compounded annually?

2 £1000 is invested for 5 years at an annual rate of interest of 12%.

The amount is rounded at the end of each year.

What is the effect of rounding on the final amount?

3 A bank operates an account where £10 000 is invested for 10 years at an annual interest rate of 6%.

What is the effect of truncation on the final amount?

There are 50 000 such investments. How much money could the bank make by truncating the results of the calculations?

4 £12 000 is invested at an annual rate of interest of 8.4%. The interest is compounded quarterly.

What effect does rounding have after 2 years?

5 The projected value of an investment of £40 000 at 5% per annum is quoted as £65 000 after 10 years.

Comment on this prediction.

6 The amount currently invested to repay an endowment mortgage is £15 000.

The projected annual interest rate is estimated as being between 4% and 6%.

How likely is it that this will be enough to repay a mortgage of £35 000 in 10 years' time?

7 Sam decides to invest £160 000 he has inherited by buying a house. The current rate of increase in house prices is 4% per annum.

What is the range of values of the house (given that he spends the whole amount on it) in 5 years' time?

 HINT

2.1% each quarter

PERCENTAGES

Percentage means 'out of' (per) 100 (cent).

$10\% = 10$ out of $100 = \frac{10}{100} = \frac{1}{10} = 0.1$

$34\% = 34$ out of $100 = \frac{34}{100} = 0.34$

26 out of $40 = \frac{26}{40} = 0.65 = \frac{65}{100} = 65\%$

Percentage change

If something was worth £30 last year and it is now worth £45 you can say:

> The price has gone up by £15. That is an increase of £15 over the original £30.

so $\frac{15}{30} = \frac{1}{2} = \frac{50}{100} = 50\%$

percentage increase $= \frac{\text{actual increase}}{\text{original amount}} \times 100 = \frac{15}{30} \times 100 = 50\%$

Similarly,

percentage decrease $= \frac{\text{actual decrease}}{\text{original amount}} \times 100$

> The price is now $\frac{45}{30} = \frac{3}{2} = \frac{150}{100} = 150\%$ of what it was last year.

In this case $\frac{150}{100} = \frac{3}{2} = 1.5$ is called the **multiplier**.

It's what you multiply the old cost by to get the new cost.

That means $30 \times 1.5 = 45$

multiplier $= \frac{\text{new amount}}{\text{old amount}}$

Multipliers allow you to work out a percentage increase or decrease in just one step on your calculator.

WORKED EXAMPLE

Last year Subhash paid £350 for a computer tablet. This year he could have bought the same one for £295. What percentage has the price reduced by?

SOLUTION

$$\text{percentage decrease} = \frac{\text{actual decrease}}{\text{original amount}}$$

$$= \frac{55}{350}$$

$$= 0.157 \quad \longleftarrow \boxed{\times 100 \text{ to convert it to a percentage}}$$

$$= 15.7\,\%$$

The price of the tablet has gone down by 15.7%

Alternatively, you can say the new price is $\frac{295}{350} = 0.843 = \frac{84.3}{100} = 84.3\%$ of the old price.

Notice: 84.3% = 100% − 15.7%

Comparing two quantities using percentages

Sometimes a saving or price increase can seem quite a large amount. Using percentages to compare the prices allows you to evaluate them better.

WORKED EXAMPLE

Subhash paid £350 for a computer tablet. He then saw it on the internet for £295. How much would he have saved if he had bought it on the internet? What percentage of the price he paid is this?

SOLUTION

The working for this problem is identical to the question above.

The percentage he would have saved was 15.7%.

Increasing/decreasing by a percentage

Either

amount of increase = percentage increase × old amount

new amount = old amount + amount of increase

Or

new amount = multiplier × old amount

WORKED EXAMPLE

In 2015, Simon had 150 followers on Twitter. In 2016 his number of followers had increased by 120%. How many followers did he have in 2016?

SOLUTION

Increase amount = percentage increase or decrease × old amount

$$= 120\% \times 150$$

$$= \frac{120}{100} \times 150$$

$$= 1.2 \times 150$$

$$= 180$$

New amount = old amount + increase amount

New amount = 150 + 180

$$= 330$$

Simon has 330 followers in 2016.

The increase in the example above is greater than 100%, which means that the new amount is more than double the old amount.

When an amount has been increased, the multiplier is 100% + the percentage increase.

When an amount has been decreased, the multiplier is 100% – the percentage decrease.

So the multiplier for the question above would be 100% + 120% = 220%

$= \frac{220}{100} = 2.2$.

You can check your answer with $150 \times 2.2 = 330$.

WORKED EXAMPLE

A car depreciates in value by 27% during the first year. Its price when new was £15 650. What is it worth after one year?

SOLUTION

Using the multiplier method,

multiplier = 100% – the percentage decrease

$$= 100\% - 27\% = 73\%$$

$$= \frac{73}{100} = 0.73$$

New amount = multiplier × old amount

$$= 0.73 \times £15\,650$$

$$= \frac{73}{100} \times 15\,650$$

$$= 11\,424.5$$

The car is worth £11 424.50 after one year.

Repeated percentage increase/decrease

Using a multiplier allows these calculations to be done in one step on a calculator. See Chapter 3.16 for more examples.

new amount = (multiplier)^n × old amount

where n = number of repeats

WORKED EXAMPLE

A car depreciates in value by 27% each year. Its price when new was £15 650. What is it worth after three years?

SOLUTION

Multiplier = 100% − 27% = 73% = 0.73

New amount = $(0.73)^3 \times 15\,650$

$$= 6088 \text{ (to the nearest whole number)}$$

The car is worth £6088 to the nearest pound.

Original value problems

These are problems where you know the final amount after an increase and need to know the original amount before the increase.

We have already seen that

new amount = multiplier × old amount

Rearranging,

old amount = $\dfrac{\text{new amount}}{\text{multiplier}}$

So to find the original amount **divide** the new amount by the multiplier.

WORKED EXAMPLE

A computer game costs £54.60 including VAT at 20%. What was the price before VAT?

SOLUTION

The original cost was increased by 20%.

So the multiplier = 100% + the percentage increase = 100% + 20% = 120% = 1.2

Old amount = $\frac{\text{new amount}}{\text{multiplier}}$

$= \frac{54.6}{1.2}$

$= 45.5$

The game cost £45.50 before VAT was added.

QUESTIONS

1   Raphael earns £1100 a month. He spends £450 on rent, £85 on services and £110 on council tax. Work out the percentage of his earnings he spends on rent, services and council tax.

2   Gregor had 20 tickets to a concert. He gave away 6 tickets. Calculate the percentage of his tickets he gave away.

3   A moped depreciates by 23% every year. A moped was bought for £1995. Calculate its value after 2 years.

4   Gemina and Gita are arguing about percentages. Gemina says percentages cannot be bigger than 100%. Is Gemina correct?

Give an example where a percentage cannot be more than 100%.

Give an example where a percentage can be more than 100%.

5   Each year a house increases in price by 3%. In 2016 it was worth £200 000.

   a   What is it worth after one year?

   b   What is it worth after three years?

6   a   A dining room table costs £57 + VAT at 20%. How much does Sam have to pay for the table?

   b   A different table costs £65 including VAT. How much did this table cost before VAT?

7   A pair of shoes were originally £49. In the sale they cost £35. What percentage had they been reduced by?

8   Simon sold a bike for £564. It included VAT at 20%. How much was the VAT?

9   A bank account pays interest at 0.4% per year. Su-ling invests £500 in the account for 4 years. Each year the interest is paid into the account. How much does she have after 4 years?

INTEREST

If you have some money in some sort of savings account, you can expect to be paid interest as a reward for making your money available for the bank or building society to use.

Interest may be paid to the saver ('paid away') rather than being added on to the savings in the account. This is called **simple interest**. If the interest is added to the savings, increasing them so that more interest is paid next year, this is called **compound interest**.

Simple interest

WORKED EXAMPLE

You have £500 invested at an annual simple interest rate of 1.2%.

What is the interest after one year?

What is the interest after eight years?

SOLUTION

$$\text{Interest after one year} = £500 \times \frac{1.2}{100}$$
$$= £500 \times 0.012$$
$$= £6$$

If you keep the money in the account for eight years at the same rate of interest, the total amount of interest you get is $8 \times £6 = £48$

This can be done in one step:

Total interest $= £500 \times 0.012 \times 8 = £48$

Compound interest

With coumpound interest the interest is added to the money in the account instead of being 'paid away'.

WORKED EXAMPLE

You have £500 invested at an annual compound interest rate of 1.2%.

What is the interest after one year?

What is the interest after eight years?

SOLUTION

At the end of the first year

$$\text{amount in the account} = £500 \times \frac{101.2}{100}$$
$$= £500 \times 1.012$$
$$= £506$$

So the interest earned after one year is £6, as with simple interest.

To find the amount in the account at the end of the second year, multiply by 1.012 again. Repeat the multiplication by 1.012 for each additional year.

So after eight years, the amount in the account after eight years is £500 multiplied by 1.012 eight times. Amount in account after eight years = £500 × 1.012^8 = £550.07 to the nearest penny

So the interest earned on the account after eight years is £550.07 − £500 = £50.07, which is a bit more than was earned in simple interest.

Compound interest is the standard method used for saving and borrowing.

Interest is often calculated and compounded once a year, but it can be done at more or less frequent intervals.

WORKED EXAMPLE

£350 is put into a savings account with an annual interest rate of 1.8%.

Interest is compounded quarterly.

How much will be in the account at the end of the year?

SOLUTION

An interest rate of 1.8% per year is 0.45% interest per quarter.

At the end of the year, the amount will be £350 ×1.0045^4 ≈ £356.34, giving you £6.34 in interest.

If the interest were compounded only at the end of the year, the amount would be £350 ×1.018 ≈ £356.30.

So you get an extra four pence with the quarterly compounding!

The quarterly compounded interest is $\frac{6.34}{350} \times 100\% \approx 1.81\%$ of the original £350, which is very slightly higher than the 1.8% quoted.

This higher rate is called the **AER (Annual Equivalent Rate)**. All savings accounts are required to quote the AER so that potential customers can compare like with like.

The AER can be worked out as shown above, but you may prefer to use a formula:

$$r = \left(1 + \frac{i}{n}\right)^n - 1$$

where r is the AER, i is the annual rate (both expressed as decimals) and n is the number of times per year the interest is compounded.

The corresponding term for borrowing is **APR (Annual Percentage Rate)**. Again, those offering loans are required to quote the APR. APRs on short-term borrowing, such as pay day loans, can be astonishingly high (see practice question 7).

APR

WORKED EXAMPLE

A credit card company charges 3% per month interest on money outstanding on your credit card bill.

If you owe £250 and make no further purchases or repayments, what is the amount owing at the end of the year?

SOLUTION

Amount owing at the end of the year is $£250 \times 1.03^{12} \approx £356.44$

So the APR is $\frac{356.44 - 250}{250} \times 100\% \approx 42.6\%$

This example presents a simplification of the usual situation, when people normally pay back a loan over a period of time, reducing the debt throughout the loan period.

To work with such situations, it is helpful to use the following formula.

The annual percentage interest rate (APR) is given by

$$C = \sum_{k=1}^{m} \left(\frac{A_k}{(i+i)^{t_k}}\right)$$

where $£C$ is the amount of the loan, m is the number of repayments, i is the APR expressed as a decimal, $£A_k$ is the amount of the kth repayment, t_k is the interval in years between the start of the loan and the kth repayment.

It may be assumed that there are no arrangement or exit fees.

Source: AQA specification

This looks quite daunting, but the formula becomes more manageable if we consider a loan paid back by means of just a few instalments.

WORKED EXAMPLE

A loan of £1000 is taken out at an APR of 6%, to be paid back in two equal annual instalments.

Calculate the size of the equal payments.

SOLUTION

Using the formula above,

$$1000 = \frac{A}{1.06} + \frac{A}{1.06^2}$$
$$= A\left(\frac{1}{1.06} + \frac{1}{1.1236}\right)$$
$$\approx 1.83339A$$
$$\therefore A \approx \frac{1000}{1.83339} \approx 545.44$$

Each payment is £545.44.

QUESTIONS

1 Calculate how much simple interest you will earn over five years on a sum of £75 put into an account paying 0.8% interest annually.

2 Set up a spreadsheet to show the simple interest on £75 invested for five years at different annual rates of interest.

3 Work out the compound interest earned on £350 invested for five years at an annual rate of 2.1%.

4 If £2000 is invested in an account with an annual interest rate of 4.3% (compounded annually), how long will it take for the amount in the account to double?

5 Calculate the AER when £1000 is put in a savings account at an annual rate of 3.6% compounded every month.

6 Calculate the APR if a credit card company charges:

  a 2.9% interest per month

  b 0.1% interest per day.

7 A pay day loan has a daily interest rate of 0.8%. Find the annual interest rate if:

  a the interest is not compounded

  b the interest is compounded daily.

8 A loan of £3000 at an APR of 9% is to be paid back in three equal monthly instalments. How much is paid back each month?

CHAPTER

> **3.17**

GRAPHS WITH A FINANCIAL CONTEXT

Here are three phone tariffs:

Tariff 1
SIM only
£7.50 a month
5000 texts
500 MB of data
250 minutes then
25p per minute.

Tariff 2
Free phone
£10 a month
Unlimited texts
500 MB of data
500 minutes then
35p per minute.

Tariff 3
Free phone
£25 a month
Unlimited texts
2 GB of data
500 minutes then
30p per minute.

David is upgrading his phone before starting university. Which tariff should he select?

Use the graph below to support your argument.

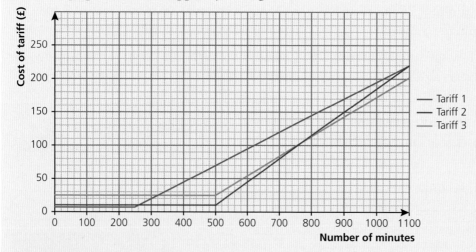

SOLUTION

Tariff 1 is the cheapest up to 250 minutes.

After 250 minutes, tariff 2 is the cheapest up to about 800 minutes, at which point both tariff 2 and tariff 3 have the same cost.

After 800 minutes tariff 3 is the cheapest.

Tariff 1 will suit David the most if he is happy to keep his current mobile phone and will make no more than 250 minutes of calls in a month.

If David needs more minutes than tariff 1 can offer, and he would like a new mobile phone, then he should select tariff 2.

David might select tariff 3 if he feels he will be using a higher amount of data than provided in tariffs 1 and 2, and if he feels 800 minutes of calls a month will not be sufficient.

QUESTIONS

1   A local taxi firm is struggling for business. It is looking to update its charges to boost income.

> **Current model**
>
> Fixed charge of £2 per journey followed by £0.50 per mile travelled.
>
> **New model**
>
> Fixed charge of £1 followed by £1 per mile travelled.

The average distance travelled by the firm's customers is 5 miles.

Should the taxi firm update its charges to the new model?

Use linear graphs to support your solution.

2   Asha has been given £500 for her 18th birthday, which she is going to save in order to pay for a car. Here are two bank accounts that Asha is considering:

> **Account 1**
>
> Interest paid at 1.25% AER

> **Account 2**
>
> Free £100 cash after one year*
> Interest paid at 0.75% AER
>
> *If no withdrawals have been made during the one year period.

Asha plans to buy her car in either two or three years' time. She is going to try not to make any withdrawals from her bank account.

Which bank account would you advise Asha to use to save her money?

Use linear graphs to support your solution.

3   The graph shows an electrician's charges.

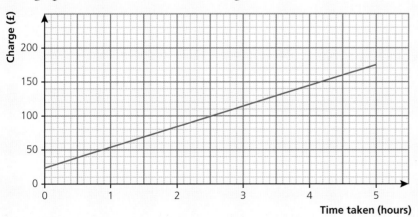

HINT

The *y*-intercept is where a graph cuts the *y*-axis.

HINT

The gradient of a line tells you how much the *y*-value increases for every one unit increase in *x*.

Explain what the *y*-intercept and gradient mean in this context.

TAX

Income tax

This is deducted from your earnings by your employer and passed to the Government. How much you pay depends on the amount you earn.

Income tax rates and bands 2016/2017

Band	Taxable income (per year)	Tax rate
Personal allowance	Up to £11000	0%
Basic rate	£11001 to £43000	20%
Higher rate	£43001 to £150000	40%
Additional rate	over £150000	45%

WORKED EXAMPLE

Sadiq has just started a new job. His annual salary is £45000. How much tax will he need to pay?

SOLUTION

Find the amounts of salary in each tax band.

£45000 comes into the higher rate band.

Amount of salary subject to tax at the higher rate is £45000 − £43000 = £2000

Amount of salary subject to tax at the basic rate is £43000 − £11000 = £32000

No tax is paid on the first £11000

Calculate the tax to be paid.

Basic rate: 20% of £32000 = £6400

Higher rate: 40% of £2000 = £800

Total tax to pay = £6400 + £800 = £7200

National Insurance (NI)

This is also deducted from your earnings by your employer and passed to the Government. How much you pay depends on the amount you earn.

National Insurance Rates 2016/2017

Your pay	Class 1 National Insurance rate
£155 to £827 a week (£672 to £3583 a month)	12%
Over £827 a week (£3583 a month)	2%

WORKED EXAMPLE

Sadiq has just started a new job. His annual salary is £45 000. How much National Insurance will he need to pay for the year?

SOLUTION

Find the amount of salary at each rate.

Sadiq's monthly salary is £45 000 ÷ 12 = £3750

He will pay 2% on £3750 − £3583 = £167 per month

He will pay 12% on £3583 − £672 = £2911 per month

Calculate the NI to be paid over the year.

2% of £167 × 12 = 2% of £2004 = £40.04

12% of £2911 × 12 = 12% of £34 932 = £4191.84

Total NI to be paid = £4191.84 + £40.04 = £4231.88

Value Added Tax (VAT)

VAT is a tax on the purchase of goods and services. The rate of VAT is applied to the price the seller wishes to charge.

Rate	% of VAT	What the rate applies to
Standard rate	20%	Most goods and services
Reduced rate	5%	Some goods and services, for example, children's car seats and home energy
Zero rate	0%	Zero-rated goods and services, for example, most food and children's clothes

WORKED EXAMPLE

Kira has her car serviced. The service cost £84 for labour plus parts costing £22.99. Her final bill included VAT. How much did she have to pay?

SOLUTION

Total costs = £84 + £22.99 = £106.99

VAT = 20% of £106.99 = £21.40

Total bill = £106.99 + £21.40 = £128.39

WORKED EXAMPLE

A television set is priced at £499.99 including 20% VAT. How much is the VAT?

SOLUTION

VAT adds 20% to the net price (price before VAT) of the TV.

Therefore, 120% of the net price is £499.99.

So 1% of net price = $\frac{£499.99}{120}$.

VAT is 20% of the net price.

So VAT is $\frac{£499.99}{120} \times 20 = £88.33$.

QUESTIONS

1 The price of a new car is £14 455 + VAT. What is the price when 20% VAT is added?

2 Amina earns £500 per week. How much income tax and NI will she pay in one year at the rates current in 2016/2017?

3 Jess renovates furniture. He sells a table for £135 + VAT. The table cost him £30. His costs including wages were £73.
How much VAT did his customer pay?

4 Greg earns £20 000 per year. How much does he earn after deductions each year at the rates current in 2016/2017?

5 Mira wants to buy a new computer. Model A costs £459 + VAT. Model B costs £487 + VAT. How much more will Mira pay if she buys model B rather than model A?

6 An e-reader costs £59.99 including 20% VAT. What is the net price?

7 Naela works part-time. She earns £320 a week. How much will she earn in one year after deductions with the tax and National Insurance rates current in 2016/2017?

8 The charge for three months' gas usage is £126 including VAT at 5%. Calculate the cost of the gas before VAT is added.

CHAPTER 3.19

INDICES AND CURRENCY

The **Retail Price Index** (RPI) and the **Consumer Price Index** (CPI) are measures of inflation – the rate at which prices rise or fall.

The RPI was introduced during the First World War, and the CPI in 1996.

Since 2013, the CPI has been used as the main national inflation measure as it meets international statistical standards and is comparable with what other countries do, but the RPI is still used for some purposes.

Many payments, such as pensions, benefits and index-linked savings, vary according to inflation as measured by the CPI or the RPI.

Both RPI and CPI are calculated by considering the current cost of a 'basket' of goods representing typical household spending and comparing the total with that in previous years. The major difference between the RPI and the CPI is that the RPI includes housing costs and the CPI does not. The CPI is generally more stable than the RPI and usually has a lower value, as shown in the graph.

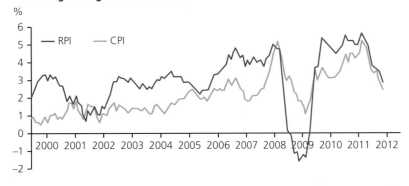

Inflation 2000–2012
Percentage change over 12 months

Source: Office for National Statistics.

The following table shows the CPI over the 10 years from 2006 to March 2016.

YEAR	JAN	FEB	MAR	APR	MAY	JUN	JUL	AUG	SEP	OCT	NOV	DEC	AVE.
2016	236.916	237.111	238.132										
2015	233.707	234.722	236.119	236.599	237.805	238.638	238.654	238.316	237.945	237.838	237.336	236.525	237.017
2014	233.916	234.781	236.293	237.072	237.900	238.343	238.250	237.852	238.031	237.433	236.151	234.812	236.736
2013	230.280	232.166	232.773	232.531	232.945	233.504	233.596	233.877	234.149	233.456	233.069	233.049	232.957
2012	226.665	227.663	229.392	230.085	229.815	229.478	229.104	230.379	231.407	231.317	230.221	229.601	229.594
2011	220.223	221.309	223.467	224.906	225.964	225.722	225.922	226.545	226.889	226.421	226.230	225.672	224.939
2010	216.687	216.741	217.631	218.009	218.178	217.965	218.011	218.312	218.439	218.711	218.803	219.179	218.056
2009	211.143	212.193	212.709	213.240	213.856	215.693	215.351	215.834	215.969	216.177	216.330	215.949	214.537
2008	211.080	211.693	213.528	214.823	216.632	218.815	219.964	219.086	218.783	216.573	212.425	210.228	215.303
2007	202.416	203.499	205.352	206.686	207.949	208.352	208.299	207.917	208.490	208.936	210.177	210.036	207.342
2006	198.300	198.700	199.800	201.500	202.500	202.900	203.500	203.900	202.900	201.800	201.500	201.800	201.600

WORKED EXAMPLE

Use the data in the table to find the percentage increase in prices from March 2006 to March 2016. If a restaurant meal cost £15 in March 2006, how much would it cost at a similar restaurant in March 2016?

SOLUTION

In March 2006, the CPI was 199.800; in March 2016, it was 238.132.

So, between these two dates, prices went up by an average of

$$\frac{238.132 - 199.800}{199.800} \times 100\% \approx 19.185\%$$

In March 2016, the meal would cost £15 × 1.19185 ≈ £17.88.

Currency exchange

If you are going on holiday abroad, you will need some foreign currency. You can get it from the Post Office or a bank or travel agent before you go. Alternatively, you can get it at the ferry port or airport (usually more expensive) or from a bank, bureau de change, hotel or cash machine while you are abroad.

WORKED EXAMPLE

On a day when the official exchange rate was £1 = €1.2658, a travel agent was selling €1.2382 for £1. How many euros would you get for £150?

SOLUTION

If you are changing £150 to euros, you may expect to get

$$€1.2658 \times 150 = €189.87$$

But from the travel agent you would actually get

$$€1.2382 \times 150 = €185.73$$

So you have paid the travel agent €4.14, or about £3.27, for the exchange. The travel agent might claim to charge 'no commission', but they are unlikely to offer the exchange service for nothing.

QUESTIONS

1 In 1918, the average annual salary was £133. In 2015 it was £25 027. What percentage increase is this?

2 The price of a pint of milk in 2015 was 49p.

   a What would it have cost in 2006, according to the CPI?

   b Estimate how long it will be until a pint of milk costs £1.

3 The average salary of the chief executives of FTSE 100 companies went up from £4.29 million to £4.50 million from 2012 to 2013.

   a Is this in line with the CPI?

   b In the same period, overall annual average earnings went up from £24 452 to £24 743. Does this match the CPI?

4 In 2016, the official exchange rate for US dollars is given as $1 = £0.69. How many dollars is equivalent to £300?

5 Choose a world currency and draw a graph to convert between British pounds and your chosen currency.

6 Find the current interbank pound-to-euro exchange rate and the rate offered by some banks and travel agents to the general public. Set up a spreadsheet to draw graphs that compare how many euros you get for different numbers of pounds at the interbank exchange rate and from different agencies.

MATHS HELP

p 359 Chapter 3.15 Percentages

CHAPTER

〉 3.20

ESTIMATING

Estimating gives a rough idea of the 'answer' to a calculation.

In this section the numbers to use are estimated as well.

Each problem can be broken down into separate amounts that are estimated and then used in the final calculation.

Estimating the numbers to use can be done by comparing them with known amounts or choosing the nearest power of 10, for example, 1, 10, 100 and so on.

WORKED EXAMPLE

Estimate how many soft toys are on this Christmas tree.
Is your estimate reasonable?
How could the estimate be improved?

SOLUTION

Assume the soft toys are only on the surface.
Assume the tree is the shape of a cone.
Estimate the width of the base: 4 metres
Estimate the height of the tree: 20 metres
Estimate the area covered by one soft toy: 0.1 square metres
The curved surface area of a cone is $\pi r l$ where l is the slant height of the cone.
The tree is tall and thin so the slant height is also about 20 m.
Area $\approx 3.14 \times 2 \times 20$
$\qquad \approx 120 \text{ m}^2$
Number of toys $\approx 120 \div 0.1$
$\qquad\qquad \approx 1200$ soft toys
Calculating the answer in a different way may give a different estimate.

1200 seems the right order of magnitude.

The estimate could be improved by getting a more accurate figure for the number of toys in a square metre.

WORKED EXAMPLE

How many eggs are consumed in a year in the UK?

SOLUTION

Eggs are eaten both as a dish in their own right and as ingredients in baked items.

These are the assumptions or estimates for a 'power of 10' estimate.

Each person eats 1, 10, 100 eggs per month – say 10.

Each person eats 1, 10, 100 eggs as ingredients per month – between 10 and 100 so say 10.

Number of months in a year – 12 for this estimate as we know this.

Number of people in the UK, 10 million or 100 million – 100 million

Number of eggs $\approx (10 + 10) \times 12 \times 100\,000\,000$

$$\approx 20 \times 12 \times 100\,000\,000$$

$$\approx 24\,000 \text{ million}$$

An estimate should be easy to calculate, so it is appropriate to round so that the answers are easy to work out. You are looking for an estimate of the right order of magnitude. Is it measured in tens, hundreds or thousands?

You should state any assumptions you use to make it easier to evaluate your answer.

You should show clearly the calculations you do to get to your answer.

There were in fact 2000 soft toys in the tree, and the UK egg consumption for 2015 was 12.2 billion. Each of these is out by about a factor of 2 and that is reasonable for an estimate of this kind.

QUESTIONS

1 Estimate how much money you could make walking other people's dogs in a year.

2 Estimate how many Easter eggs are consumed each year in the UK.

3 Estimate the area of the hide on a reindeer.

4 Estimate the cost of furnishing a one-bedroom flat.

5 Estimate how many books you will read in your lifetime.

6 Estimate how many hours you spend on the internet in a year.

NORMAL DISTRIBUTION

Features of a normal distribution

A normal distribution has a classic, symmetrical bell shape.

The mean is at the line of symmetry.

Approximately **two-thirds** of the data falls within **one standard deviation** of the mean.

Approximately **95%** of the data falls within **two standard deviations** of the mean.

Almost all of the data falls within three standard deviations of the mean.

WORKED EXAMPLE

The heights of sunflowers are normally distributed with a mean of 2.3 m and standard deviation of 0.5 m.

a   What percentage of sunflowers will grow to a height of between 1.8 m and 2.8 m?

b   In a field of 200 sunflowers, how many sunflowers would you expect to have a height of between 2.3 m and 3.3 m?

SOLUTION

a   A height of 1.8 m lies one standard deviation below the mean and 2.8 m lies one standard deviation above the mean. Therefore we are looking for the region that lies within one standard deviation of the mean. For a normal distribution we know that this corresponds to $\frac{2}{3}$ of the data or 66.7% of sunflowers.

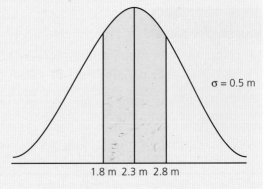

b A height of 3.3 m lies two standard deviations above the mean. We are looking for the region that is between the mean and this value. We know that 95% of the data lies within two standard deviations of the mean, so we want half this amount, which corresponds to 47.5%.

47.5% of 200 = 95 sunflowers.

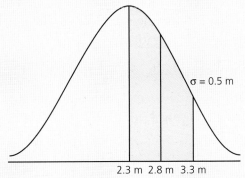

σ = 0.5 m

2.3 m 2.8 m 3.3 m

Note

You can find normal distribution tables on AQA's website

http://www.aqa.org.uk/subjects/mathematics/aqa-certificate/mathematical-studies-1350/statistical-tables

The standardised normal distribution

$N(\mu, \sigma^2)$ means a variable is normally distributed with a mean of μ and a standard deviation of σ.

$N(0, 1)$ means a normally distributed variable with mean 0 and standard deviation 1.

z is often used to denote the standardised normal variable.

Values of the standardised normal distribution can be looked up in statistical tables, or found using calculators and spreadsheets.

WORKED EXAMPLE

For $N(0,1)$ find the proportion that lies between $z = -0.3$ and $z = 0.9$.

SOLUTION

The diagram is a sketch of the standardised normal distribution $N(0,1)$. We will use the normal distribution tables to help us here. The normal distribution table gives the probability that a data item drawn from the standardised normal distribution $N(0, 1)$ lies in the shaded region under the curve. That is, it gives you the probability $P(Z < z)$ that the data item is less than or equal to z.

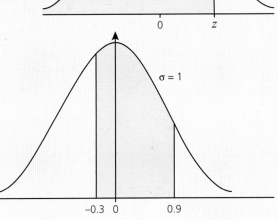

This diagram is a sketch of the region we are seeking.

When $z = 0.9$ this corresponds to a value of 0.81594 in the probability table, which is the probability that z is **less than** 0.9.

Since the tables only represent positive values of z, when $z = -0.3$, we have to use the symmetry of the normal distribution and look up $z = 0.3$ instead.

This corresponds to a value of 0.61791 in the table.

However, the correct region for our situation is obtained using $1 - 0.61791 = 0.38209$.

This tells us the probability that z is **less than** -0.3.

To find the proportion that lies between $z = -0.3$ and $z = 0.9$, we need to start with the probability that z is less than 0.9 and subtract the probability that z is less than -0.3 from this result.

$0.81594 - 0.38209 = 0.43385$

Therefore 43.4% of the distribution lies between $z = -0.3$ and $z = 0.9$.

QUESTIONS

1 The heights of adult males are normally distributed with a mean of 178 cm and a standard deviation of 8.5 cm.

   In a random selection of 1000 men, how many would you expect to be smaller than 169.5 cm?

2 In a maths competition, the top 2.5% of competitors win a prize.

   The mean mark is 52 and the standard deviation for the marks is 12.

   What score do you need to win a prize?

3 The sketch shows two curves, A and B, which are normally distributed.

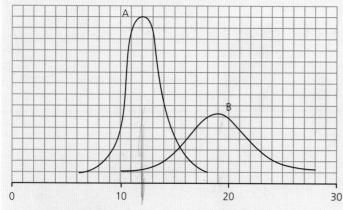

   Estimate the mean and standard deviation of each distribution.

4 For the distribution N(0,1):

   a find the proportion that lies below $z = 1.4$

   b find the proportion of the distribution that lies above $z = 1.4$.

5 Find:

   a $P(Z > 1.2)$

   b $P(Z < -0.7)$

   c $P(0.9 < Z < 1.8)$.

6 Find the value for z such that:

   a $P(Z < z) = 0.77035$

   b $P(Z < z) = 0.11507$.

HINT

z is often used to denote a variable of the normal distribution.

HINT

It is helpful to draw a sketch of the normal distribution for the given situation. This helps you to identify which area you are finding.

NORMAL DISTRIBUTION – FINDING PROBABILITIES

Standardising

To standardise a result, use the formula standardised result $= \dfrac{\text{result} - \text{mean}}{\text{standard deviation}}$.

We use the symbol μ to represent the mean and the symbol σ to represent the standard deviation.

WORKED EXAMPLE

In a maths test, Jay scored 56% and Poppy scored 72%.

The mean score for Jay's class was 52% with a standard deviation of 2%.

The mean score for Poppy's class was 68% with a standard deviation of 4%.

Who did better in relation to their class?

SOLUTION

We need to standardise the scores for Jay and Poppy before we can compare them.

Jay: $\dfrac{56 - 52}{2} = 2$

Poppy: $\dfrac{72 - 68}{4} = 1$

Jay's score is 2 standard deviations better than the average for his class whereas Poppy's score is only 1 standard deviation above the average for her class, therefore Jay did better.

Finding probabilities

The probabilities for a standardised variable can be looked up in tables.

Looking at the table below, find P(Z < 0.42).

z	0.00	0.01	0.02	0.03	0.04
0.0	0.50000	0.50399	0.50798	0.51197	0.51595
0.1	0.53983	0.54380	0.54776	0.55172	0.55567
0.2	0.57926	0.58317	0.58706	0.59095	0.59483
0.3	0.61791	0.62172	0.62552	0.62930	0.63307
0.4	0.65542	0.65910	0.66276	0.66640	0.67003
0.5	0.69146	0.69497	0.69847	0.70194	0.70540
0.6	0.72575	0.72907	0.73237	0.73565	0.73891
0.7	0.75804	0.76115	0.76424	0.76730	0.77035
0.8	0.78814	0.79103	0.79386	0.79673	0.79955

Go down the page for the first decimal place then across for the second decimal place.

P(Z < 0.42) = 0.66276.

It looks as if many values are missing, but you can use the symmetry of the distribution to work these out. A diagram is particularly useful in these cases.

You can also find the probabilities using a calculator. Graphing calculators are more likely to have this facility. You will need to look up how to do this in the calculator manual as the process varies between models.

WORKED EXAMPLE

The contents of jars of jam follow a normal distribution with a mean weight of 360 g and a standard deviation of 8 g.

a   Find the probability that the weight of jam in a jar is more than 366 g.

b   The label on a jar states that the minimum weight of jam in the jar is 340 g. Can this claim be justified?

c   Find the weight of jam below which 95% of the population of jars lie.

→

SOLUTION

a

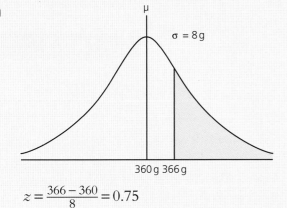

$$z = \frac{366 - 360}{8} = 0.75$$

Using the probability tables, or your calculator, a z-score of 0.75 tells us that the probability that a randomly selected item from the standard normal distribution is less than 0.75, is 0.77337.

This means that the probability that it is greater than 0.75 is $1 - 0.77337 = 0.22663$.

Therefore the probability that the weight of jam in a jar is more than 366 g is 0.22663 or approximately 22.6%.

b

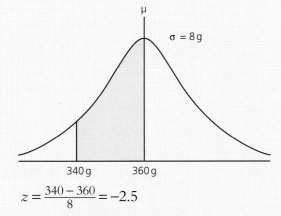

$$z = \frac{340 - 360}{8} = -2.5$$

Using the probability tables, or your calculator, a z-score of 2.5 tells us that the probability that a randomly selected item from the standard normal distribution is less than 2.5, will be 0.99379.

This means that the probability that it is greater than 2.5 will be $1 - 0.99379 = 0.00621$.

Using the symmetry of the normal distribution, the probability that a randomly selected item is greater than 2.5 is the same as the probability of it being less than -2.5.

Therefore, the probability that the weight of jam in a jar is less than 340 g is 0.00621 or approximately 0.6%.

Since this is such a small probability, the claim on the jar could be accepted.

c   Use the probability tables or your calculator to find the z-score that gives 95% or 0.95000.

The closest z-score for 95% is between 1.64 and 1.65. It is slightly closer to 1.64.

We now need to find the value which is 1.64 standard deviations above the mean.

The required weight of jam is $360 + (1.64 \times 8) = 373.1 \, \text{g}$.

Therefore 95% of the population of jam jars will weigh below 373.1 g.

QUESTIONS

1 Two athletes are practising the high jump in preparation for an upcoming competition.

Athlete A has a mean jump height of 1.8 m with a standard deviation of 0.3 m.

Athlete B has a mean jump height of 1.9 m with a standard deviation of 0.5 m.

During the last practice session both athletes believe they jumped 1.85 m.

Who is more likely to be wrong

2 Fred sat his mock exam for biology and scored 79. The results are normally distributed with a mean of 64 and a standard deviation of 5.8. ~~Bio 2.5~~

Fred's standardised score for his physics mock was 1.8.

Comment on Fred's performance in his mock exams.

3 Annabel is measuring worms for a biology experiment. The lengths of worms is normally distributed with a mean of 7.5 cm and a standard deviation of 2 cm.

Annabel is currently comparing two worms. One is 9.2 cm and the other is 5.3 cm.

Comment on the lengths of the two worms.

4 Given that $X \sim N(20, 5^2)$, find the following probabilities:

a $P(X < 26.3)$

b $P(X > 33.5)$

c $P(X < 12.2)$

d $P(18 < X < 22)$.

> **HINT**
>
> $N(20, 5^2)$ means the standard deviation is 5.

5 After the birth of a calf, the production of milk by dairy cows follows a normal distribution. The amount of milk produced is normally distributed with a mean of 27 litres per day and a standard deviation of 4 litres per day.

a Calculate the probability that a dairy cow will produce less than 33 litres of milk per day.

b Calculate the probability that a dairy cow will produce between 24 litres and 33 litres of milk per day.

6 Resting heart rates are assumed to be normally distributed. Their distribution is approximately given by $N(70, 144)$.

a Find the probability that a person has a resting heart rate of over 100.

b Find the probability that a person has a resting heart rate of below 60.

c Find the heart rate below which 95% of the population lie.

> **HINT**
>
> To find a z-score use the formula $z = \frac{x - \mu}{\sigma}$, where x is the given value, μ is the mean and σ is the standard deviation.

> **HINT**
>
> It is helpful to draw a sketch of the normal distribution for the given situation. This can help with identifying which area you are trying to find.

CONFIDENCE INTERVALS FOR THE MEAN

When we try to find information about, say, the heights of a large population such as all the adult males in the country, we are unlikely to be able to get data from everyone, but we can use a sample.

The interesting thing about samples is that the distribution of the means of the samples is close to a normal distribution. The more samples you take and, more importantly, the larger the size of the samples (30 or more is a good size), the more closely the distribution of the sample means will match a normal distribution. This is true even if the distribution from which the samples are taken is not normal.

The mean height of the sample of adult males is called a **point estimate** of the mean height of the population.

We use the formula $s = \frac{\sigma}{\sqrt{n}}$, where s is the standard deviation of the sample means and n is the sample size, to find confidence intervals for the point estimate of the mean.

This tells us that the larger the sample size, the smaller the standard deviation of the sample means.

A 90% **confidence interval** for the mean is a range of values within which we are 90% confident that the mean of the population lies. This is because in 90% of the samples the mean falls in this interval.

> **HINT**
>
> The estimate $s = \frac{\sigma}{\sqrt{n}}$ is called the **standard error**.

WORKED EXAMPLE

We have a sample of the heights of 100 adult men. The mean height of the sample is 1.68 m. The population has a standard deviation of 0.31 m. Find a 90% confidence interval for the mean height of the population.

➡

SOLUTION

A 90% confidence interval includes everything except for 5% at each end of the distribution. So we need to use the table to find the value of z below which 95% of the data lies. This value is 1.64 (see extract from table below).

The symmetry of the graph means that the value of z below which 5% of the data lies is -1.64.

So, for $N(0,1)$, the 90% confidence interval is $-1.64 < z < 1.64$.

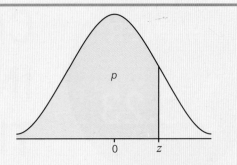

You can use tables of the standard normal distribution to find the probability of certain outcomes.

z	0.00	0.01	0.02	0.03	0.04	0.05
0.0	0.50000	0.50399	0.50798	0.51197	0.51595	0.51994
0.1	0.53983	0.54380	0.54776	0.55172	0.55567	0.55962
0.2	0.57926	0.58317	0.58706	0.59095	0.59483	0.59871
0.3	0.61791	0.62172	0.62552	0.62930	0.63307	0.63683
0.4	0.65542	0.65910	0.66276	0.66640	0.67003	0.67364
0.5	0.69146	0.69497	0.69847	0.70194	0.70540	0.70884
0.6	0.72575	0.72907	0.73237	0.73565	0.73891	0.74215
0.7	0.75804	0.76115	0.76424	0.76730	0.77035	0.77337
0.8	0.78814	0.79103	0.78389	0.79673	0.79955	0.80234
0.9	0.81594	0.81859	0.82121	0.82381	0.82639	0.82894
1.0	0.84134	0.84375	0.84614	0.84849	0.85083	0.85314
1.1	0.86433	0.86650	0.86864	0.87076	0.87286	0.87493
1.2	0.88493	0.88686	0.88877	0.89065	0.89251	0.89435
1.3	0.90320	0.90490	0.90658	0.90824	0.90988	0.91149
1.4	0.91924	0.92073	0.92220	0.92364	0.92507	0.92647
1.5	0.93319	0.93448	0.93574	0.93699	0.93822	0.93943
1.6	0.94520	0.94630	0.94738	0.94845	(0.94950)	0.95053

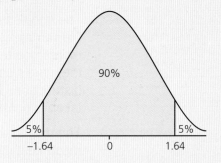

The tables give the probability of a random data item from $N(0, 1)$ being less than z.

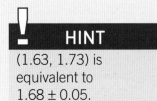

HINT

(1.63, 1.73) is equivalent to 1.68 ± 0.05.

Now we need to find the confidence interval for the male heights based on a sample with mean 1.68 m from a population with standard deviation 0.31 m.

The estimate of the mean of the population is 1.68 m and its standard deviation is $\frac{0.31}{\sqrt{100}} = 0.031$.

The z-scores for the upper and lower bounds of the confidence interval are equal to 1.64 and -1.64.

So, if the upper bound is H cm, $1.64 = \frac{H - 1.68}{0.031}$

So $H = 1.64 \times 0.031 + 1.68 \approx 1.73$

Similarly, if the lower bound is h cm, $-1.64 = \frac{h - 1.68}{0.031}$

So $h = -1.64 \times 0.031 + 1.68 \approx 1.63$

So we can be 90% confident that the mean height of the population of males lies in the interval (1.63 m, 1.73 m).

Instead of using the table of the normal distribution table 'backwards' (as shown above) to find the values with the required probabilities, you can choose to use the table showing percentage points.

Look for $p = 0.95$ in the column and row headings of the table. There are two places to find $z = 1.6449$ as it is accurate to 2 decimal places, and the tables allow for the probability to be stated accurate to 3 decimal places above $p = 0.9$.

Note

You can find normal distribution tables on AQA's website

http://www.aqa.org.uk/subjects/mathematics/aqa-certificate/mathematical-studies-1350/statistical-tables

p	0.00	0.01	0.02	0.03	0.04	0.05	0.06
0.5	0.0000	0.0251	0.0502	0.0753	0.1004	0.1257	0.1510
0.6	0.2533	0.2793	0.3055	0.3319	0.3585	0.3853	0.4125
0.7	0.5244	0.5534	0.5828	0.6128	0.6433	0.6745	0.7063
0.8	0.8416	0.8779	0.9154	0.9542	0.9945	1.0364	1.0803
0.9	1.2816	1.3408	1.4051	1.4758	1.5548	1.6449	1.7507

p	0.000	0.001	0.002	0.003	0.004	0.005	0.006
0.95	1.6449	1.6546	1.6646	1.6747	1.6849	1.6954	1.7060
0.96	1.7507	1.7624	1.7744	1.7866	1.7991	1.8119	1.8250

QUESTIONS

1 Find a 95% confidence interval for the mean height of the males in the worked example on pages 385–6.

2 The weight of a type of banana is normally distributed with a standard deviation of 17 g. A sample of 100 of these bananas has a mean of 113 g.

Work out a 95% symmetric confidence interval for the mean of the population.

3 In a sample of 1200 credit card users in the UK, the mean debt was £1983 with standard deviation £1875. Find a 90% confidence interval for the mean credit card debt in the UK.

4 Work through question 3 with a sample size of 120. Explain what effect the change in sample size has on the confidence interval.

MATHS HELP

p 381 Chapter 3.22
Normal distribution
– finding probabilities

CORRELATION

Bivariate data can be plotted on a **scatter graph** using the two data items as coordinates for a point representing, for example, a person or country.

It may show **correlation**, which may be **positive** or **negative**.

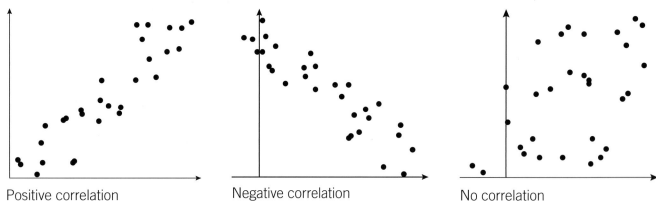

Positive correlation   Negative correlation   No correlation

The closer the points are to a straight line the stronger the correlation.

Correlation on a scatter graph does not mean that the two quantities measured are linked in any way. Further work is required to explore whether the correlation is from both quantities being affected by a third measure or whether there is any cause-and-effect relationship. Correlation does not imply causation.

Points that are away from the general trend of the data are called **outliers**. They may be unusual items of data, or errors in collecting or recording the data. Further exploration and thought is required to decide which. Appropriate action can then be taken to exclude or retain those items of data.

The strength of any correlation can be calculated using a variety of measures. In this qualification, we use the **product moment correlation coefficient** or **pmcc**.

The pmcc has a value between −1 (perfect negative correlation) and +1 (perfect positive correlation). A positive value suggests that when one measure rises so does the other. A negative value suggests that when one measure rises the other falls. A value near zero suggests that there is no linear association between the two measures.

The pmcc can be calculated 'by hand' or by using a calculator. It is expected that you will use a calculator to work it out. You will need to consult the manual for your calculator to find out how it calculates the pmcc.

HINT

You do not need to know the formula for calculating pmcc, but it is:

$$\text{pmcc} = \frac{S_{xy}}{\sqrt{S_{xx}S_{yy}}}$$

where

$$S_{XX} = \sum x_i^2 - \frac{\left(\sum x_i\right)^2}{n}$$

$$S_{yy} = \sum y_i^2 - \frac{\left(\sum y_i\right)^2}{n}$$

$$S_{xy} = \sum x_i\, y_i - \frac{\sum x_i \sum y_i}{n}.$$

The pmcc only indicates possible linear relationships. Data having a pmcc = 0 may still have a relationship. Examples include exponential and quadratic relationships.

WORKED EXAMPLE

The graph represents the population of countries winning 10 or more medals at the 2012 Olympics. Describe the correlation and identify any outliers.

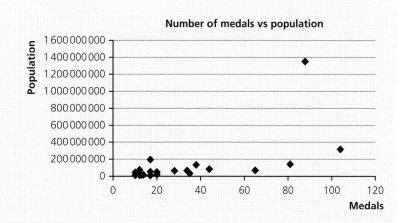

SOLUTION

Typically, the number of medals won increases as the population increases so the correlation is positive, but it is not very strong. There is one outlier with a population of 1.4 billion, which is a population at least 3 times larger than any other country, but its medal count is only the second highest.

WORKED EXAMPLE

The table opposite gives the data for the top ten medal winners (as used for the graph in the previous example). Calculate the pmcc for this data and comment on how it relates to the answer for the previous example.

Number of medals	Population
104	313 382 000
88	1 347 350 000
81	143 056 383
65	62 262 000
44	81 831 000
38	127 650 000
35	22 880 619
34	65 350 000
28	48 580 000
28	60 776 531

SOLUTION

Using either your calculator or a spreadsheet, enter the data using the number of medals as the x-variable and the population as the y-variable. The result is 0.5703 to 4 decimal places. This value indicates a positive correlation of medium strength. A value of 1 would indicate a perfect linear relationship, a value close to 0 would indicate no linear relationship and a value of -1 would indicate a perfect negative linear relationship.

QUESTIONS

1 The marks scored by a group of students in a mock examination are shown below.

Maths	Physics	Biology
51	41	83
50	50	79
40	55	79
64	59	73
46	59	74
70	69	69
55	71	74
66	79	79
60	80	79
81	84	70

Calculate the pmcc between maths and physics and the pmcc between physics and biology. Use the values obtained to describe and compare the correlation between the marks in these pairs of subjects. Without calculating the pmcc, how would the maths marks relate to the biology marks?

2 Provisional data from the Office of National Statistics gives the following details on the number of live births and number of deaths in 2015 by region in England.

Region	Live births	Deaths
North East	28400	28075
North West	85838	71299
Yorkshire and The Humber	63858	51690
East Midlands	53641	45016
West Midlands	69806	54563
East	72505	56407
London	129615	50543
South East	102703	81049
South West	58033	56667

Calculate the pmcc for the number of births and deaths. Discuss whether there is a correlation between births and deaths, and if that means that one is a consequence of the other.

3 For each of graphs A, B and C, match these pmcc values to the graph:

0.385   0.981   −0.247

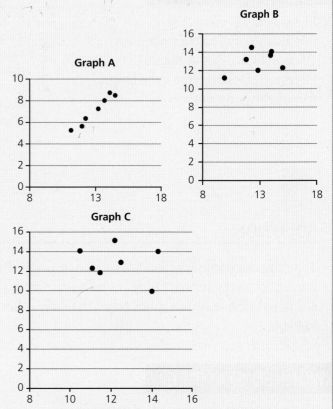

4 Juliet measured the height of a stone thrown up from a balcony and produced this scatter graph of height versus time.

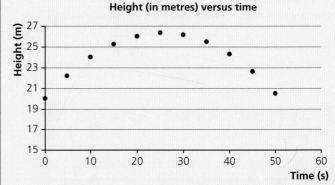

She calculated the pmcc between height and time to be 0.07. When Romeo sees the results, he says there is no relationship between the height and the time. Is Romeo correct?

CHAPTER

3.25

REGRESSION

Drawing a line of best fit through the mean point by eye

The steps are:

Calculate the mean of each data set to work out the mean point.

Plot the mean point.

Draw a straight line that:

- passes through the mean point
- has a gradient similar to the trend of the points
- has roughly equal numbers of points on each side.

WORKED EXAMPLE

The scatter graph opposite shows the weight versus age plots for a group of adults.

Mean of ages = 35 years

Mean of weights = 76 kg

Draw a line of best fit.

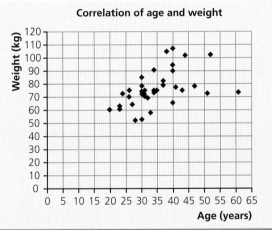

SOLUTION

Plot the mean point on the graph.

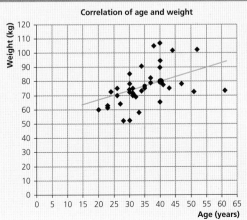

Adjust the position of your straight edge or ruler so that it passes through the mean point and has a positive gradient to match the trend of the points.

Now check that there are roughly equal numbers of points on each side of the ruler, adjusting the position of the ruler if you need to.

Draw the line of best fit on the diagram.

We can now interpret the direction of the line of best fit and conclude that there is weak positive correlation between age and weight. In other words, as people get older they tend to weigh more.

A line of best fit as described above is sufficient for many applications.

When greater accuracy is required, a regression line is calculated as this gives a 'better' line of best fit.

Plotting a regression line from its equation

The equation of the regression line can be worked out on paper or by using a calculator, as shown in the next part of this chapter.

WORKED EXAMPLE

For the same set of data:

a   Plot the regression line on the graph.

b   Use the regression line to predict the average weight of a 50-year-old.

c   Use the regression line to predict the average weight of a 5-year-old.

d   Comment on your predictions.

> **! HINT**
>
> You can read the values from the graph. The values will not be sufficiently accurate to get the equation in the solution but they should be close.

SOLUTION

a   Using a calculator, we have the equation of the regression line for the weight versus age data:

$y = 0.75x + 50$

The graph of this equation has a vertical axis intercept at 50, the point $(0, 50)$, and a gradient of 0.75.

We can find another point on the line by substituting a sensible value into the equation:

Substitute $x = 40$: $y = (0.75 \times 40) + 50 = 80$; this gives the point $(40, 80)$.

We can now draw an accurate regression line for this data using these points. This is shown by the thick black line.

The dotted portion of the line shows where the plotted regression line would intersect the vertical axis.

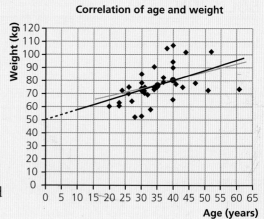

Correlation of age and weight

The blue line shows the positon of the line of best fit drawn by eye.

The regression line shows the average weight-for-age relationship for this data.

b   The average weight of a 50-year-old person is approximately 88 kg.

c   The average weight of a 5-year-old person is approximately 53 kg.

d   The prediction for a 50-year-old is within the range of the data and is sensible.

However, the prediction for a 5-year-old is far too high. It has been extrapolated beyond the range of the data, so the relationship may not apply (as is the case here).

Finding the equation of a regression line of y on x using a calculator

This requires some substantial number crunching and it is expected that you will use a calculator to do that for you. This means that the formulae will not be given and the number of marks will reflect the time needed to use a calculator.

Using a calculator requires you to:

- input the data
- select the regression line option $y = a + bx$
- note down the values of a and b.

You will need to consult the manual for the exact sequence of key presses for the particular model of calculator you have.

WORKED EXAMPLE

The table shows the number of people at a swimming pool over five days and the maximum air temperature for those days.

a   Find the equation of the regression line.

b   Estimate the number of people at the pool on a day when the air temperature is 24°C.

c   Estimate the number of people at the pool on a day when the air temperature is 4°C.

d   Comment on your estimates.

Air temperature	Number of people
22°C	37
31°C	104
15°C	15
27°C	82
13°C	5

SOLUTION

a   We can use the following steps to find the equation of the regression line for this data:

Step 1: Find $X \times Y$ and X^2 (shown in the table below).

Step 2: Find the sum of every column.

Air temperature (X)	No. of people (Y)	$X \times Y$	X^2
22	37	814	484
31	104	3224	961
15	15	225	225
27	82	2214	729
13	5	65	169
Sum (Σ) 108	243	6542	2568

Step 3: The equation of the regression line is in the form $y = a + bx$.

You can use the following formula to find the values of a and b; they may be familiar from earlier work but they will not be given in the exam.

This is the part where you can use your calculator to get a and b:

$$b = \frac{S_{xy}}{S_{xx}} = \frac{n.\sum XY - \sum X.\sum Y}{n.\sum X^2 - \left(\sum X\right)^2} = \frac{5.6542 - 108.243}{5.2568 - (108)^2} \approx 5.50$$

$$a = \text{mean of } = y - (b \times \text{mean of } x) = \bar{y} - b\bar{x} = \frac{243}{5} - 5.4983.. \times \frac{108}{5} \approx -70.16$$

> **! HINT**
> You are expected to use a calculator for this.

Therefore, the regression line equation is $y = 5.50x - 70.16$.

b   Estimate = $5.50 \times 24 - 70.16 = 62$ people.

c   Estimate = $5.50 \times 4 - 70.16 = -48$ people.

d   The second estimate is beyond the range of the data so it is an extrapolation. The relationship may not hold outside the range and it clearly doesn't here. The first estimate is an interpolation and so is likely to be valid.

QUESTIONS

1   The graph below shows the winning time for the marathon at the Olympic Games from 1948 to 2012.

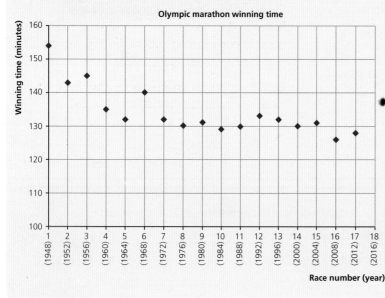

Olympic marathon winning time

Winning time (minutes) / Race number (year)

a   The mean of the race number values is 9 and the mean winning time is 134 minutes. Use this information to draw a line of best fit (by hand) for this data on a copy of the graph.

b   What does the line of best fit indicate about the change to the winning times over time?

The equation for the accurate regression line of y on x for this data is:

$y = -1.132x + 144$

c   Explain how you would use this equation to determine where to draw an accurate regression line.

d   Now draw in the accurate regression line on your copy of the graph above.

e Use the regression line to predict what you think the winning time was in the 2016 Olympics.

f What has been done to the graph above to make the correlation between winning time and year seem more dramatic?

2 The table shows the marathon times for the data in question 1.

Race number (*x*)	Time (minutes) (*y*)
1	154
2	143
3	145
4	135
5	132
6	140
7	132
8	130
9	131
10	129
11	130
12	133
13	132
14	130
15	131
16	126
17	128

a Show that the equation of the regression line of *y* on *x* for this data is $y = 144.4 - 1.13x$.

b Explain the significance of the gradient of this regression line having a negative value.

3 The table below shows the marks that the same class of students scored in two different exams.

The teacher wants to see if there is a correlation between performances in the two exams.

	Exam 1 (/100) (*x*)	Exam 2 (/100) (*y*)
Student 1	72	70
Student 2	55	45
Student 3	89	93
Student 4	43	62
Student 5	67	72
Student 6	74	77
Student 7	77	75
Student 8	81	79
Student 9	50	43
Student 10	36	41
Student 11	67	60
Student 12	75	67

a Show that the equation of the regression line of *y* on *x* for this data is $y = 0.864x + 8.716$.

b Does this line show a positive or negative correlation?

c What does the type of correlation represented by this equation indicate about performances in the two tests?

d Student 13 missed Exam 2 and gained 53 marks in Exam 1. Use the regression line to estimate their score in Exam 2.

e Student 14 also missed Exam 2 and gained 18 marks in Exam 1. Use the regression line to estimate their score in Exam 2.

f Explain why the estimate for Student 13 is likely to be more accurate.

CRITICAL PATH ANALYSIS

Critical path analysis puts planning a project on a more formal, rigorous basis. The tasks are identified with approximate durations, and then organised into a precedence table. The **activity network diagram** is then drawn, the **critical path** identified and a **Gantt chart** is drawn to identify the resources needed at each stage.

The **critical path** is so called because any delay in the activities on this path will delay the whole project.

Activities not on the critical path have spare time (called **float time**) associated with them because there is some flexibility in when they are done.

WORKED EXAMPLE

The table shows the activities needed for a loft conversion.

	Activity	To be done after (predecessor activities)	Time needed (days)
A	Make staircase off-site		2
B	On-site preparatory work		0.5
C	Install roof windows	B	1
D	Install floorboards	B	1
E	Install floor covering	D	0.5
F	Install staircase	A, B	1
G	Decorate	C, D, F	1.5
H	Clear up	E, G	0.5

Draw an activity network diagram, identify the critical path and draw a Gantt chart.

SOLUTION

The activity network diagram shows the time needed for each activity in black.

Working from left to right across the diagram, we fill in the earliest possible starting time for each activity in red.

A and B have earliest start times of 0 as they have no preceding activities.

F has earliest start time of 2 as that is the greater of the durations of A and B that are before it.

C and D have earliest start times of 0.5 as they both only follow B.

E has an earliest start time of 1.5 = 0.5 + 1, as it follows D.

G has earliest start time 3 as that is the greatest value of 2 + 1 (F), 0.5 + 1 (C) and 0.5 + 1 (D).

H has an earliest start time of 4.5 as it follows G (3 + 1.5) and that is greater than 1.5 + 0.5 for E.

Working from right to left back through the diagram, we fill in the latest possible finishing times in green.

The project takes 5 days so that is the latest finishing time for H.

E and G both precede H and so have latest finishing times of 4.5 = 5 − 0.5.

C and F both precede G and so have latest finishing times of 3 = 4.5 − 1.5.

D precedes E and G so has latest finishing time that is the lesser of 3 = 4.5 − 1.5 (G) and 4 = 4.5 − 0.5 (E)

B has a latest finishing time of 2 = 3 − 1 from each of its following activities (if they were different you would select the earliest of these).

A precedes F so has the latest finishing time of 2 = 3 − 1.

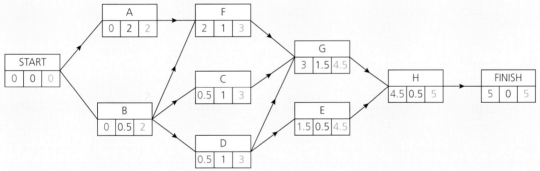

The critical path is the one linking the activities where the starting time plus the time for the activity equals the finishing time, that is A − F − G − H.

The project activities can also be shown in a Gantt chart:

The activities shown in red are on the critical path; the activities not on the critical path (shown in black) have dashed 'float time' so there is flexibility as to when they are done.

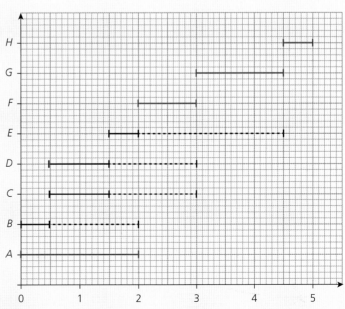

QUESTIONS

1 Here is an adapted activity list for the loft conversion of the previous example, which now includes the installation of electric lights and sockets.

 a Redraw the activity network diagram to include this.

 b Is the critical path different?

 c Which activities have float time now?

	Activity	To be done after	Time needed (days)
A	Make staircase off-site		2
B	On-site preparatory work		0.5
C	Install roof windows	B	1
D	Install floorboards	I	1
E	Install floor covering	D	0.5
F	Install staircase	A, B	1
G	Decorate	C, D, F	1.5
H	Clear up	E, G	0.5
I	Do electrical work	B	1.5

2 Charlotte plans to make a patchwork cushion. The table shows the steps, the times and which steps have to be completed before others can happen.

	Task	Predecessors	Time (minutes)
A	Design the patchwork	–	60
B	Choose fabrics	A	20
C	Cut out fabric	B	15
D	Choose backing material	–	10
E	Cut out backing material	D	10
F	Sew patches together	C	80
G	Insert zip	E, F	15
H	Press front and back of cushion	G	5
I	Sew front to back	H	10
J	Insert cushion pad	I	5

 a Construct an activity network diagram.

 b Fill in the earliest and latest times.

 c List the critical path and state the duration of the complete task.

 d Charlotte is working alone to make the cushion. How does this affect the time needed to complete the task?

3 The staff in a hotel have a limited time to clean rooms and get them ready for the next guest.

This table shows the activities that need to be done and their approximate durations.

Task	Activity	Immediate predecessors	Duration (minutes)
A	Remove used bed linen	–	1
B	Make bed	A	2
C	Remove used towels	–	1
D	Clean bathroom	–	2
E	Replace towels and toiletries	C, D	1
F	Wash/replace cups	–	2
G	Replenish tea/coffee	–	1
H	Put tidied tray away	F, G	2
I	Clean carpet	A	2
J	Tidy room	B, H, I	1
K	Check heater settings	–	1
L	Leave room	E, J, K	1

 a Draw the activity network diagram.

 b Fill in the earliest start times and latest finish times.

 c Identify the critical path and the duration.

 d Draw a Gantt chart.

 e How many people need to be involved for the job to take the minimum possible time?

4 Carole is planning to assemble a herb planter. She identifies the activities, their durations and whether some activities need to be completed before another starts. You may assume she has help.

This is a table of the tasks and the times associated with them:

Task	Activity	Duration (minutes)	Earliest start time	Latest finish time
A	Choose planter (pot)	10	0	20
B	Select herbs	20	0	20
C	Mix the compost	20	10	30
D	Decide on location	5	20	30
E	Move planter to its location	5	30	35
F	Fill with compost	10	35	45
G	Plant the herbs	15	45	60
H	Water the plants	5	60	65

a Draw a Gantt chart to show this information.

b Identify the critical path.

c How long should the project take?

d Which activity has the most float time?

5 A charity plans to publish a book to raise money. The table shows the steps in the project.

a Draw the activity network diagram.

b What are the critical activities?

c Draw the Gantt chart.

Task	Activity	Immediate predecessors	Duration (weeks)
A	Assemble writing team	–	3
B	Share ideas	A	2
C	Develop template	B	1
D	Write a sample chapter	C	3
E	Feedback on sample chapter	D	1
F	Revise sample chapter	E	2
G	Samples go to designer	F	2
H	Writers write their allocation	F	8
I	Editor collates and edits	H	3
J	Feedback to designer	G	1
K	Finalise design	J	2
L	Chapters to designer	I, K	4
M	First proofs to editor	L	3
N	Amendments to designer	M	2
O	Final proofs	N	1
P	Print	O	2

CALCULATING PROBABILITIES

Probability of single events

The probability of a single event can be calculated theoretically by considering **equally likely outcomes**.

$$\text{Probability} = \frac{\text{number of outcomes that satisfy the condition}}{\text{total number of possible outcomes}}$$

A set of outcomes is called **exhaustive** if it covers all of the possible outcomes. The probabilities add up to 1.

WORKED EXAMPLE

A fair die is thrown.

a   Work out the probability that it lands showing a square number.

b   Work out the probability that it does not land showing a square number.

SOLUTION

a   There are only two square numbers on a die, 1 and 4.

There are six possible outcomes.

Therefore, P(showing a square number) $= \frac{2}{6}$

This may be simplified to $\frac{1}{3}$

b   This probability covers all of the other outcomes.

$$\text{Probability} = 1 - \frac{1}{3}$$
$$= \frac{2}{3}$$

Sometimes this type of analysis is not possible and it is necessary to rely on experimental data to estimate the probability of an event.

WORKED EXAMPLE

When 100 of a certain make of mobile phone are dropped in water for ten minutes to test if they are waterproof, 38 of them fail to work afterwards. What is the probability that one of these phones will fail to work after being dropped in water for ten minutes?

→

SOLUTION

The probability is simply the fraction of phones that fail to work after being in water for ten minutes. This is $\frac{38}{100}$, which simplifies to $\frac{19}{50}$ or 38%.

When a probability is found from experimental trials as in this case, it is known as the **relative frequency**.

Combined events

The probability of combined events is calculated using the probabilities of the individual events.

The probabilities are multiplied if the events are **independent**.

Events are independent if what happens in one event does not affect what happens in the other event. For example, when two coins are thrown, the outcome for one is independent of the outcome for the other.

When events are not independent they are described as **dependent**. For example, a ball is removed from a bag containing three red and five blue balls. The ball is not returned to the bag and a second ball is removed. The probability for the second event depends on what happened in the first event.

WORKED EXAMPLE

Two fair coins are tossed. What is the probability of getting two heads?

SOLUTION

Each head has a probability of $\frac{1}{2}$ of appearing. P(2 heads) $= \frac{1}{2} \times \frac{1}{2} = \frac{1}{4}$

Tree diagrams are a useful way to represent problems involving combined events.

WORKED EXAMPLE

Natalie drives to work on a route with two sets of traffic lights.

The first set has a probability of 0.4 of showing green. The second set has a probability of 0.35 of showing green.

What is the probability that Natalie gets stopped by the lights at least once on her journey to work?

SOLUTION

First draw a tree diagram.

G is the event that the lights are green.

G′ is the event that the lights are not green.

Natalie is stopped on any branch where there is G′.

The only time she is not stopped is the top branch, so

P(GG) = 0.4 × 0.35 = 0.14

The probability that she is stopped = 1 − 0.14

= 0.86

So there is a 0.86 probability that Natalie will be stopped by traffic lights at least once.

We could have worked out the probability of each of the other three branches and added them together in order to get the answer.

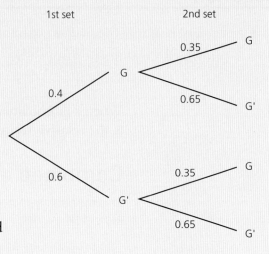

Tree diagrams are useful for calculating the probabilities when the events are dependent.

WORKED EXAMPLE

Clare is organising an event to raise money for charity. If it rains there is a probability of 0.3 of meeting her target of £500. If it does not rain the probability is 0.9. The probability of rain is 60% according to the weather forecast.

What is the probability of raising £500?

SOLUTION

Draw a tree diagram

P(£500) = 0.6 × 0.3 + 0.4 × 0.9

= 0.54

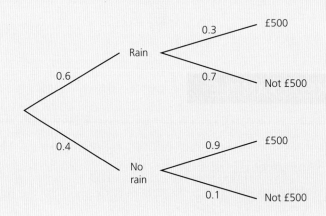

QUESTIONS

1   A bag contains 5 white balls and 7 red balls. Two balls are removed from the bag in succession. Work out the probability that both of the balls are red.

2   The probability of passing the driving test first time is 65%. The probability of passing on the second attempt is 0.5.
What is the probability that someone passes on the first or second attempt?

3   A game involves throwing a die and tossing a coin.
Throwing a 6 is a win.
If you do not throw a 6 you toss the coin.
Throwing tails wins.
What is the probability of winning this game?

4   If a disease is carried by 5% of the population and the test for it is 95% accurate, what is the probability that those who test positive will actually have the disease?

CHAPTER 3.28

VENN DIAGRAMS

Venn diagrams are a way to organise information and represent it visually in order to solve problems.

WORKED EXAMPLE

The table opposite shows the sports played by a group of students.

Draw a Venn diagram to show the information.

SOLUTION

The rectangle represents the whole population that we are working with.

Each set is represented by a circle.

H is those who play hockey.

F is those who play football.

O is those who play something else.

	Football	Hockey	Other
Anna	✓	✓	
Joe			✓
Angela	✓		
Raewyn		✓	
Brendan			
Marc	✓	✓	✓
Elizabeth			
Sameer	✓		
Rory	✓	✓	
Louis	✓	✓	
Alex			✓
Joanna	✓		✓

Sport choices

```
F                              H

    Angela      Rory
                Anna
    Sameer      Louis        Raewyn

             Marc

        Joanna

                    Alex
 Brendan        Joe
 Elizabeth
                         O
```

Hint

Marc is in each of the three sets F, H and O and so is in the overlap between the three circles. He counts as a member of each one because he is within each circle.

Using the Venn diagram to determine probabilities

WORKED EXAMPLE

From the information in the previous example, work out the probability of selecting a student who plays hockey.

SOLUTION

Out of the 12 names shown on the Venn diagram, there are 5 students who play hockey (the names in the shaded red circle).

So, the probability of selecting the name of a student who plays hockey is:

$$P(H) = \frac{5}{12} \text{ or } 41.7\%$$

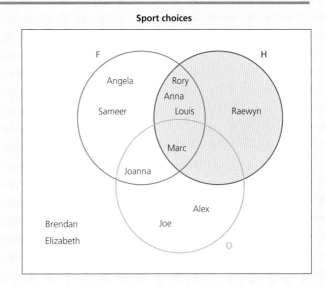

Sport choices

Complement of an event

The **complement of A, A´**, is the probability of the event that is **not** A.

WORKED EXAMPLE

Work out the probability of a student not playing hockey.

SOLUTION

5 out of the 12 students play hockey, so the number of students who don't play hockey is 7.

This is shown in the yellow shaded space on the diagram opposite.

We can work out the probability of selecting a student who does not play hockey in two ways:

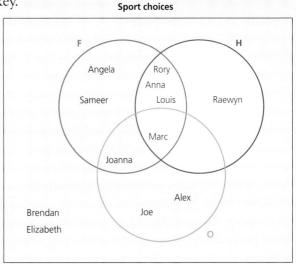

Sport choices

Method 1

$$P(H') = \frac{\text{number of students who do not play hockey}}{\text{total number of students}}$$

$$= \frac{7}{12}$$

Method 2

The probability of all possible outcomes for an event add up to 1.

This means that $P(H) + P(H') = 1$.

So $P(H') = 1 - P(H) = 1 - \frac{5}{12} = \frac{7}{12}$

Union of events

This is all of the outcomes in A or B and includes those in both A and B.

The probability of it is written $P(A \cup B)$.

WORKED EXAMPLE

Work out $P(H \cup O)$.

SOLUTION

The shaded space on the Venn diagram shows the union of H and O, or $H \cup O$.

This is the space on the Venn diagram that shows the students who play either hockey or another sport or both hockey and another sport.

$P(\text{Hockey} \cup \text{Other}) = \frac{8}{12}$

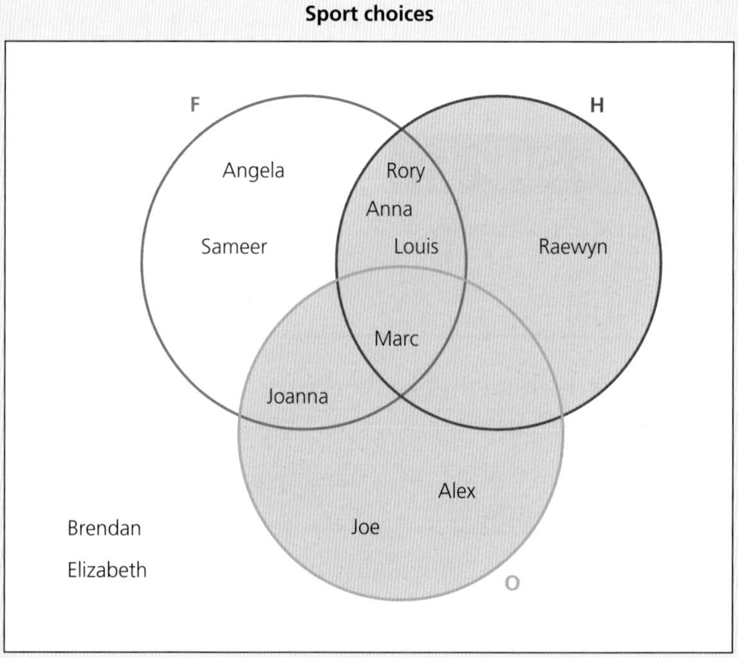

Sport choices

Intersection of events

This is all the outcomes in both **A** and **B**, that is, each event must be in both sets. It is the overlap between the sets.

We use the notation P(A ∩ B) to show that we are finding the probability of outcomes that appears in both sets.

WORKED EXAMPLE

Work out P(H ∩ F).

SOLUTION

The shaded space on the Venn diagram shows the **intersection** of the two groups.

This gives the students who play both football and hockey.

$$P(F \cap H) = \frac{4}{12}$$

Sport choices

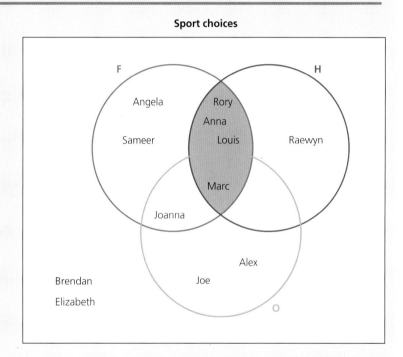

QUESTIONS

1 Using the information from the sport choices Venn diagram, work out:

  a P(O)

  b the probability that someone doesn't play any sport

  c P(H ∪ F′)

  d P(O ∩ F).

2 A hotel offers treatments in its spa. On one day 50 clients attended the spa.

The Venn diagram shows the number who chose a manicure (M) and the number who chose to have a leg wax (W).

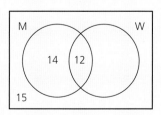

Work out:

a how many clients chose to have a leg wax and not a manicure

b P(W′)

c P(M ∩ W′).

3 There are 30 people working in an office. 12 like to drink coffee and 17 like to drink tea. There are 5 who do not like either coffee or tea.

a Draw a Venn diagram to show this information.

b Work out the probability that someone chosen randomly does not like tea.

c Work out the probability that someone chosen randomly does not like tea but does like coffee.

4 50 people were surveyed and the results are shown in the Venn diagram.
I represents people who shop online.

D represents people who drive a car.
T represents the people who live in an urban area.

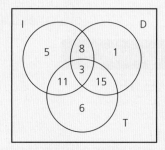

a Work out how many people are in (I ∪ D ∪ T)′.

b Calculate P(D ∩ T′).

c Work out the probability of picking someone who drives a car and is not living in an urban area.

d One of the people who drives a car is picked at random. What is the probability that they shop online?

USING PROBABILITY FOR EXPECTED AMOUNTS

Probability can help us to calculate the possible payback on games, investments, business decisions, and so on.

WORKED EXAMPLE

In a card game, you get 10 points for picking an ace, 5 points for a picture card, and 0 points for any other card.

What is the expected gain for this game?

SOLUTION

The probability of picking an ace out of a standard pack of 52 playing cards is $\frac{4}{52}$, so in $\frac{4}{52}$ picks you will win 10 points.

The expectation is therefore $\frac{4}{52} \times 10 = \frac{40}{52}$

Similarly, the expectation from picking a picture card is $\frac{12}{52} \times 5 = \frac{60}{52}$

The expectation from the other cards is $\frac{36}{52} \times 0 = 0$

So the total expectation $= \frac{40}{52} + \frac{60}{52} + 0 = \frac{100}{52} \approx 1.92$

So the mean number of points expected is 1.92.

This leads on to the idea of expected costs and benefits.

You convert probabilities to amounts in order to help make decisions about likely losses or gains. This enables you to introduce control measures, such as buying insurance or contracting more people to work on a job. This may involve extra expense, but the calculation helps you decide whether it is worth it.

WORKED EXAMPLE

A tourist attraction has the following expected ticket sales based on the outdoor temperature in summer:

Temperature, t	$t \geq 20\,°C$	$12\,°C \leq t < 20\,°C$	$t < 12\,°C$
Expected ticket sales (£)	£25 000	£12 000	£3000
Probability	0.1	0.65	0.25

The probability of each temperature category is shown in the table.

Calculate the expected value of ticket sales per day for the tourist attraction.

SOLUTION

We need to multiply each probability by each ticket sales value for all three temperature categories:

For $t \geq 20\,°C$, $0.1 \times £25\,000 = £2500$

For $12\,°C \leq t < 20\,°C$, $0.65 \times £12\,000 = £7800$

For $t < 12\,°C$, $0.25 \times £3000 = £750$

Because these events are mutually exclusive (they cannot happen at the same time), we add the individual expected values together.

So the expected value of ticket sales per day = £2500 + £7800 + £750 = £11 050.

WORKED EXAMPLE

A summer fair costs £1200 to put on. If it rains on the day, takings are estimated to be £500; if not, the takings are estimated at £2000. The likelihood of rain is 40%. The organisers consider taking out insurance for £1000 at a cost of £250 in case of rain. What would you advise?

SOLUTION

The expected takings are $0.4 \times 500 + 0.6 \times 2000 = £1400$.

This is greater than the £1200 cost, but if it does rain, the takings will only be £500 (a loss of £700).

With the extra £250 cost of insurance, the expected takings will be $0.4 \times 1500 + 0.6 \times 2000 = £1800$.

This is greater than the increased costs of £1450 and still gives a slight profit if it rains.

They should buy the insurance.

QUESTIONS

1 In a dice game, you get 0 points for throwing a one or a two, 1 point for a three, four or five, and 5 points for a six. Work out the expected number of points per game.

2 A manufacturer produces approximately 3200 car parts per day.
The probability of a part being faulty is 0.025. Calculate the expected number of faulty parts per day.

3 The probability that it snows on any given day in December is 0.06.
Calculate the expected number of days when it snows in December.

4 The probability that a football team wins a match is 0.45.
The probability that the team loses a match is 0.25.
If the team plays 60 matches this season, calculate the expected number of matches they will draw.

5 Paige's business adviser estimates that in the year ahead she has a 10% chance of making £20 000, a 50% chance of making £15 000, a 20% chance of making £10 000, a 10% chance of breaking even and a 10% chance of losing £10 000. What is the expected amount of money Paige will make in the next year?

6 1000 tickets are sold for a raffle in which there is a £100 prize, a £50 prize and five £10 prizes. Darren has five tickets. How much should he expect to win in the raffle?

7 The probability of it raining on any given day is 0.55.
The table shows the average number of car accidents in a city in a day when it has rained and not rained:

	Average number of accidents
On a day when it has rained	24
On a day when it has not rained	7

Calculate the expected number of car accidents in the city per day.

8 In a referendum, the probability that an individual will vote 'Yes' is 48%. If approximately 33 million people vote in the referendum, calculate the expected number of 'Yes' votes.

9 An event costs £1200 to stage. Advance tickets sales are worth £900, and the organisers expect between £300 and £700 in further sales. There is a 20% chance that the event will have to be cancelled because of illness and, in that case, 50% of the ticket price (for those already sold) is refunded. Insurance of £1500 is available at a cost of £500 if the event is cancelled. Is it worth it?

CHAPTER

❯ 3.30

GRAPH SKETCHING AND PLOTTING

Plotting graphs involves finding pairs of values for x and y that satisfy the equation of the function.

This is usually done by selecting a range of values for x and calculating the corresponding value for y by substituting into the equation.

WORKED EXAMPLE

a Plot the graph of $y = x^3 - 3x^2 - 6x + 8$ for $-3 < x < 5$

b Write down the coordinates of the two turning points from the graph.

SOLUTION

a First calculate the coordinates of some points

x	−3	−2	−1	0	1	2	3	4	5
y	−28	0	10	8	0	−8	−10	0	28

Notice that −3 and 5 are not included in the values for x. This is because they require a much larger range of values for y, as can be seen from the table of values.

Plot the points on a pair of axes such that the coordinates can be plotted so $-3 < x < 5$ and $-12 < y < 12$. Join with a smooth curve.

b Turning points are (−0.7, 10.4) and (2.7, −10.4) to 1 decimal place read from graph.

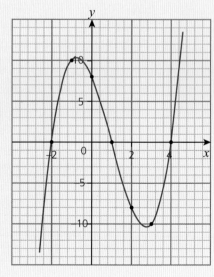

WORKED EXAMPLE

a Plot the graph of $y = 32 \times 0.5^x$ for $0 \leqslant x \leqslant 6$.

b Use your graph to estimate the value of x when $y = 10$.

SOLUTION

a To plot a graph we need to find the coordinates of some points on the curve. So, we need to construct a table of values, in this case for $0 \leqslant x \leqslant 6$.

We find the corresponding values of y by substituting the values of x into $y = 32 \times 0.5^x$
For example, when $x = 6$, $y = 32 \times 0.5^6 = 0.5$. ◀

x	0	1	2	3	4	5	6
y	32	16	8	4	2	1	0.5 ◀

We plot the graph using pairs of values from the table as coordinates, such as $(0, 32)$.

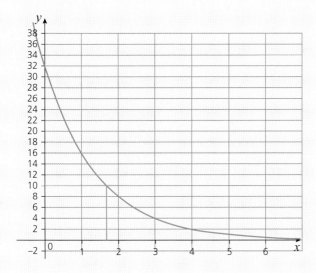

b We need to draw a horizontal line at height $y = 10$ until it connects with the curve.

Then we read down to the x-axis to find the value of x.

Therefore, when $y = 10$, $x \approx 1.7$.

The solution to an equation can be found using a graph. The graph may come from a pair of simultaneous equations, but only the x-values are required. The expression on one side of the equals sign is one function, and this is equated to y to form one graph. The same thing happens to the expression on the other side of the equals sign.

The solution can be read off the graph where the two graphs intersect.

WORKED EXAMPLE

The graph of $y = x + 5$ is shown on the axes opposite.

By plotting a suitable graph, find the solution to the equation $x + 5 = -x^2 + x + 8$.

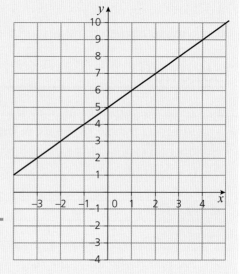

SOLUTION

We already have a graph for the left-hand side of the equation, so we need to plot one for the right-hand side. So we will plot the graph of $y = -x^2 + x + 8$.

We start by constructing a table of values for the same range of x as the graph we already have, as follows:

x	−3	−2	−1	0	1	2	3
y	−4	2	6	8	8	6	2

We plot each pair of points and connect them with a smooth curve.

To solve the equation, we find the coordinates of the intersections of the curve and the line. These are shown as the red dots on the diagram. We then read down to the x-axis to find the solutions. So the solutions to the equation are $x \approx 1.7$ and $x \approx -1.7$.

We do not need the y-coordinates as the equation is in x.

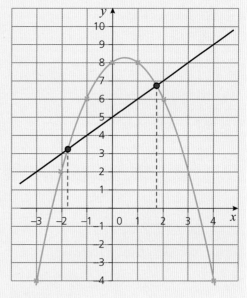

It may not always be necessary to **plot** the graph of a function. A **sketch** is sufficient to show the key aspects, such as the shape, intercepts with the axes and turning points.

WORKED EXAMPLE

Sketch the graph of $y = 4^x + 1$.

SOLUTION

This graph is an exponential shape because the variable x is used as a power.

When $x = 0$, $y = 4^0 + 1 = 1 + 1 = 2$ so the y-intercept will be at $(0, 2)$.

When $x = 1$, $y = 4^1 + 1 = 4 + 1 = 5$ so the graph will go through $(1, 5)$.

The graph is as shown opposite.

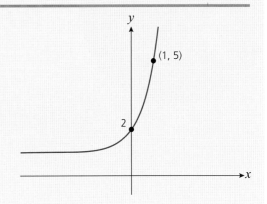

QUESTIONS

1   **a** Plot the graph of $y = x^2 - 5x$ for $0 \leqslant x \leqslant 6$.

   b Use your graph to find the value of x where y is a minimum.

2   **a** Plot the graph of $y = 0.25 \times 2^x$ for $0 \leqslant x \leqslant 6$.

   b Use your graph to estimate the solution to the equation $0.25 \times 2^x = 5$.

3   The volume (V) of a container is modelled by the equation $V = \frac{1}{2}h^3 - 4$, where h represents the height of the container in centimetres.

   a Plot the graph of $V = \frac{1}{2}h^3 - 0.5$ for $1 \leqslant h \leqslant 4$.

   b Use your graph to estimate the height when $V = 20$.

4   The path of a projectile is modelled by the equation $h = -t^2 + 2t + 10$, where h represents the height of the projectile in metres after t seconds.

   a Plot a graph of the path of the projectile over the first 5 seconds.

   b Use your graph to estimate the value of t when $h = 8$.

   c Estimate how long the projectile is in flight before it hits the ground.

   d What was the launch height of the projectile?

5   **a** Plot the graph of $y = -x + 3$ for $0 \leqslant x \leqslant 3$.

   b On the same axes plot the graph of $y = 2^x - 3$ for $0 \leqslant x \leqslant 3$.

   c Use your graphs to solve the equation $2^x - 3 = -x + 3$.

6   Jo wants to solve the equation $x^3 - x = 3^x$.

   State the equations of two graphs that Jo could plot to find the solution.

7   **a** Plot the graph of $y = x^2 + x - 6$ for $-3 \leqslant x \leqslant 3$.

   b By plotting a suitable second graph on the same axes, find the solution(s) to the equation $x^2 + x - 6 = 2x - 3$.

Sketch the graphs of the following equations. You must show the coordinates of any points where the graph crosses an axis.

8   $y = 20 - x^2$.

9   $y = 500 + 100^x$.

10 $y = 100x - 20$.

11 $y = x^3 - 10$.

RATES OF CHANGE

CHAPTER 3.31

Gradient and rate of change of a straight line

The gradient of a line is a measure of the rate of change of the dependent variable (y) as the independent variable (x) increases. This is constant for a straight line.

A straight line is used to model relationships where there is direct proportionality.

The gradient, or steepness, of the line is $\dfrac{\text{vertical change}}{\text{horizontal change}}$

The **gradient** shows the **rate of change**.

The graph is a solid line if it models continuous data.

WORKED EXAMPLE

Many long–distance runners measure their speed using the number of minutes that it takes to run 1 mile (or 1 kilometre).

A runner runs at an average speed of one mile every 7 minutes.

a   Work out how long they take to run 3 miles.

b   Work out how far they have run after 14 minutes.

c   Draw a distance–time graph for the first 35 minutes.

d   Work out the gradient of the graph.

e   What does the gradient represent?

>
> **Hint**
>
> We refer to average speed because a runner's speed varies, even if their pace is consistent. Here we assume that the runner ran every mile at the same speed, whatever the terrain.

SOLUTION

a   One mile takes 7 minutes, so 3 miles take 3×7 minutes
    $= 21$ minutes.

b $14 = 2 \times 7$ so 14 minutes is the time to run 2 miles.

c

Running speed -minutes per mile

(graph: Distance (miles) vs Time (minutes), straight line through origin)

(28 min, 4 miles)
(21 min, 3 miles)
(14 min, 2 miles)
1 mile
7 minutes
(7 min, 1 mile)

Time (minutes)

d The gradient or steepness of the line is:

$$\frac{\text{vertical change}}{\text{horizontal change}} = \frac{1}{7} \text{ miles per minute.}$$

e The gradient shows the speed, or rate of change of the runner's distance with respect to time, and it tells us that every 7-minute change in time corresponds to a 1-mile change in distance.

Gradient of a curve

The gradient of a curve changes.

It is only possible to calculate the average rate of change over a period of time, or the instantaneous rate of change at a point.

The average rate of change is calculated by selecting two points and working out the gradient of the straight line joining them.

WORKED EXAMPLE

A football is kicked into the air. The table and graph below show the height of the ball.

Time (seconds)	0	1	2	3	4	5	6
Height of ball (metres)	0	5	8	9	8	5	0

a Draw a graph to show the motion of the football.

b Describe and interpret the rate of change of the gradient when $t = 0.5$ s.

c Describe and interpret the rate of change of the gradient when $t = 1.5$ s.

d Describe and interpret the rate of change of the gradient when $t = 3$ s.

e Work out the average rate of change in the first second.

f Work out the average rate of change between 4 and 6 seconds.

SOLUTION

a

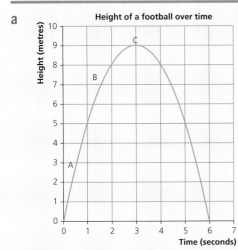

Height of a football over time

Time (seconds)

b At A, when $t = 0.5$ s, the steepness of the curve gives the rate of change of the ball's height. Initially, the gradient is very steep because the height of the ball increases quickly over a small period of time.

c At B, when $t = 1.5$ s, the gradient is less steep than at A. This is because the ball's height is changing at a slower rate.

d There is a turning point of the graph at C, when $t = 3$ s, showing this is a turning point of the graph showing the maximum height of the ball.

Here the gradient is zero because the ball stops instantaneously, so the rate of change, or speed, is zero.

After the maximum point, the height of the ball decreases as the ball falls to the ground.

e Average gradient between 0 and 1 seconds =

$$\frac{\text{vertical change}}{\text{horizontal change}} = \frac{5}{1}$$

This tells us that the ball travelled at an average speed of $5\,\text{m}\,\text{s}^{-1}$ during this time.

f Average gradient between 4 and 6 seconds =
$$\frac{8-0}{4-6} = \frac{8}{-2} = -4$$

This tells us that the ball's speed **decreased** at a rate of $4\,\text{m}\,\text{s}^{-1}$ during this time period.

Drawing a tangent

The instantaneous rate of change at a point can be estimated by drawing a tangent.

A tangent is a line drawn that touches the curve at a point so that its gradient is close to that of the curve at that point. Some judgement is required to place the tangent, hence it is an estimate.

Alternatively, you can find the gradient of a chord as before, but choose two points that straddle the required point and are very close to it. You can read off values or calculate values of the function to do this.

WORKED EXAMPLE

The graph shows the height of a football during the 6 seconds after it is kicked.

Work out the instantaneous rate of change of the height 4 seconds after the ball is kicked. You will need to use the graph from the previous worked example.

SOLUTION

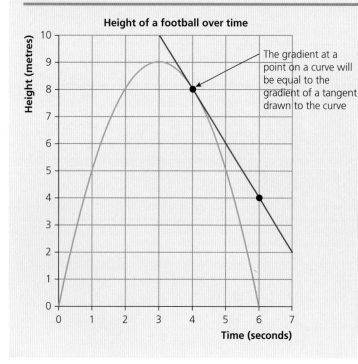

The gradient at a point on a curve will be equal to the gradient of a tangent drawn to the curve

We can use the fact that the tangent touches the curve at the point $(4, 8)$ and also passes through the point $(6, 4)$ to work out the gradient of the tangent (hence the gradient of the curve at that point).

Tangent gradient $= \frac{4-8}{6-4} = \frac{-4}{2} = -2$ so the ball was travelling at a speed of 2 metres per second at exactly 4 seconds.

QUESTIONS

1 Use the information about the runner in the first worked example.

   a How far will an athlete running at this pace have run after 35 minutes?

   b How long will it take to run 15 miles at this pace?

   c Explain why using the points (14 min, 2 miles) and (28 min, 4 miles) to work out the gradient gives the same gradient as using the points (14 min, 2 miles) and (21 min, 3 miles).

2 A water tank has a leak. The graph below shows the amount of water in the tank over time.

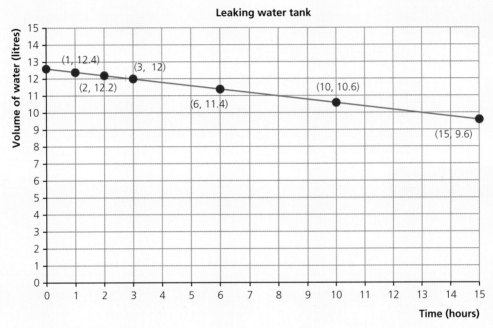

Leaking water tank

a How much water was there in the tank before the leak started?

b Why is the graph decreasing?

c Work out the gradient of the graph and explain what this gradient tells us about the relationship between the volume of water in the tank and the amount of time that has elapsed.

d Explain why the graph is a straight line.

e How long will it take for the water in the tank to run out?

3 Use the information in the worked example about a football.

   a Work out the average rate of change of the ball's height between 0 and 2 seconds.

b Work out the average rate of change of the ball's height between 3 and 6 seconds.

c Explain why the answer in b only gives an estimate of the rate of change of the ball's height over this time period.

4 Use the information in the worked example about a football.

   a Work out the speed of the ball at exactly 1 second.

   b Show that the ball was travelling at a slower speed at 2 seconds than at 5 seconds.

   c Show that the ball had zero speed at exactly 3 seconds.

DISTANCE, SPEED AND TIME

Average speed

To calculate the average speed for a journey, divide the total distance travelled by the total time taken. If the distance is in metres and the time in seconds, you are calculating how far you go on average in one second, so your speed is in metres per second. If the distance is in miles and the time in hours, you are calculating how far you go on average in one hour, so your speed is in miles per hour.

WORKED EXAMPLE

You drive 50 miles in one hour ten minutes. What is your average speed?

SOLUTION

The average speed is the total distance divided by the total time:

$$\text{Average speed} = \frac{50 \text{ miles}}{\left(1+\frac{10}{60}\right)} \approx \frac{50}{1.17} \text{ mph} \approx 42.9 \text{ mph}$$

Graphs of distance and velocity against time

The gradient of a **distance–time** graph represents speed.

When you work out the gradient, you divide distance by time so the units are metres per second or miles per hour, depending on the units that the distance and time are measured in.

These are the units of speed.

The gradient of a **velocity–time** graph represents acceleration.

When you work out the gradient you divide velocity by time so the units are metres per second per second or miles per hour per hour, depending on the units that the velocity and time are measured in.

These are the units of acceleration.

WORKED EXAMPLE

The Earth moves round the Sun at an average speed of approximately 110 000 kilometres per hour.

The graph shows the time taken against the distance travelled.

The gradient of the graph represents the speed of the Earth – how many kilometres the Earth travels each day.

Work out the speed of the Earth.

SOLUTION

The graph shows that the Earth travels 800 million km in 300 days.

So, in one day, it travels $\frac{800 \times 10^6}{300}$ km $\approx 2.67 \times 10^6$ km.

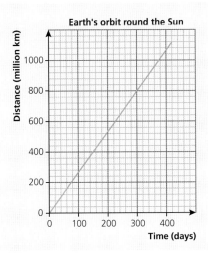

QUESTIONS

1 The world record for the 100 metres was set by Usain Bolt at 9.58 seconds in 2009.

 a What was Usain Bolt's average speed?

 b Could Usain Bolt run 1000 metres in 95.8 seconds?

2 You travel 323 miles at an average speed of 54 mph. How long does the journey take?

3 A car journey of 250 miles was done at an average speed of 60 mph. On the return journey, heavy traffic reduced the average speed to 38 mph.

 What was the average speed for the round trip?

4 This distance–time graph shows the orbit of Mercury.

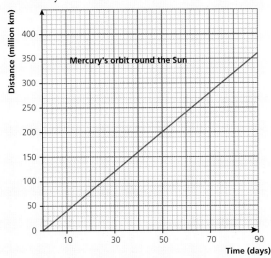

Calculate the speed at which Mercury is moving.

5 The graph shows the motion of a golf ball.

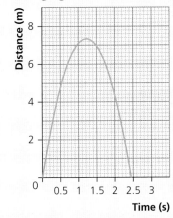

 a Estimate the velocity of the ball at half-second intervals and draw a velocity–time graph for the ball.

 b Calculate the gradient of the velocity–time graph.

 c Interpret your result in b.

6 Draw a distance–time graph showing a journey of half an hour at an average speed of 23 mph followed by an hour and a half at an average speed of 32 mph.

 Work out the total distance travelled.

421

EXPONENTIALS

You need to be confident with using your calculator for this topic.

Find the x^y, e^x and ln buttons on your calculator. The ln x function is the inverse of e^x, and so they are usually on the same button (but one requires use of the shift key).

x^y means 'to the power of' so 3 x^y 4 means '3 to the power of 4' or 3^4.

$e \approx 2.718$.

e^x means e 'to the power of' whatever number you choose to substitute for x.

ln (number) means $\log_e$ (number).

Logarithms, or logs, are another word for power or index, with the extra idea that a logarithm is a function. When you use logarithms you use the same base number for your powers or indices.

Your calculator uses a base of 10 or e for its log functions.

WORKED EXAMPLE

Calculate $3^{1.5}$.

SOLUTION

Use the x^y calculator button.

$3^{1.5} = 5.196$ (to 3 decimal places).

 Hint

It's wise to estimate your answer.

It should be between 3 and 9 because $3^1 = 3$ and $3^2 = 9$.

HINT

$\ln a^x = x \ln a$

So $x \ln a = \ln b$

and so $x = \dfrac{\ln b}{\ln a}$

Solving equations that involve exponents (powers) such as $a^x = b$ and $e^{kx} = b$ may be done by trial and improvement or by plotting a graph.

However, it is much more efficient to use logarithms.

Logarithmic functions are inverses of exponential functions and so this idea is used to simplify expressions involving logs or exponentials. In this qualification we use the natural logarithm, which is written ln x and is the inverse of the exponential function e^x.

Applying the natural logarithm function to both sides of this equation allows you to simplify it so that the equation is more easily solved.

$a^x = b$

$\ln a^x = \ln b$

DID YOU KNOW?

$e^{\ln a} = a$, so $(e^{\ln a})^x = a^x$.

The laws of indices tell us that $(e^{\ln a})^x = e^{x \ln a}$

So $a^x = e^{x \ln a}$

So $\ln a^x = x \ln a$

WORKED EXAMPLE

Find the value of x when $2^x = 12$.

SOLUTION

We could draw or **sketch** the graph of $y = 2^x$ and use it to estimate the solution.

From this the value is approximately 3.5.

Algebraic method

$2^x = 12$

Apply the natural logarithm function to both sides

$\ln (2^x) = \ln 12$

This can be rewritten, using the rule that $\ln a^x = x \ln a$, as

$x \ln 2 = \ln 12$

$x = \dfrac{\ln 12}{\ln 2}$

$x = 3.585$ (to 3 decimal places).

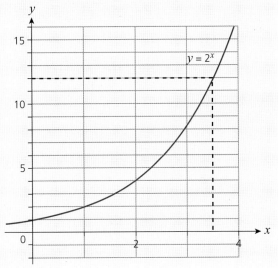

HINT

The ln and exponential functions cancel each other out.

The same method is used when the equation has the form $e^{kx} = b$.

$e^{kx} = b$

$\ln e^{kx} = \ln b$

$kx = \ln b$

so $x = \dfrac{\ln b}{k}$

WORKED EXAMPLE

Find the value of x when $e^{2x} = 24$.

SOLUTION

$\ln (e^{2x}) = \ln 24$

$2x = \ln 24$

$x = \dfrac{\ln 24}{2}$

$x = 1.589$ (to 3 decimal places).

One of the most significant facts about e is that the gradient of $y = e^x$ at any point is equal to the y-coordinate at that point.

WORKED EXAMPLE

Find the gradient of the curve given by the equation $y = e^x$ at the point with x-coordinate 4.

SOLUTION

The gradient at any point is equal to the y-value at that point.

So, at $x = 4$, the y-value is $e^4 = 54.598$ and this is also the gradient of the curve at that point.

QUESTIONS

1 Use your calculator to find these values, giving your answer to 3 decimal places.

  a $3^{3.1}$

  b e^2

  c e^{-2}

  d $4e^{0.5}$

  e $3.5^2 + 2^{3.5}$

2 Write down the value of x that satisfies the equation $3^x = \dfrac{1}{3}$.

3 Use your calculator to solve these equations, giving your answer to 3 decimal places.

  a $3^x = 12$

  b $4.5^x = 21$

  c $e^{2x} = 30$

  d $e^{3.5x} = 45$

  e $e^x = 30$

4 a A curve is represented by the equation $y = e^x$. Calculate the gradient of the curve at $x = 2.1$.

  b A curve is represented by the equation $y = e^x$. Calculate the gradient of the curve at $x = 3$.

  c A curve is represented by the equation $y = e^x$. Calculate the gradient of the curve at $x = 3.5$.

  d A curve is represented by the equation $y = e^{-x}$. Calculate the gradient of the curve at $x = 4$.

EXPONENTIAL GROWTH AND DECAY

Formulating equations of the form $y = Ca^x$

We can use the following exponential growth formula to represent the growth of some money in an account over time, where $a = 1 + r$.

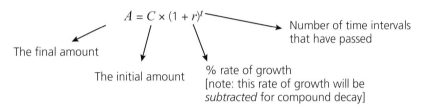

$$A = C \times (1 + r)^t$$

The final amount

The initial amount

% rate of growth
[note: this rate of growth will be *subtracted* for compound decay]

Number of time intervals that have passed

WORKED EXAMPLE

A person deposits £1000 into a bank account. This money will earn interest at a compound rate of 0.5% per month.

Substitute the given values into this equation:

$A = 1000 \times (1 + 0.5\%)^t$

Simplify the values inside the bracket by converting the percentage value to a decimal and then adding the 1:

$A = 1000 \times (1.005)^t$

Use this formula to work out the amount of money in the account after a certain period of time. For example:

t	Working	Answer
$t = 1$	$A = 1000 \times (1.005)^1 = 1000 \times 1.005$	£1005.00
$t = 2$	$A = 1000 \times (1.005)^2 = 1000 \times 1.010025$	£1010.03
$t = 12$	$A = 1000 \times (1.005)^{12}$	£1061.68

If the scenario involves a doubling, tripling or halving rate instead of a percentage growth or decay rate, then a slightly different formula can be used:

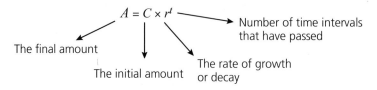

$A = C \times r^t$ — Number of time intervals that have passed

The final amount

The initial amount

The rate of growth or decay

WORKED EXAMPLE

A tennis tournament starts with 128 players.

After each round, half of the players are eliminated.

a   Write down an equation for the number of players (A) in the tournament after t rounds.

b   Work out how many players remain after the first round.

c   Work out how many players remain after the fifth round.

SOLUTION

a   We can represent this scenario using the formula above as follows:

$$A = 128 \times \left(\frac{1}{2}\right)^t$$

b   Number of players in the tournament after 1 round:

$$A = 128 \times \left(\frac{1}{2}\right)^t = 128 \times \left(\frac{1}{2}\right)^1 = 128 \times \left(\frac{1}{2}\right) = 64$$

c   Number of players in the tournament after 5 rounds:

$$A = 128 \times \left(\frac{1}{2}\right)^5 = 128 \times \frac{1}{32} = 128 \times \frac{1}{32} = 4$$

This would be referred to as the semi-finals.

Formulating equations of the form $y = Ce^{kx}$

The formula $y = Ce^{kx}$ represents a more general form than the formulae used above, with e representing the standard base for exponential functions.

When the increase is continuously happening through the year an increase of $r\%$ per year can be modelled by $e^{\left(\frac{r}{100}t\right)}$.

This formula applies when interest on an account is compounded continuously rather than at specific time intervals. It also applies when growth or decay happens continuously, as will usually be the case.

WORKED EXAMPLE

A recycling centre currently processes 235 000 tonnes of recyclable materials every year. It has the capacity to grow by 5% per year.

Represent this using the formula:

$A = 235\,000 \times e^{0.05t}$

Work out how many tonnes it will process in the first year and the fifth year.

SOLUTION

When $t = 1$: $A = 235\,000 \times e^{0.05(1)} = 235\,000 \times e^{0.05} = 247048.71 \approx 247\,000$ tonnes.

When $t = 5$: $A = 235\,000 \times e^{0.05(5)} = 235\,000 \times e^{0.25} = 301745.97 \approx 302\,000$ tonnes.

WORKED EXAMPLE

An adult takes a 20 ml dosage of flu medicine.

The medicine decreases in the person's system at a rate of 25% per hour.

Represent this scenario using the general formula as follows:

$A = 20 \times e^{-0.25t}$

Work out the amount of the medicine in the person's system after 1 hour and after 3 hours.

SOLUTION

Rewrite the equation to account for the negative exponent as follows:

$A = \dfrac{20}{e^{0.25t}}$

When $t = 1$: $A = \dfrac{20}{e^{0.25(1)}} = \dfrac{20}{e^{0.25}} \approx 15.58$ ml

When $t = 3$: $A = \dfrac{20}{e^{0.25(3)}} = \dfrac{20}{e^{0.75}} \approx 9.45$ ml

QUESTIONS

1 The formula below shows the change in the number of a particular type of bacteria:

$A = 5 \times 3^t$

a Work out how many bacteria there were to start with.

b Does this formula represent growth or decay? Explain your answer.

c The rate of growth or decay is given per hour.

How fast per hour is the bacteria growing or decaying?

d Work out how many bacteria there will be after 3 hours.

e Work out how many bacteria there will be after 1.5 hours.

2 The original price of a car is £5000.

a Does the value of the value of the car increase or decrease over time?

Explain your answer.

b The value of this car initially depreciates (decreases) at a rate of approximately 10% per year.

Write down a formula that can be used to estimate the value of the car.

c Use the formula to estimate how much the car is worth after 1 year.

d Use the formula to estimate how much the car is worth after 4 years.

3 A recycling centre currently processes 235 000 tonnes of recyclable materials every year. It has the capacity to grow by 5% per year.

We can model this using the formula:

$A = 235\,000 \times e^{0.05t}$

Work out how many tonnes of recyclable materials the recycling centre should be able to process after 10 years.

4 A child takes a 5 ml dosage of flu medicine.

We can model the amount of medicine left in their body using the general formula as follows:

$A = 5 \times e^{-0.25t}$

Work out how many hours it will take for the medicine in their body to be less than 1 ml.

5 The current world population is estimated at 7.4 billion people. The population growth is modelled by the formula:

$A = 7.4 \text{ (billion)} \times e^{0.013t}$

Work out what the approximate population will be in 20 years' time.

6 A cup of coffee contains around 95 mg of caffeine. The rate of decrease of the caffeine in the body is modelled by:

$A = \dfrac{95}{e^{0.1t}}$

Work out how much caffeine there is in the body after 3 hours.

INDEX